Reactions of Organosulfur Compounds

This is Volume 37 of
ORGANIC CHEMISTRY
A series of monographs
Editors: ALFRED T. BLOMQUIST and HARRY H. WASSERMAN

A complete list of the books in this series appears at the end of the volume.

Reactions of Organosulfur Compounds

Eric Block

Department of Chemistry
University of Missouri—St. Louis
St. Louis, Missouri

1978

ACADEMIC PRESS New York San Francisco London
A Subsidiary of Harcourt Brace Jovanovich, Publishers

ACADEMIC PRESS, INC.
111 Fifth Avenue, New York, New York 10003

United Kingdom Edition published by
ACADEMIC PRESS, INC. (LONDON) LTD.
24/28 Oval Road, London NW1 7DX

Library of Congress Cataloging in Publication Data

Block, Eric.
Reactions of organosulfur compounds.

(Organic chemistry: a series of monographs; v.)
Bibliography: p.
Includes index.
1. Organosulphur compounds. 2. Chemical reactions. I. Title. II. Series.
QD412.S1B56 547'.06 77-25040
ISBN 0-12-107050-6

PRINTED IN THE UNITED STATES OF AMERICA

To my wife, Judy,
and children, David and Melinda

Contents

List of Tables

Preface

It is my hope to capture in these pages the flavor of research in organosulfur chemistry during the past decade. It is also my hope that this volume will provide a timely introduction to the basics of organosulfur chemistry for students and practitioners of chemistry and biochemistry as well as a source of new ideas and useful information for those already experienced in the area. The particular perspective that I bring reflects my own graduate education at Harvard University in the mid 1960s during which time in the "Corey Group" we witnessed the coming of age of dimethyl sulfoxide, dithiane, and sulfur ylides as useful synthetic reagents and the application of these and related reagents to the total synthesis of sesquiterpenes, prostaglandins, and other natural products. During these same exciting years the concept of orbital symmetry was being enthusiastically espoused and was finding important applications in diverse areas including sulfur chemistry.

In presenting a current view of the field of organosulfur chemistry I considered two organizational schemes: a traditional approach based on organosulfur functional groups, and an approach based on intermediate types in reactions of organosulfur compounds. An orderly procession through structural types, outlining structure, spectra, reactions, synthesis, natural occurrence, etc., has considerable merit and has in fact been widely used. However, a mechanistic approach has the advantages of permitting a continuous comparison of properties of different structural types, a simultaneous treatment of applications in organic synthesis based on specific mechanistic conjectures and anticipated properties of intermediates and reaction products, and finally, a clearer appreciation of where new research is needed.

A prime consideration in the ultimate selection of the second, mechanistic approach is that this approach seems better suited for classroom use since it can easily be incorporated piecemeal within the framework of typical courses in advanced organic chemistry, organic synthesis, or even general organic chemistry, or it can form the basis for a more advanced "special topics" course. (I have used the material in all these ways during the past ten years.) Of course, through the use of the index the various characteristic reactions and properties of any desired organosulfur functional group can also be determined.

I have placed heavy emphasis on the area of applications of organosulfur compounds in organic synthesis since this area has dominated the field of sulfur chemistry during this past decade and should be a subject of great general interest. I have also attempted to make this book a useful source of general information on organosulfur compounds through incorporation of tabulations of spectral data in Appendix B and through other compilations located throughout the text. To keep down the length of this volume, I have shortened to seven a substantially longer list of potential chapters, reluctantly omitting extensive treatment of such topics as substitution reactions of sulfur (sulfur as a chiral center; reactions of disulfides, organosulfur halides, and esters), oxidation and reduction reactions of sulfur (including the chemistry of the recently discovered hypervalent tetra- and hexacoordinate systems), the chemistry of sulfur-containing heterocycles (a vast area in itself), and photochemical reactions of sulfur compounds. Such topics might be suitable for a future effort.

Acknowledgments

Work on this volume was begun in 1974 during the tenure of a visiting professorship at Harvard University, and was nourished in the ensuing years by financial support from the University of Missouri–St. Louis and the Petroleum Research Fund administered by the American Chemical Society (supporting some of the research described herein). Every effort has been made to provide thorough yet current (through August 1977) references on all subjects covered. An unfortunate consequence of a fetish for current references is that on occasion significant early contributions may be slighted. To these colleagues I express my apologies. It is a pleasure to acknowledge the counsel and constructive criticism of Professors Alfred G. Hortmann, Rudolph E. K. Winter, Saul Wolfe, David Harpp, and Frank Davis, the expert technical assistance of Michael Ennis, Gordon Sze, and Theresa Orso, and the generous use of library facilities by the Monsanto Corporation. The staff at Academic Press is applauded for their speedy and precise conversion of a jumble of pages into the present volume. I am particularly grateful to my wife, Judy, and children, David Michael and Melinda Gayle, for their encouragement and understanding during the long process of transforming an idea into a reality.

Abbreviations

The following abbreviations have been used:

Ac	acetyl
Ar	aryl
Bz	benzoyl
CIDNP	chemically induced dynamic nuclear polarization
DABCO	1,4-diazabicyclo [2.2.2] octane
DCC	dicyclohexylcarbodiimide
DME	1,2-dimethoxyethane
DMF	*N*,*N*-dimethylformamide
DMSO	dimethyl sulfoxide
ESR	electron spin resonance
HMPA	hexamethylphosphoric amide
MCPBA	*meta*-chloroperbenzoic acid
MO	molecular orbital
NBS	*N*-bromosuccinimide
NCS	*N*-chlorosuccinimide
NMR	nuclear magnetic resonance
Ph	phenyl
PNB	*para*-nitrobenzoate
Py	pyridine
Ra Ni	Raney nickel
SCF	self-consistent field
THF	tetrahydrofuran
THP	tetrahydropyranyl
TMEDA	*N*,*N*,*N'*,*N'*-tetramethylethylenediamine
Ts	*para*-toluenesulfonyl

Chapter 1 | Introduction

1.1 Prologue

The chemistry of sulfur is, in a sense, a study in contrasts. Elemental sulfur was well known to the ancients, yet current research on the element and its compounds continues apace. Sulfuric acid, sulfur dioxide, and sulfur, when found in "acid rain," polluted urban air, or coal, respectively, are the subjects of worldwide concern as environmental hazards to be eliminated at all costs; these same substances are valued bulk commodities, with sulfuric acid annually topping the list in tonnage of all synthetic chemicals. Organosulfur compounds are found in such diverse locations as deep interstellar space (Oppenheimer and Dalgarno, 1974); the hot (89°C), acidic, sulfur-rich volcanic areas near Naples (De Rosa *et al.*, 1975); and the oceanic depths off Hawaii (Moore, 1971), as well as the backyard garden patch. In the form of certain essential amino acids and enzymes, organosulfur compounds occur in the bodies of all living creatures. Simple organosulfur compounds such as the alkanethiols are repugnant to man and most higher animals even at incredibly low concentrations and are used as defensive secretions by a variety of species; related sulfur compounds found in such foods as garlic, onions, radishes, asparagus, chives, cabbage, leeks, mustard, truffles, coffee, tea, and pineapple are sources of olfactory and gustatory delight. Sulfur-containing mustard gas [bis(β-chloroethyl) sulfide] was a much feared chemical warfare agent, while another sulfur compound, penicillin, is a much valued antibiotic. Divalent as well as higher valent sulfur shows a remarkable ability to stabilize adjacent carbanions by electron acceptance or charge dispersal, yet sulfur is also well able to stabilize by electron donation to adjacent carbon cations, free radicals, and carbenes. Sulfur in some forms shows good insulating properties (sulfur-based foams show promise as insulation under roads in arctic regions); polymeric sulfur nitride [polythiazyl or $(SN)_x$] and complexes of tetrathiofulvalenes (see Chapter 6) are of great current interest because of their metallike ability to conduct electricity. While some sulfur compounds are so inert that they find application in aerospace technology (e.g., polysulfones and certain rubber analogs) or as atmospheric chemical markers (sulfur hexafluoride), other compounds have lifetimes measured in seconds or fractions thereof (thioformaldehyde, sulfene, thiirene).

While we may note the extremes in properties, we should also point to the vast number of less notorious organosulfur compounds possessing a variety of useful properties as synthetic reagents, intermediates, and solvents; as drugs, biochemicals, natural products, and insecticides; and as substances of interest for mechanistic, structural, or spectroscopic studies. These organosulfur compounds can contain a wide variety of sulfur functional groups in which sulfur can vary from a coordination number of two to six and in which sulfur may bond to itself (catenate). Organosulfur compounds are known to undergo a remarkably broad range of transformations on exposure to heat, light, ionizing radiation, metals, acids and bases, oxidizing and reducing agents, free radicals and carbenes, and enzymes. This book will attempt to portray many of the characteristics of organically bound sulfur, with an emphasis on reactions, particularly those of synthetic utility. More extensive treatment of certain features of organic and inorganic sulfur chemistry is to be found in the various volumes listed in the bibliography (Appendix A).

1.2 A Word on Odor

The first organosulfur compound to be prepared was ethanethiol (Zeise, 1834; Liebig, 1834). The offensive odor of this compound [as little as one part in fifty billion of ethanethiol in air may be detected by odor (Connor, 1943)] and other low molecular weight thiols and dithiols† is taken by some‡ as the most distinguishing characteristic of organosulfur chemistry. There is certainly no denying the unappealing aroma of certain sulfur compounds. Indeed, the sensitivity of the human nose to these small sulfur molecules and the fact that many animals also find their smell offensive supports speculation that natural selection developed this olfactory sensitivity as a form of protection for the organism against the ingestion of decaying food (low molecular weight organosulfur compounds are also products of biological decay).

It is sometimes pointed out by defensive sulfur chemists that only a small fraction of the known organosulfur compounds have unpleasant odors (most higher-valent organosulfur compounds and higher molecular weight divalent

† A particularly notorious compound is propane 2,2-dithiol (previously thought to be monomeric thioacetone). Early German researchers noted that it "has an offensive nauseating odor, much worse than that of other volatile sulfur compounds" and that its preparation "caused disturbances in the neighboring streets in Freiburg and precipitated a storm of protests . . ." (Fromm and Baumann, 1899).

‡ For example the chapter on sulfur-containing compounds in one text on organic chemistry (Gutsche and Pasto, 1975) begins with the quotation from Coleridge: "I counted two and seventy stenches / All well defined, and several stinks."

compounds have little, if any, odor). While this is true, it is also a fact that simple low molecular weight organosulfur compounds are readily available, inexpensive, and of great synthetic utility so that, like it or not, most research involving organosulfur compounds will at one time or another involve the use of obnoxious-smelling substances. It should be borne in mind that the lower mercaptans are used at low concentrations as odorants for natural gas (odorless by itself). Thus, accidental release of vapors of these compounds, even after dilution with substantial volumes of air on passage through a fume hood, can result in some mischief (entire office buildings have been known to be evacuated following use of small amounts of mercaptan in an adjacent laboratory building). It is therefore advisable to vent vapors from potentially troublesome reactions through oxidizing solutions such as alkaline peroxide or hypochlorite. With good laboratory technique, multimolar reactions involving these odorous molecules can be conducted as easily as any other common organic reaction.

1.3 Sulfur in the Natural Environment

Sulfur is widely distributed in nature. As free sulfur and inorganic sulfides and sulfates it is found in mineral deposits and in particularly high concentrations in areas of volcanic or geothermal activity; in the form of various divalent organosulfur compounds it is found in petroleum deposits and coal; as sulfur oxides, sulfur acids (H_2S, H_2SO_4), and certain low molecular weight organic sulfur compounds it is found in the atmosphere, particularly under conditions of pollution. Simple sulfur compounds such as H_2S, OCS, H_2CS, CS, and SO have even been detected in interstellar space. A rich variety of organic sulfur compounds are found in living systems. A list of biochemically notable organic sulfur compounds (see Chart 1-1) would include, among others, the essential amino acids *cysteine* and *methionine*, peptides such as *glutathione*, polycyclic peptide antibiotics such as *pencillin*, *cephalosporin*, *bacitracin*, and *gliotoxin*, cofactors and vitamins such as *thiamine*, *biotin*, *coenzyme A*, *lipoic acid*, a biological alkylating agent, *S-adenosylmethionine*, sulfur-containing alkaloids such as *cassipourine*, sulfur-containing bases found in bacterial transfer-RNA such as 4-*thiouracil* and the potent carcinogen *ethionine*. Mention should also be made of iron–sulfur proteins (e.g., *ferridoxin* and *rubredoxin*) which are ubiquitous biological redox agents. Certain of these small proteins contain a remarkable cysteine-bound "inorganic cubane" structural unit at the active site.

A remarkable variety of simpler organosulfur compounds are widely distributed throughout the plant kingdom. These compounds often make an important contribution to the odor and flavor of many of the common

Chart 1-1. Some Biochemically Notable Organic Sulfur Compounds

$HSCH_2CH(NH_2)COOH$
Cysteine

$CH_3SCH_2CH_2CH(NH_2)COOH$
Methionine

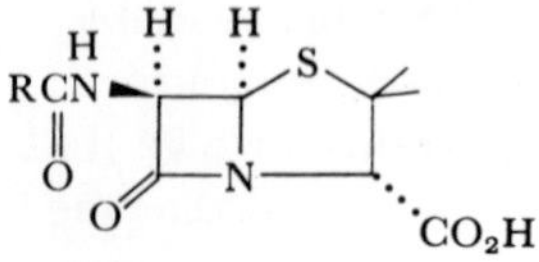

Penicillin

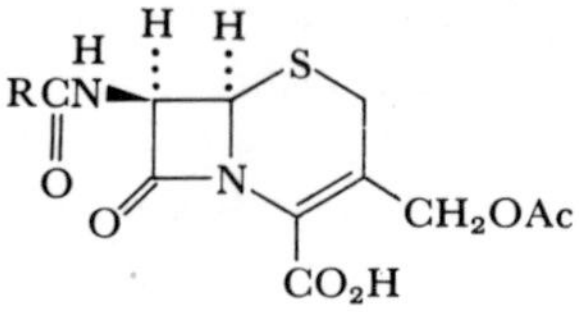

Cephalosporin

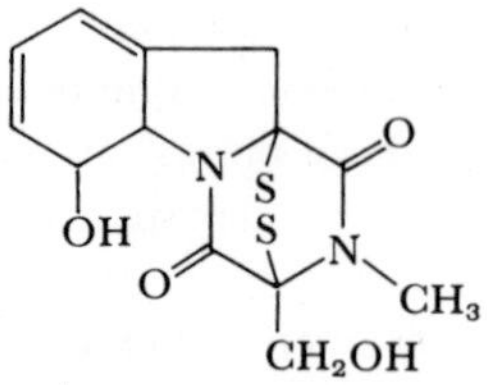

Gliotoxin

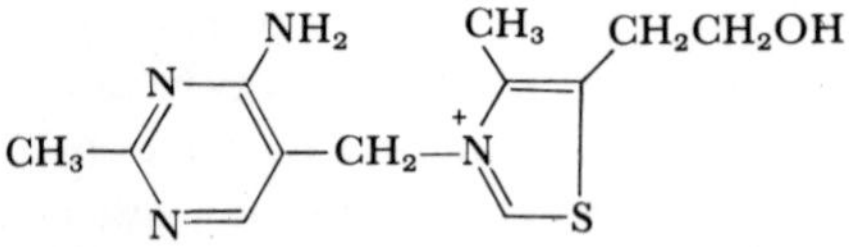

Thiamine (a B vitamin)

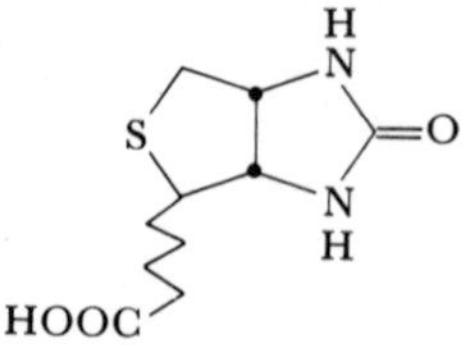

Biotin

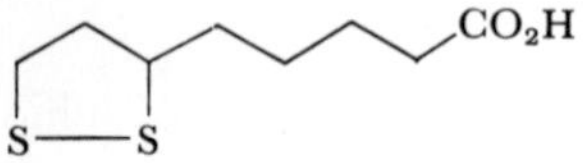

α-Lipoic acid

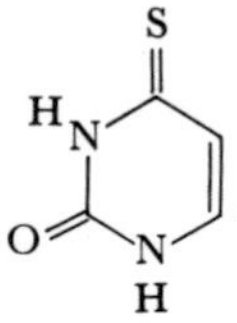

4-Thiouracil

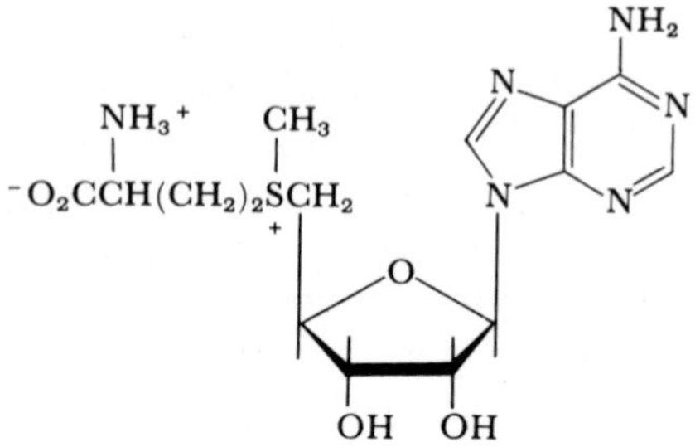

S-Adenosyl-L-methionine

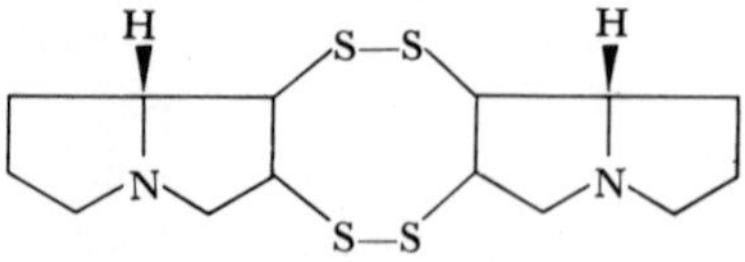

Cassipourine

$C_2H_5SCH_2CH_2CH(NH_2)COOH$

Ethionine

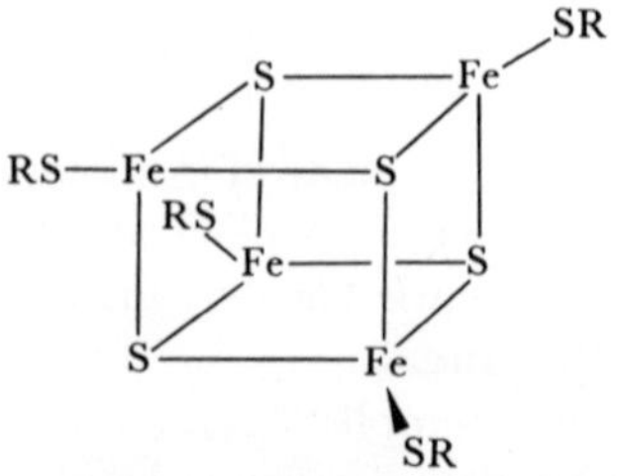

A portion of the active site of a ferrodoxin
(RS = cysteinyl group)

Chart 1-2. Some Smaller Organosulfur Molecules of Natural Origin[a]

$CH_3CH{=}CHS(O)CH_2CH(NH_2)COOH \longrightarrow C_2H_5CH{=}S{=}O$

(Onion)

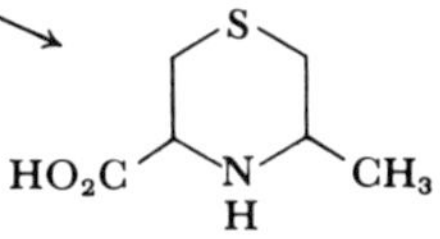

$\searrow$

S

HO_2C N CH_3

H

$n\text{-}C_3H_7S(O)CH_2CH(NH_2)COOH \longrightarrow n\text{-}C_3H_7SS\text{-}n\text{-}C_3H_7,\ n\text{-}C_3H_7SO_2S\text{-}n\text{-}C_3H_7$

(Onion)[b]

$n\text{-}C_3H_7SSS\text{-}n\text{-}C_3H_7$

$CH_2{=}CHCH_2S(O)CH_2CH(NH_2)COOH \longrightarrow [CH_2{=}CHCH_2SOH] + (CH_2{=}CHCH_2S)_2$

(Garlic)

$\downarrow$

$CH_2{=}CHCH_2S(O)SCH_2CH{=}CH_2$

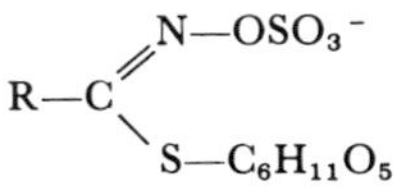

$N{-}OSO_3^-$

$R{-}C \longrightarrow RN{=}C{=}S$

$S{-}C_6H_{11}O_5$

$R = CH_3S(O)CH{=}CHCH_2CH_2$

(Radish)

$CH_2{=}CHCH_2SCN$ (Penny cress)

$CH_3SCH_2CH_2COOCH_3$ (Pineapple)

$CH_3SCH_2SCH_3$ (Truffle)

$(CH_3)_2C{=}CHCH_2SC(O)CH(CH_3)_2$

(Agathosma)[c]

$S(O)CH_2CH(NH_2)COOH$

CH_2

$SCH_2CH(NH_2)COOH$

(Djenkol bean)

$CH_3SCH(C_2H_5)SSCH{=}CHCH_3$

(Asafetida)[d]

$(C_2H_5)_2NC(S)SSC(S)N(C_2H_5)_2$

(The mushroom *Coprinus atramentarius*)

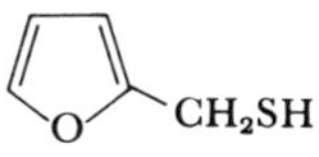

O CH_2SH

(Coffee)

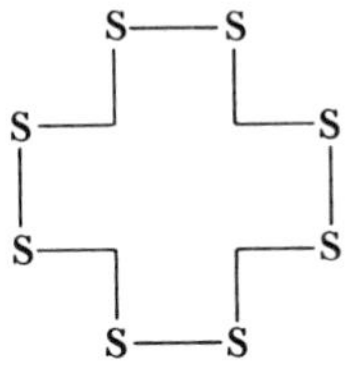

S—S, S, S, S, S, S—S

(Red algae *Chondria californica*)[e]

X

S

S—S

Y

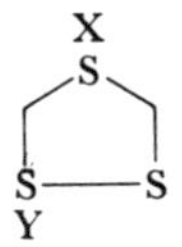

X = Y = lone pair

X = lone pair, Y = O

Y = lone pair, X = O

(Red algae *Chondria californica*)[e]

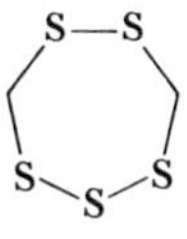

S—S

S S

S

(Shiitake mushroom)

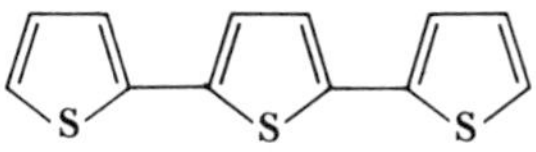

S S S

(Marigold)

COOH

SH SH

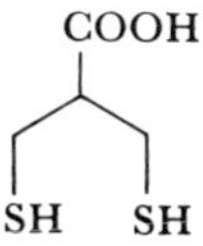

(Asparagus)

(continued)

Chart 1-2 (*continued*)

$(n\text{-}C_8H_{17}C(O)CH_2CH_2SS\text{-})_2$ (Hawaiian algae *Dictyopteris plagiogramma*)[f]	$CH_3SO_2CH_3$ (Various algae and lichen)	CH_3SO_2H (Cauliflower)
$(CH_3)_2\overset{+}{S}CH_2CH_2CO_2^-$ (Various marine algae)	CH_3SSSCH_3 (Ponerine ants)[g]	*trans*-$CH_3CH{=}CHCH_2SH$ (Skunks)[h]
$CH_3C(O)C(O)CH_2CH_2SCH_3$ (Striped hyena)[i]	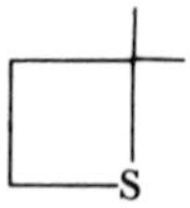 [thietane structure] (Mink)[j]	CH_3SSCH_3 (Hamsters)[k]

[a] Unless otherwise noted information was obtained from Kjaer (1977), Ohloff (1969), or Richmond (1973).
[b] Boelens (1971).
[c] Rivett (1974).
[d] Naimie (1972).
[e] Wratten (1976).
[f] Moore (1971).
[g] Casnati (1967).
[h] Andersen (1975).
[i] Wheeler (1975).
[j] Schildknecht *et al.* (1976).
[k] Singer *et al.* (1976).

comestibles. In some instances these sulfur compounds may also serve by their odor and taste to repel predators or to act for the plant as resistance factors against infection by microorganisms. In many cases these low molecular weight substances occur naturally in an odorless combined form (e.g., as peptides or glycosides) and are released enzymatically when the plant tissue is injured, i.e., during cooking or attack by a predator. Certain of these sulfur compounds, reportedly of natural origin, may in fact be artifacts formed during the isolation procedures through reactions of unstable precursors.

A partial list of plants and vegetables with conspicuous sulfur content reads like a grocery list: onion, garlic, leek, chive, asparagus, cabbage, cress, turnip, radish, horseradish, mustard, truffle, pineapple, coffee, tea, and various algae and mushrooms (Richmond, 1973). A number of animals and insects have also been found to possess unusual small organosulfur molecules. These compounds found in certain species of ants, skunks, hamsters, mink, and hyenas may serve as defensive secretions or sex attractants. Chart 1-2 indicates some of the structurally diverse types of organic sulfur compounds found in nature, a natural source and where appropriate the stable precursor found in undamaged tissue. Sulfur-rich plants have attracted the attention of biologists working in diverse areas such as chemotaxonomy and ethnobiology. The chemotaxonomist, aided by the powerful analytical tool of coupled gas chromatography–mass spectrometry, has used the natural distribution of organosulfur compounds in establishing relationships among plants. The ethnobiologist, in studying folk medicine in primitive cultures, has noted that certain of the sulfur-rich plants, particularly garlic and onion, were objects of reverence and were used to ward off disease and evil spirits (and werewolves!). That the folk medicine was not without a certain merit is suggested by recent interest in the physiological effects of garlic and onions, for example, in controlling hyperglycemia and hyperlipemia (Jain *et al.*, 1973; Bordia and Bansal, 1973).

1.4 Nomenclature and Synthesis of Organosulfur Compounds

The nomenclature of organosulfur compounds can be bewildering both for the neophyte and for the seasoned sulfur chemist. The difficulties stem not only from the myriad of known structural types but also from disagreement among the "experts" on the proper choice of names. For example, compounds of type **1-1** have been variously named sulfimides (IUPAC), sulfilimines (*Chemical Abstracts*), iminosulfuranes, sulfinimines, and sulfimines while compounds of type **1-2** are termed sulfimides (*Chemical Abstracts*), *N*-sulfonylamines (IUPAC), and *N*-sulfonyl imides (ACS; Fletcher, 1974). Table 1-1

$$R_2S{=}NR' \qquad RN{=}SO_2$$
1-1 **1-2**

indicates the various names used for the fifty most common organosulfur structural types. This table is certainly not complete, for it omits various derivatives of the listed compounds such as certain S-oxides of disulfides, polysulfides, and thioacetals; hydrodisulfides (RSSH); sulfenic, sulfinic, and sulfonic anhydrides, peroxides, and other derivatives; esters, anhydrides, halogen derivatives, etc., of thiocarboxylic acids; organosulfur compounds containing elements other than oxygen, nitrogen, and halogen. References in Appendix A should be consulted for information on the nomenclature of these and other organosulfur compounds.

Another sometimes troublesome facet of organosulfur nomenclature is the designation of valence states and coordination numbers. Andersen *et al.* (1970) has suggested an unambiguous nomenclature in which the number of ligands define the coordination number and the valence (the number of bonds to sulfur plus the formal charge on sulfur) is enclosed in parenthesis. Electron pairs are not counted as ligands. To illustrate the system with

Table 1-1. Nomenclature of Organosulfur Structural Types

Elements present and entry	Class name[a]	Specific example	Name(s)[a]	General reference[e]
CHS				
1	Disulfides	CH_3SSCH_3	Dimethyl disulfide,[b,c] (methyldithio)methane,[b,c] dimethyldisulfane[c]	Field (1977), Kice (1971)
2	Polysulfides	$CH_3S_4CH_2CH_3$	Ethyl methyl tetrasulfide,[b] ethylmethyltetrasulfane,[c] (methyltetrathio)ethane[b,c]	Field (1977)
3	Sulfonium salts	$Ph_3S^+Br^-$	Triphenylsulfonium bromide[b,c]	Marino (1976), Stirling (1977)
4	Sulfuranes	$(C_6F_5)_4S$	Tetrakis(pentafluorophenyl)sulfur,[c] tetrakis(pentafluorophenyl)sulfurane	Martin and Perozzi (1974)
5	Thioethers	$CH_3SCH_2CH_2CH_3$	Methyl propyl sulfide,[b] 1-(methylthio)propane[b,c]	Brandsma and Arens (1967), Tagaki (1977)
6	Thioacetals [mercaptals]	$CH_3CH_2CH(SC_2H_5)_2$	1,1-Bis(ethylthio)propane[b]	Brandsma and Arens (1967)
7	Thioaldehydes, thioketones	$CH_3C(S)CH_3$	2-Propanethione,[b,c] dimethyl thioketone[b,c] [thioacetone][b]	Ohno (1977)
		$CH_3C(S)CH_2$–C_6H_4–COOH	*p*-(2-Thioxopropyl)benzoic acid	
8	Thiols [mercaptans]	$C_6H_5CH_2SH$	Phenylmethanethiol[b,c] [α-toluenethiol][b] [benzyl mercaptan], 2,3-dimercaptobutanedoic acid	Patai (1974), Ohno and Oae (1977)
CHSO				
9	Oxysulfonium salts	$(C_2H_5)_2S^+—OCH_3\ Cl^-$	Diethyl(methoxy)sulfonium chloride	Marino (1976)
10	Oxysulfoxonium salts	$(CH_3)_2\overset{+}{S}(=O)—O$–$C_6H_4$–$CH_3$	Dimethyl(*p*-tolyloxy)sulfoxonium tetrafluoroborate[d]	Durst and Tin (1971), Chalkley *et al.* (1970)

11	Sulfates	$(CH_3O)_2SO_2$	Dimethyl sulfate[b,c]	Kaiser (1977)
12	Sulfenate esters	$CH_3S—O—C_2H_5$	Ethyl methanesulfenate	Kühle (1973)
13	Sulfenes	$(CH_3)_2C{=}SO_2$	2-Propanethione *S*,*S*-dioxide	King (1975)
14	Sulfinate esters	$CH_3CH_2S(O)OCH_3$	Methyl ethanesulfinate	Oae and Kunieda (1977)
15	Sulfines	$CH_2{=}SO$	Thiomethanal *S*-oxide, thioformaldehyde *S*-oxide, sulfine	Block *et al.* (1976a)
16	Sulfites	$(C_2H_5O)_2S{=}O$	Diethyl sulfite	Oae and Kunieda (1977)
17	Sulfonate esters	$C_6H_5SO_2OCH_3$	Methylbenzene sulfonate	Kaiser (1977)
18	Sulfones	$C_2H_5SO_2C_2H_5$	(Ethylsulfonyl)ethane, diethyl sulfone	Truce *et al.* (1977)
19	Sulfoxides	$C_2H_5S(O)C_2H_5$	(Ethylsulfinyl)ethane, diethyl sulfoxide	Oae (1977)
20	Sulfoxonium salts [oxosulfonium salts]	$(CH_3)_3S^+{=}O\ I^-$	Trimethylsulfoxonium iodide[d]	Kamiyama *et al.* (1973)
21	Sulfoxylates	$C_2H_5OSOC_2H_5$	Diethyl sulfoxylate[c]	Thompson (1965)
22	Sulfurane oxides	$(C_6H_5)_2(CH_3O)_2S{=}O$	Dimethoxydiphenylsulfurane oxide; dimethoxydiphenylsulfur oxide	Perozzi and Martin (1974)
23	Thiocarbonates, dithiocarbonates, trithiocarbonates [thionocarbonates][e]	$CH_3OC(S)SCH_3$	*O*,*O'*-Dimethyl thiocarbonate[b] [dimethyl thionocarbonate]	
		$CH_3OC(O)SCH_3$	*O*,*S*-Dimethyl thiocarbonate[b]	
24	Thiosulfinates	$C_2H_5S(O)SCH_3$	Methyl ethanethiosulfinate	Block and O'Connor (1974), Isenberg and Grdinic (1973), Kice (1971)
CHSO (acids)				
25	Sulfenic acids	C_2H_5SOH (and/or $C_2H_5S(O)H$)	Ethanesulfenic acid	Kühle (1973), Block and O'Connor (1974)
		$HOS—C_6H_4—COOH$	4-Sulfenobenzoic acid	

(continued)

Table 1-1 (*continued*)

Elements present and entry	Class name[a]	Specific example	name(s)[a]	General reference[e]
26	Sulfinic acids	$C_2H_5S(O)OH$	Ethanesulfinic acid	Oae (1977)
		$HOS(O)-C_6H_4-COOH$	4-Sulfinobenzoic acid	
27	Sulfonic acids	$C_2H_5SO_3H$	Ethanesulfonic acid	
		$HO_3S-C_6H_4-COOH$	4-Sulfobenzoic acid	
28	Thiocarboxylic acids	$CH_3C(O)SH$	Ethanethioic *S*-acid[b,c] [thioacetic acid]	Janssen (1969)
		$CH_3C(S)OH$	Ethanethioic *O*-acid[b,c] [thionoacetic acid]	
		$CH_3C(S)SH$	Ethanedithioic acid[b,c]	
		$C_6H_5C(O)SH$	Benzenecarbothioic *S*-acid[b,c] [thiobenzoic acid]	
CHSHal				
29	Halosulfonium salts	$(CH_3)_2S^+—Cl\ Cl^-$	Chlorodimethylsulfonium chloride[d]	Marino (1976), Magee (1971)
30	Organosulfur halides	CH_3SCl_3	Methylsulfur trichloride,[b] trichloro(methyl)sulfur[d]	Marino (1976)
		$(C_6H_5)_2SF_4$	Tetrafluorodiphenylsulfur[d]	
31	Sulfenyl halides	CH_3SCl	Methanesulfenyl chloride,[b] methylsulfur monochloride[b]	Marino (1976), Kühle (1973), Danehy (1971), Russ and Douglass (1971)
CHSOHal				
32	Chlorosulfoxonium salts	$(CH_3)_2S^+(O)Cl\ Cl^-$	Chlorodimethylsulfoxonium chloride[d]	Durst and Tin (1971)

33	Sulfinyl halides	$CH_3S(O)Cl$	Methanesulfinyl chloride	Russ and Douglas (1971), Magee (1971)
34	Sulfonyl halides	$CH_3-C_6H_4-SO_2Br$	4-Toluenesulfonyl bromide[c] [tosyl bromide]	Danehy (1971), Magee (1971), Russ and Douglass (1971)
CHSN				
35	Azasulfonium salts	$(CH_3)_2S^+—N(C_2H_5)_2\ Cl^-$	*N*,*N*-diethylaminodimethylsulfonium chloride	Marino (1976)
36	Isothiocyanates	C_2H_5NCS	(Thiocarbonylamino)ethane [ethyl isothiocyanate]	Reid (1966)
37	Sulfenamides	$3—O_2NC_6H_4SNH_2$	3-Nitrobenzenesulfenamide	Kühle (1973)
38	Sulfimides[b,c] Sulfilimines[d]	$(C_2H_5)_2S{=}NC_6H_5$	*S*,*S*-Diethyl-*N*-(phenyl)sulfimide[b,c] *S*,*S*-Diethyl-*N*-(phenyl)sulfilimine[d]	Yoshimura *et al.* (1976), Oae (1977), Roesky (1971)
39	Sulfur diimides[d] (variously punctuated)	$C_2H_5N{=}S{=}NC_2H_5$	Diethyl sulfur diimide	Roesky (1971), Kresze (1975)
40	*S*,*S*-Diorgano-sulfodiimides,[b,e] sulfone diimines[e]	$(C_2H_5)_2S(=NH)_2$	*S*,*S*-Diethyl sulfodiimide,[e] diethylsulfone diimine	Haake (1976), Roesky (1971)
41	Thiocyanates	C_2H_5SCN	Ethyl thiocyanate	Reid (1966)
42	Thioamides	$CH_3CH_2CH_2C(S)NH_2$ $C_6H_5C(S)NH_2$	Butanethioamide[c] Benzenecarbothiomide[c] [thiobenzamide]	Walter and Voss (1970)
CHSON				
43	Azasulfoxonium salts	$(CH_3)_2S^+(=O)—N(CH_3)_2\ BF_4^-$	(Dimethylamino)dimethylsulfoxonium tetrafluoroborate	Johnson (1973)
44	Diazasulfoxonium salts	$CH_3S^+(=O)[N(CH_3)_2]_2\ Br^-$	Bis(dimethylamino)methylsulfoxonium bromide	Schroek and Johnson (1971)

(continued)

Table 1-1 (*continued*)

Elements present and entry	Class name[a]	Specific example	and name(s)[a]	General reference[e]
45	Oxyazasulfoxonium salts	$C_3H_7S^+(=O)(OC_6H_5)-N(CH_3)_2\ Cl^-$	(Dimethylamino)(phenoxy)propylsulfoxonium chloride	Schroek and Johnson (1971)
46	Sulfinamides	$CH_3S(O)NHCH_2CH_3$	*N*-Ethylmethanesulfinamide	
47	Sulfinylamines,[b] sulfinylimides,[c] thionylimides[d]	$PhN{=}S{=}O$	*N*-Sulfinylaniline,[b,d] *N*-phenylsulfinylimide[c]	Roesky (1971)
48	Sulfonamides	$CH_3SO_2NHC_2H_5$	*N*-Ethylmethanesulfonamide	
49	Sulfonylamines,[b] sulfonylimides,[c] sulfimides[d]	$CH_3CH_2CH_2N{=}SO_2$	*N*-Sulfonylpropylamine,[b] *N*-propylsulfonyl imide,[c] propylsulfimide[d]	
50	Sulfoximides[b]	$(C_2H_5)_2S({=}O){=}NH$	Diethylsulfoximide,[b] diethylsulfoximine[d]	Kennewell and Taylor (1975), Truce *et al.* (1977), Roesky (1971)

[a] Names in brackets are no longer recommended.
[b] IUPAC nomenclature: IUPAC (1971).
[c] ACS nomenclature: Fletcher *et al.* (1974).
[d] Chemical Abstracts nomenclature: Chemical Abstracts, 1972 (and subsequent years); also see Loening (1972).
[e] Not all of these classes are illustrated by examples here.

entries from Table 1-1: sulfonium salts (entry 3) and sulfoxides (entry 19) contain tricoordinate sulfur(IV), sulfuranes (entry 4), tetracoordinate sulfur(IV), and sulfones (entry 18), tetracoordinate sulfur(VI).

Table 1-1 also includes one or more recent reference dealing with the synthesis, structure, properties, and reactions of each of the listed structural types. Extensive coverage of the earlier literature on the synthesis of organosulfur compounds is to be found in Houben–Weyl, the volumes by Reid, and the chapter by Connor, while recent advances in this area are to be found in the Chemical Society's Specialist Periodical Reports on sulfur, selenium, and tellurium (see Appendix A for all references).

Chart 1-3 indicates in summary fashion methods for the synthesis and interconversion of some of the simpler types of organosulfur compounds; the numbers are keyed for purposes of nomenclature to the entries in Table 1-1.

Chart 1-3. Synthesis of Organosulfur Compounds

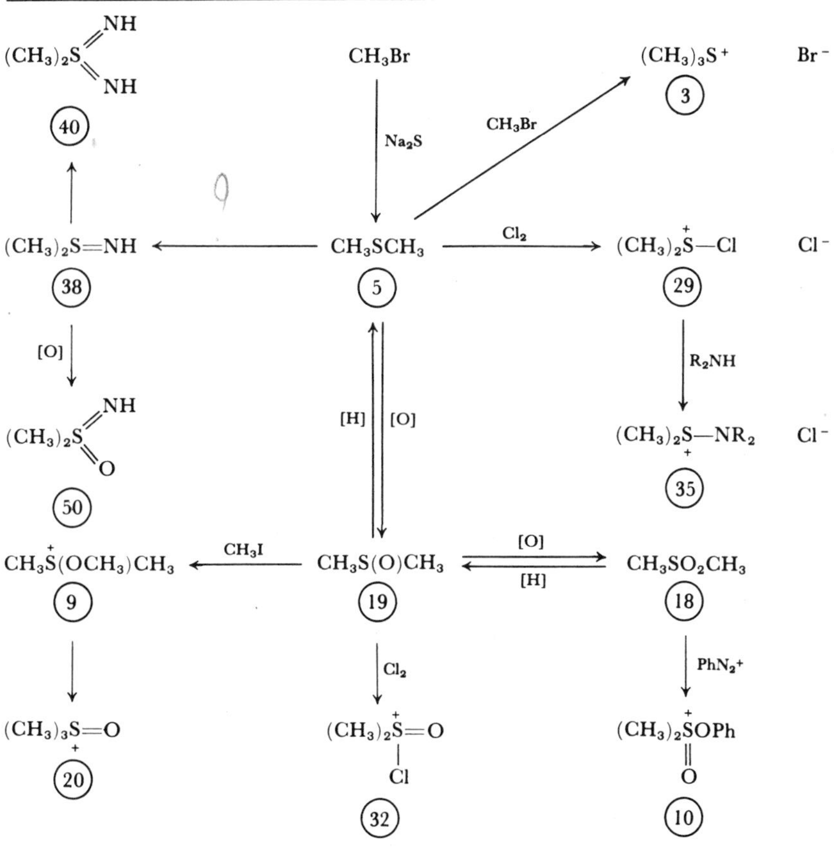

Chart 1-3 *(continued)*

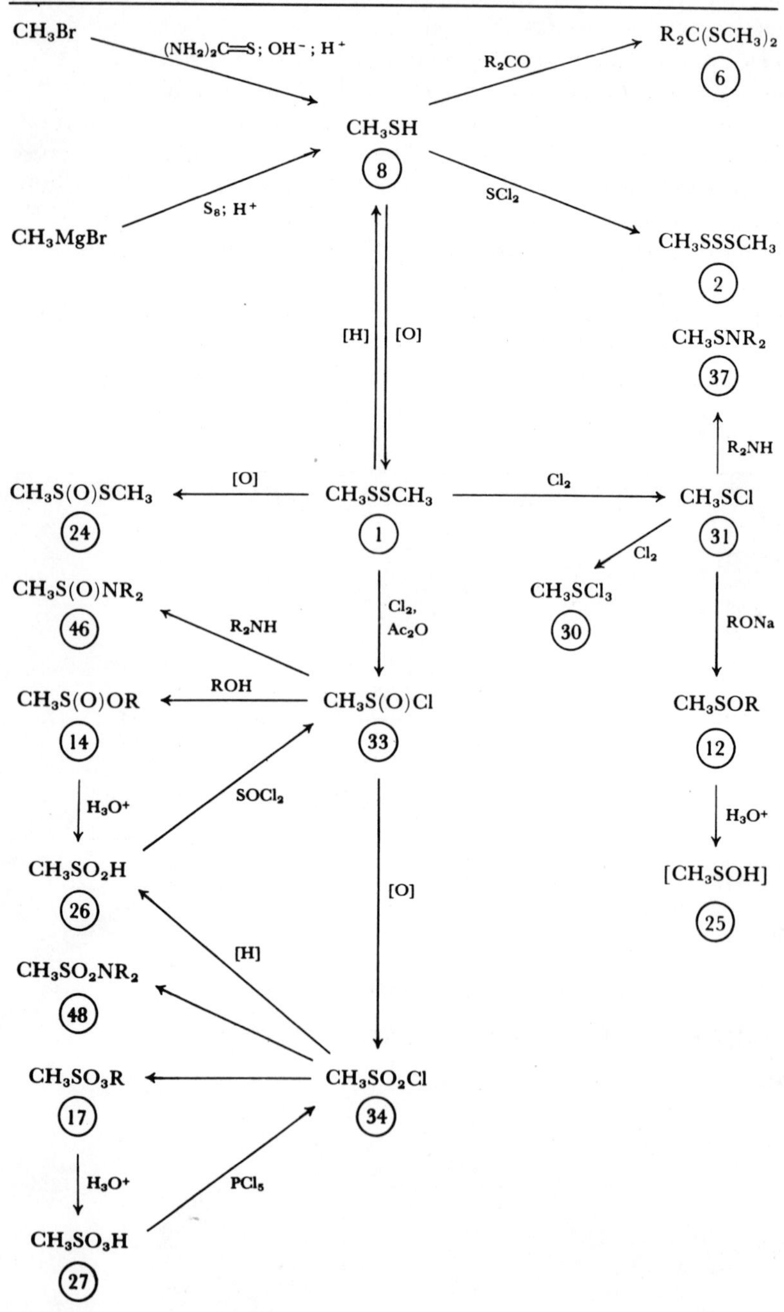

Since oxidation–reduction reactions play such an important role in many of these interconversions, a separate listing is given in Table 1-2 of some of the oxidizing and reducing (or deoxygenating) reagents commonly employed in organosulfur chemistry. The choice of a particular reagent depends on a careful consideration of such factors as the scale and economics of the reaction, the ease of isolation of product, the presence of other functional groups that might be adversely affected by the reagent, the deactivation of the reaction center by steric or electronic effects, the instability or high reactivity of the desired product, the need for stereochemical control, and the desire to introduce isotopic labeling at the reaction center.

Certain of these points may be illustrated with the sequence of reactions in Eq. 1-1 (Block, *et al.*, 1966c). Conversion of **1-4** to **1-3** required a particularly gentle reagent since both **1-3** and **1-4** are quite sensitive to acids and bases. Use of the borane–tetrahydrofuran complex at room temperature effected the

1-3 **1-4** **1-5** **1-6** **1-7** **1-9** **1-8** (1-1)

a 2 mol equiv 1 *M* THF–BH_3, 24 hr at 25°
b $PhICl_2$, CH_3CN, $(C_2H_5)_3N$, $H_2{}^{18}O$
c $PhICl_2$, Py, H_2O, −30° or MCPBA, CH_2Cl_2, 0°
d 0.67 mol equiv $KMnO_4$, $MgSO_4$, acetone, −30°
e CH_3CO_3H, $CHCl_3$, 0°
f 30 mol equiv CH_3CO_3H, 100°, 4 hr

1-10

Table 1-2. Oxidation–Reduction Reactions of Organosulfur Compounds

Transformation	Reagents	References
$R_2S \xrightarrow{[O]} R_2SO$	(a) N_2O_5, $NaIO_4$, H_2O_2/acetone, MCPBA, CrO_3/Py, O_3, PhIO, $PhICl_2$, *t*-C_4H_9OCl, *i*-C_3H_7OCl; (b) *N*-chlorotriazole; (c) $HAuCl_4$; (d) 2,4,4,6-tetrabromocyclohexadienone; (e) Br_2/OAc^-; (f) *t*-C_4H_9OOH/V(IV); (g) H_2O_2/V_2O_5/*t*-C_4H_9OH; (h) Ce(IV); (i) SO_2Cl_2/silica gel/H_2O; (j) $AcONO_2$; (k) *N*-chloro-nylon 66; (l) $^1O_2$	(a) Rigau *et al.* (1970); (b) Siegl and Johnson (1970); (c) Bordignon *et al.* (1973); (d) Calo *et al.* (1971); (e) Cox and Gibson (1973); (f) Curci *et al.* (1974); (g) Handy *et al.* (1969); (h) Ho and Wong (1972); (i) Hojo and Masada (1976); (j) Louw *et al.* (1976); (k) Sato *et al.* (1972); (l) Foote and Peters (1971)
$R_2SO \xrightarrow{[H]} R_2S$	(a) $Fe(CO)_5$; (b) $LiAlH_4$; (c) $HBCl_2$; (d) $BH_3 \cdot THF$; (e) Cl_3SiH; (f) Si_2Cl_6; (g) $NaBH_4/CoCl_2$; (h) $NaBH_3CN$–18-crown-6; (i) PBr_3/DMF; (j) CH_3COCl; (k) 2-phenoxy-1,3,2-benzodioxaphosphole; (l) $TiCl_3$; (m) $SnCl_2$/HCl; (n) $NaHSO_3$; (o) R_3P, $(C_2H_5O)_2PS_2H$, RCS_2H; (p) dithiothreitol; (q) $h\nu$/HOAc; (r) 1,3,2-benzodioxaborole (catecholborane); (s) $K_3W_2Cl_9$	(a) Alper and Keung (1970); (b) Anastassiou *et al.* (1975); (c) Brown and Ravindran (1973); (d) Block *et al.* (1976c); (e) Oae *et al.* (1972a); (f) Chan *et al.* (1969); (g) Naumann *et al.* (1969); (h) Chaser (1971); (i) Durst *et al.* (1974); (j) Kukolja *et al.* (1976); (k) Numata and Oae (1973); (l) Dreux *et al.* (1974); (m) Ho and Wong (1973a, b); (n) Johnson *et al.* (1972); (o) Oae (1972a, b, c); (p) Polzhofer and Ney (1971); (q) Gurria and Posner (1973); (r) Kalbalka *et al.* (1977); (s) Nuzzo *et al.* (1977).
$R_2SO \xrightarrow{[O]} R_2SO_2$	(a) O_2/Ir or Rh catalyst; (b) H_2O_2/HOAc, CrO_3; (c) $KMnO_4/MgSO_4$, CH_3CO_3H.	(a) Henbest and Trocha–Grimshaw (1974); (b) Connor (1943); (c) Block (1976c)
$R_2SO_2 \xrightarrow{[H]} R_2SO$ or R_2S	(a) (*i*-$C_4H_9)_2AlH$; (b) $LiAlH_4$; (c) S.	(a) Gardner *et al.* (1973); Anastassiou *et al.* (1975); (b) Siegl and Johnson (1970); (c) Connor (1943)
$R_2S(OR')_2 \xrightarrow{[O]} R_2S(O)(OR')_2$	RuO_4	Perozii, 1972.

$RSH \xrightarrow{[O]} RSSR$	(a) cobalt maleonitriledithiolate; (b) $K_2Fe(CN)_6$; (c) $K_3Fe(CN)_6$, H_2O_2/KI, I_2/R_3N or OH^-; (d) polymer-SH; (e) dithiothreitol/NH_3; (f) Mn(III)(acac)$_3$; (g) $FeCl_3$; (h) DMSO; (i) dithiobis(thioformate); (j) $CuSO_4$	(a) Dance *et al.* (1974); (b) Walti and Hope (1971); (c) Field and Khim (1972); (d) Gorecki and Patchornik (1973); (e) Meienhofer *et al.* (1971); (f) Nakaya *et al.* (1970); (g) Rae and Pojer (1971); (h) Sandler and Karo (1980); (i) Stout *et al.* (1974); (j) Connor, (1943); general reference, Capozzi and Modena (1974)
$RSH \xrightarrow{[O]} RSO_2H$	MCPBA	Filby *et al.* (1973)
$RSH \xrightarrow{[O]} RSO_3H$	$KMnO_4$, CrO_3, Br_2/H_2O, HNO_3, H_2O_2	Sandler and Karo (1968)
$RSSR \xrightarrow{[H]} RSH$	(a) Dithiothreitol, $NaBH_4$, H_3PO_2, $LiAlH_4$, R_3P, glucose-OH^-, Sn/H^+, $Zn/H^+Na_2S_2O_4$, Na–Hg, $NaSO_3$; (b) Ph_3P/H_2O	(a) Field (1977); (b) Overman *et al.* (1974)
$RSSR \xrightarrow{[O]} RS(O)SR$	(a) H_2O_2, $KMnO_4$, CrO_3, KIO_4; (b) CH_3CO_3H, $O_2/h\nu$/sensitizer, $(PhO)_3PO_3$	(a) Field and Khim (1972); (b) Block and O'Connor (1974); Murray *et al.* (1971)
$RSSR \xrightarrow{[O]} RSO_2SR$	$KIO_4/I_2/i\text{-}C_3H_7OH$	Field and Khim (1972)
$RSCN \xrightarrow{[H]} RSH$	$LiAlH_4$	Wilson *et al.* (1971)
$RSCN \xrightarrow{[O]} RSO_2CN$	MCPBA	Pews and Corson (1969)
$R_2S{=}NTs \xrightarrow{[H]} R_2S$	RCS_2H, $(RO)_3PS_2H$	Oae *et al.* (1972b, c)
$R_2S{=}NTs \xrightarrow{[O]} R_2S(=O)=NTs$	(a) MCPBA; (b) $H_2O_2/NaOH$; (c) $KMnO_4/Py$	(a) Cram *et al.* (1970); (b) Johnson *et al.* (1975); (c) Siegl and Johnson (1971)
$RSNR_2 \xrightarrow{[O]} RS(O)NR_2$	MCPBA	Harpp and Back (1973)
$RSO_2Cl \longrightarrow RSO_2H$	(a) Na_2SO_3; (b) H_2S; (c) Zn/H_2O	(a) Sandler and Karo (1968); (b) Schöllkopf and Hilbert (1973); (c) Connor (1943)

desired reduction in 70% yield. Reoxidation of **1-3** with $PhICl_2$ in acetonitrile/$H_2{}^{18}O$ and triethylamine gave **1-4**-^{18}O, which was required for microwave studies. Compound **1-4** could be converted into sulfone **1-7** in 96% yield by treatment with $KMnO_4/MgSO_4$ in acetone at −20°. The remarkable selectivity of this oxidation (sulfide is more easily oxidized than sulfoxide with most oxidants) may reflect coordination of sulfoxide oxygen by permanganate as shown in **1-10**. Oxidation of **1-4** with iodobenzene dichloride or metachloroperbenzoic acid (MCPBA) gave bissulfoxides **1-5** (*cis*) and **1-6** (*trans*) in ratios of 3:1 and 2:3, respectively. Treatment of sulfone **1-7** with peracetic acid at 0° gave **1-8** in 90% yield while oxidation of compounds **1-4**–**1-8** with excess peracetic acid at 100° for several hours gave bissulfone **1-9** in high yields. Stereoselectivity in the oxidation of sulfides, as seen in the formation of **1-5** and **1-6** or more dramatically in the oxidation of 4-*tert*-butylthiane **1-11** (Johnson and McCants, 1965; Barbieri *et al.*, 1968), has been attributed to thermodynamic or kinetic control (Johnson and McCants, 1965).

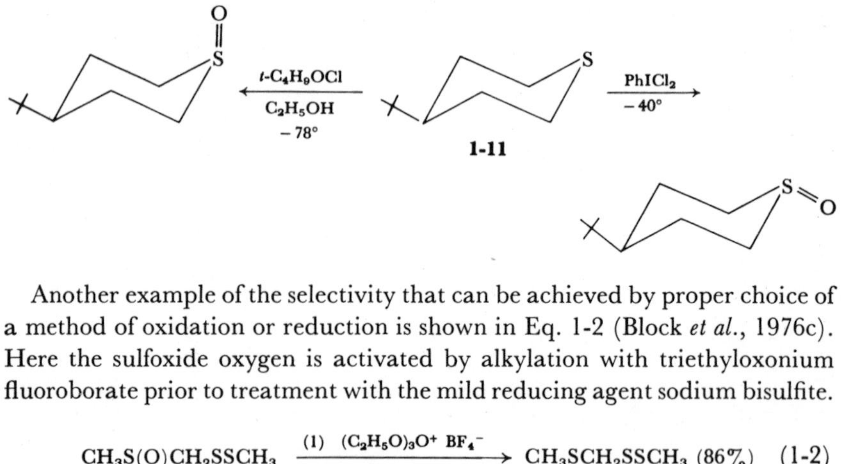

1-11

Another example of the selectivity that can be achieved by proper choice of a method of oxidation or reduction is shown in Eq. 1-2 (Block *et al.*, 1976c). Here the sulfoxide oxygen is activated by alkylation with triethyloxonium fluoroborate prior to treatment with the mild reducing agent sodium bisulfite.

$$CH_3S(O)CH_2SSCH_3 \xrightarrow[\text{(2) } NaHSO_3,\ 0°,\ 30\ \text{min}]{\text{(1) } (C_2H_5O)_3O^+\ BF_4^-} CH_3SCH_2SSCH_3\ (86\%) \quad (1\text{-}2)$$

1.5 3d Orbital Effects

Sulfur can form stable compounds such as SF_4 and SF_6 involving higher valences than seen with oxygen. Following Pauling (1939), the concept of *octet* or *valence shell expansion* has been invoked to describe the bonding in these "violators" of the Lewis–Langmuir octet principle. Thus, utilization of the unfilled sulfur $3d_{z^2}$ orbital allows trigonal bipyramidal sp^3d hybridization for SF_4 while use of both the $3d_{z^2}$ and $3d_{x^2-y^2}$ orbitals of sulfur gives the hybrid sp^3d^2 orbitals with the necessary octahedral symmetry for SF_6. The postulated hybridization schemes for SF_4 and SF_6 involve dσ bonding (see **1-12**). For molecules such as sulfoxides and sulfones, which in their neutral resonance forms are also "violators" of the octet principle, dπ bonding can be illustrated by overlap of an oxygen $2p_y$ orbital with a sulfur $3d_{xy}$ orbital in dimethylsulfoxide (**1-13**).

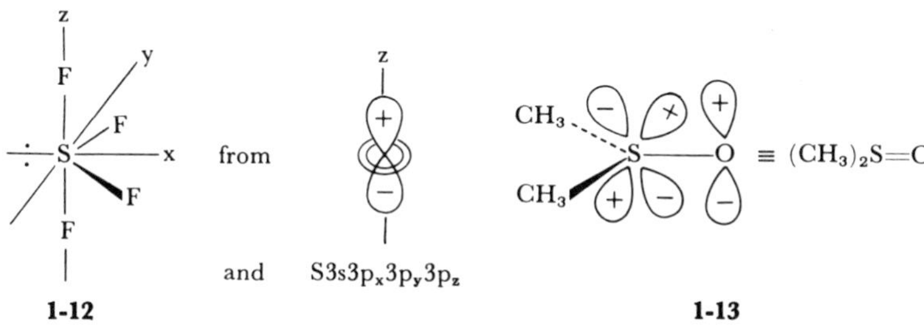

There are difficulties associated with the postulation of bonding involving outer d orbitals for these and related compounds. Thus, if the 3d orbitals are energetically too far above the 3s and 3p orbitals, they will not mix to any great extent to form either dσ or dπ type bonds. Furthermore, if the radial maxima of the d orbitals are large with respect to the bond lengths, the d orbitals can make only small contributions to atom–atom overlap in the bonding region. However, the presence of highly electronegative substituents such as fluorine or oxygen on sulfur can increase the effective nuclear charge on sulfur and contract all orbitals and lower all orbital energies (Craig *et al.*, 1954; Coulson, 1969). The 3s- and 3p-orbital sizes seem to be far less sensitive to net nuclear charge than the 3d-orbital size (Coulson, 1969, 1973).

There is a voluminous literature in which diverse experimental facts involving both dicoordinate and higher coordinate sulfur are "explained" by participation of 3d orbitals in bonds. More recent reviews of this subject include Cilento (1960), Price and Oae (1962), Cram (1965), Salmond (1968), Coffen (1969), Mitchell (1969), Seebach (1969), Coulson (1969, 1973), Peterson (1972), Dyatkina and Klimenko (1973), Brill (1973), and Kwart and King (1977). In some instances it is thought that d orbitals function as nothing more than "finagle factors." Quoting Hoffman *et al.* (1972), "Far too often have 3d orbitals been invoked as a kind of theoretical *deus ex machina* to account for facts apparently otherwise inexplicable. Our attitude toward 3d orbitals is pragmatic. We begin by working without them. We then analyze the way they would perturb the valence orbital picture *if* they were active. . . ."

An excellent illustration of the type of approach advocated by Hoffmann is the three-center four-electron bond picture of *hypervalent* molecules developed by Rundle (1963) and by Musher (1969). Hypervalent molecules are those in which the central atom exceeds the number of valences allowed by the traditional Lewis–Langmuir theory (Musher, 1969), for example, as seen with SF_4 with its lone pair and four bonding pairs about sulfur. To illustrate the hypervalent bonding picture for this molecule, consider the hypothetical conversion of SF_2, unhybridized with 3s and 3p lone pairs, to SF_4 via addition of two new fluorine ligands *colinearly and along the axis* of the non-bonded p electron pair (**1-14**; Musher, 1969). The resulting three-center four-electron hypervalent bond involves the combination of a doubly occupied bonding and a purely ligand nonbonding orbital (see **1-15**; Koutecky and Musher, 1974). The addition of sulfur 3d orbitals has been shown to have a negligible effect here (Koutecky and Musher, 1974).

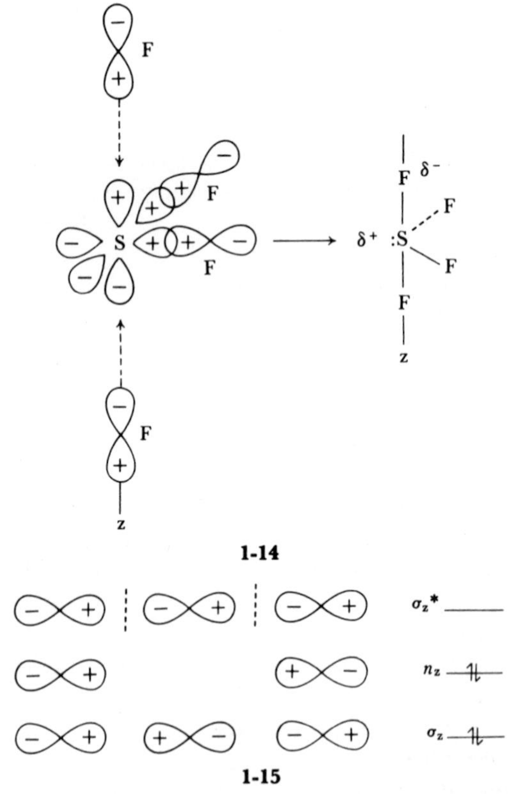

1-14

1-15

Ambiguities exist in discussing the participation of outer 3d orbitals in chemical bonding. Outer atomic orbitals, whether 3d orbitals of sulfur or 2p orbitals of hydrogen, can play the role of "polarization functions" (Dyatkina and Klimenko, 1973). That is, in describing the polarization or perturbation of a $p\pi$ orbital by a uniform field, one can approximate the wave function of the perturbed state by a superposition of some dπ orbital on the original pπ orbital (Coulson, 1969). While hybrids of this kind, involving a small amount of d character might be the easiest way of expressing small perturbations of p orbitals, these are not considered to be valid illustrations of d contributions to bonding (Coulson, 1969). A second instance of non-significant introduction of outer d-orbital character in bonding is described by Coulson (1969):

> If a large basis set of atomic orbitals is used, involving s, p, d, f, . . . orbitals on several atoms, and a variational principle is used to determine the best energy and corresponding coefficients in the expansion of the different molecular orbitals in terms of the atomic orbitals, d orbitals will certainly occur in all, or nearly all, of these MOs. Their coefficients, and so their weights, will usually be small. It seems to me that there is very little, if any, chemical significance in such members. To "throw everything into" the computer will result in some d character . . . .

An additional source of ambiguity is that the inclusion of d orbitals can have different effects on different properties of a molecule, with some calculated properties being significantly less sensitive to expansion of the basis set (Dyatkina and Klimenko, 1973). It should also be noted that certain of the simpler semiempirical methods, such as the CNDO/2 procedure ("complete neglect of differential overlap") are known to exaggerate d-orbital participation (Fabian, 1975) and are more likely to muddle than clarify the outer-orbital problem (Dyatkina and Klimenko, 1973).

Some recent studies indicate that *qualitative* features of the electronic structure of hypervalent molecules such as tetracoordinate and hexacoordinate sulfur and pentacoordinate phosphorus are not affected by d functions† (Musher, 1972; Hoffman *et al.*, 1972; Rauk *et al.*, 1972; Florey and Cusachs, 1972; Keil and Kutzelnigg, 1975; Schwenzer and Schaefer, 1975; Chen and Hoffmann, 1976; Rösch *et al.*, 1976; Arduengo and Burgess, 1977). However, other current works conclude that in order to get accurate results, 3d orbitals have to be included in calculations of binding energies and geometries of various molecules containing second-row elements (Ungemach and Shoefer, 1976; Collins *et al.*, 1976; Collins and Duke, 1976). In this book we shall carefully examine when possible the evidence for particular properties of sulfur compounds (e.g., the acidity of sulfides, sulfonium salts, etc.) being *primarily* attributable to dπ–pπ overlap as opposed to other properties of the sulfur atom.

† Indeed it is thought that compounds of the type OF_4, OF_6, and NF_5, where d-orbital involvement is out of the question, would probably exist except for the insufficient amount of space around the first row atom kernels for the large number of fluorine atoms (Rundle, 1963).

1.6 Other Bonding Properties of Sulfur

It is informative to compare sulfur with its close relative in Group VI, oxygen. We have already considered differences associated with the availability of low-lying vacant d orbitals on sulfur but not on oxygen. Other differences include the lower ionization potential and electronegativity (see Table 1–3) and greater polarizability of sulfur compared to oxygen. In terms of Pearson's theory of hard and soft acids and bases (HSAB theory; Pearson, 1973), sulfur would be considered a soft base (high polarizability, low electronegativity, easily oxidized, and possessing empty low-lying orbitals). An important qualitative rule associated with the HSAB principle is that "hard bases prefer to combine with hard acids and soft bases prefer to combine with soft acids." In accord with this rule tervalent phosphorus, mercury, silver, and a variety of other soft acids have a higher affinity for the soft base sulfur than for the hard base oxygen. As we shall see, the interaction of sulfur with the above (and other) soft acids leads to a variety of interesting and synthetically useful reactions.

While sulfur is an excellent *nucleophile*,† it is not very *basic* nor does it tend to form strong hydrogen bonds as in the case of oxygen. A corollary of the weaker basicity of RS^- compared to RO^- is the greater acidity of RSH compared to ROH.‡ It should be recalled that both bases and nucleophiles have a tendency to form covalent bonds by sharing electron pairs. Basicity is a thermodynamic property measured by an acid–base equilibrium, i.e., Eq. 1-3, while nucleophilicity is a kinetic property measured by the rate

$$RS^- + H_3O^+ \rightleftharpoons RSH + H_2O \tag{1-3}$$

Table 1-3. Selected Electronegativity and Ionization Potential Values

Element	Electronegativity[a]	Ionization potential[b]	Element	Electronegativity[a]	Ionization potential[b]
Si	1.64		H	2.79	
P	2.11	10.10	Cl	2.84	12.80
C	2.35		N	3.16	10.87
Se	2.40		O	3.52	12.61
S	2.52	10.48	F	4.00	15.77
Br	2.52	11.62			

[a] For trivalent Group V, tetravalent Group IV, divalent Group VI, and univalent Group VII (Simons *et al.*, 1976).

[b] In electron volts data are for hydrides and are taken from *Nat. Bur. Std.* No. 26 (1969).

† The high nucleophilicity of sulfur anions in solution is now thought to be purely an artifact of solidation effects (Olmstead and Brauman, 1977).

‡ For example, the estimated pK_a values for methanethiol and methanol in dimethylsulfoxide are 17.3 and 27.9, respectively (Arnett and Small, 1977).

constant, for example, for an S_N2 reaction, Eq. 1-4. The proton is a unique electrophile because of its small size and concentrated positive charge and

$$RS^- + R{-}X \xrightarrow{k} RSR + X^- \qquad (1\text{-}4)$$

requires a partner with correspondingly concentrated negative charge for the strongest bonds.

1.7 Spectroscopic Properties of Sulfur

The spectroscopic properties of an organosulfur compound reflect the electronic structure of the molecule, although generally in an indirect manner. A particularly powerful spectroscopic technique is microwave spectroscopy which, together with x-ray crystallography and electron diffraction, allows direct determination of molecular geometry. Appendix B, Table B-1 lists some representative bond lengths for sulfur compounds as obtained by these three methods. An intriguing application of microwave spectroscopy is the study of highly reactive, short-lived small organosulfur molecules generated by the technique of flash vacuum pyrolysis (FVP). For instance, Block and Penn found that 1,3-dithietane 1-oxide smoothly cleaved in a flow system at temperatures above 300° to sulfine CH_2SO and thioformaldehyde CH_2S (Eq. 1-5; Block *et al.*, 1976a). The structures of these two molecules as determined

$$\underset{\textbf{1-4}}{\text{1,3-dithietane 1-oxide}} \xrightarrow{300^\circ} H_1H_2C{=}S{=}O + H_2C{=}S \qquad (1\text{-}5)$$

Sulfine: C—H₂ 1.077, C—H₁ 1.085, C=S 1.610, S=O 1.469; angles 115.6°, 121.9°, 122.5°, 114.7°. Thioformaldehyde: C—H 1.093, C=S 1.611; angles 116.9°, 121.5°.

by microwave spectroscopy are indicated (Penn and Olsen, 1976; Johnson *et al.*, 1971). The photoelectron spectrum of sulfine could also be determined by FVP techniques, providing ionization energies that could be assigned by *ab initio* SCF calculations (Block *et al.*, 1976b). Two independent calculations (Block *et al.*, 1976b; Flood and Boggs, 1976) indicate that the best representation for sulfine is $CH_2{=}S^+{-}O^-$ with no contribution from $CH_2^-{-}S^+{=}O$. The calculations (Flood and Boggs, 1976) also reproduce the unusual 5° shift toward sulfur of the hydrogen atom (H_2 in sulfine in Eq. 1-5) *trans* to the oxygen atom, a shift said to reflect the smaller lone pair-H_2 repulsion compared to the oxygen lone pair-H_1 repulsion.

The possibility of the reaction shown in Eq. 1-5 was first suggested by the appearance of significant m/e 62(CH_2SO) and 46(CH_2S) peaks in the mass spectrum of 1,3-dithietane 1-oxide, indicating that the process of Eq. 1-5 occurs under electron impact conditions. By conducting a flash vacuum pyrolysis with a device attached directly to the mass spectrometer (which was operated at low ionizing voltage to display only parent ions), it could be shown that at temperatures above 300°, the growth in intensity of the m/e 62 and 46 peaks were coupled with the disappearance of the original parent ion and its fragment peaks. A point to be made here is the great utility of mass spectrometry in the study of organosulfur compounds. Even in the absence of exact mass measurements, the presence of a sulfur atom in an organic molecule can be readily recognized by the abundant ^{34}S isotope, which gives a characteristic peak at two mass numbers above the parent or fragment ion mass. The relative natural abundances of the ^{32}S, ^{33}S, and ^{34}S isotopes are 95.0, 0.76, and 4.2%, respectively. From the data in Appendix B, Table B-2, it can be shown that the ratio of m/e 110 to m/e 108 for 1,3-dithietane l-oxide will be about 9:100; whereas for m/e 64 to m/e 48 and m/e 48 to m/e 46 assuming formulas of CH_2SO and CH_2S, respectively, the ratios will be 4.4:100 (neglecting contributions from isotopes of C, H, and O). Another very valuable feature of the mass spectra of organosulfur compounds is that the parent ion abundance is generally significantly higher for these compounds than is the parent ion in the mass spectrum of the corresponding oxygenated compound. For example, the parent peak in the case of *n*-propyl propionate is only 0.04% of the base peak, while for *S*-*n*-propyl propanethioate ($C_3H_7C(O)SC_3H_7$) the relative intensity of the parent peak is 22% (Beynon *et al.*, 1968). This difference in relative molecular ion intensities may be attributed to the lower ionization potential of sulfur compared with oxygen, and makes sulfur compounds more easily identifiable from their mass spectra than the corresponding oxygen compounds.

A second example, also taken from the author's research, of applications of spectroscopic techniques in organosulfur chemistry involves studies on 1,3-dithietanes **1-3**–**1-9** (Eq. 1-1; Block *et al.*, 1976c). The microwave spectrum of compound **1-4** revealed that it was puckered with the angle between the two CSC planes being 39.3° with oxygen equatorial as shown in **1-4a**. The nonbonded S···S distance of 2.60 Å is surprisingly short (only 0.56 Å longer than the *covalent* S—S distance indicated in Table B-1) suggesting possible contributions from hypervalent structure **1-4b**. However, the facts that (1) the S=O distance (1.473 Å) is normal or even slightly shorter than normal (cf. Table B-1), and (2) the S···S distance in bissulfone **1-9** is even shorter (2.590 Å as determined by x-ray crystallography) argues against any significant contribution from **1-4b** (Block *et al.*, 1976c).

The ultraviolet spectrum of **1-4** (λ_{max} 266 nm, ϵ101) when compared to that of thietane (λ_{max} 270 nm, ϵ32) again seems to argue against invoking any special bonding effects as in **1-4b**. On the other hand, the ultraviolet spectrum of 1,3-dithietane (**1-3**) itself (λ_{max} 311 nm, ϵ20) is suggestive of some form of excited state interaction between the sulfur atoms (also compare the λ_{max} for the several dithioacetals listed in Appendix B, Table B-3, with the λ_{max} for the corresponding sulfides; see also Price and Oae, 1962).

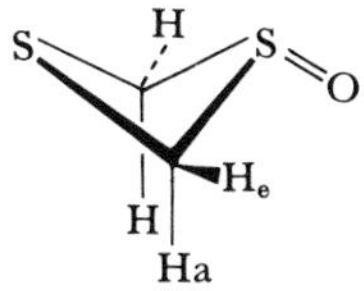

1-4a 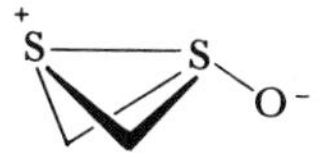 **1-4b**

The successive oxidation of 1,3-dithietane **1-3** to each of its S-oxides **1-4–1-9** could be nicely followed by infrared and NMR spectroscopy. Thus, **1-3** showed a singlet in its NMR spectrum at δ4.05($CDCl_3$), monosulfoxide **1-4** a multiplet at δ4.23($(CD_3)_2SO$), *cis*-bisulfoxide **1-5** an AA'BB'multiplet at δ4.78 and δ5.72 (CF_3COOH), *trans*-bissulfoxide **1-6** a singlet at δ4.97 (CF_3COOH), monosulfone **1-7** a singlet at δ5.27 ($(CD_3)_2SO$ or CF_3COOH), sulfoxide-sulfone **1-8** an AA'BB' pattern at δ5.17 and 5.72 ($(CD_3)_2SO$), and bissulfone **1-9** a singlet at δ6.40 ($(CD_3)_2SO$). As might be expected, the methylene protons are shifted to lower field (deshielded) as the oxidation level of the sulfur increases (see also the data in Appendix B, Table B-4, on chemical shifts for methyl group in organosulfur compounds for additional examples of this trend). As will be discussed further in Section 2.6, the α-methylene protons of sulfoxides are magnetically nonequivalent and appear in NMR spectra as AB quartets. The *degree* of nonequivalence would be expected to vary among the series of sulfoxides **1-4**, **1-5**, **1-6**, and **1-8**. Ring inversion in *trans*-bissulfoxide **1-6** interconverts the methylene protons (cf. **1-6a** ⇌ **1-6a'**) accounting for the sharp singlet observed.† In the isomeric *cis*-bissulfoxide **1-5** the "double-barreled" effect of the two flanking sulfoxide functions should attenuate the magnetic differences between the axial and equatorial protons (cf. **1-5a**) compared to monosulfoxide **1-4** or sulfoxide–sulfone **1-8**. Indeed the separation of the "A" and "B" components in the NMR spectrum of **1-5** is substantially larger than in the spectra of **1-4** and **1-8**. It is in fact possible to

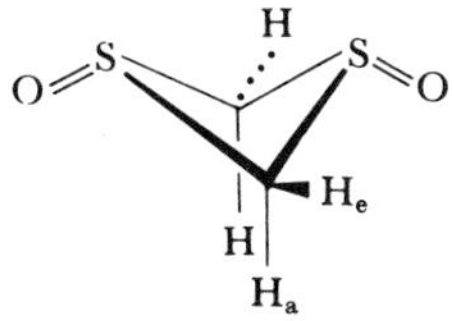

1-5a 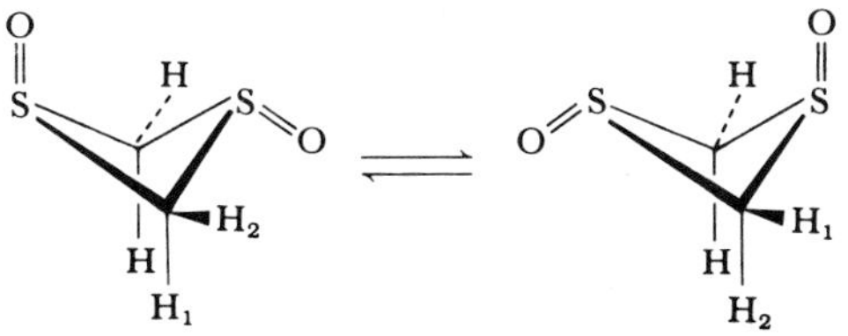 **1-6a** **1-6a'**

† If **1-6** is planar it would also show a singlet in its NMR spectrum.

magnify the chemical shift differences between the diastereotopic α-sulfinyl methylene protons using benzene as a solvent (the aromatic solvent-induced shift, or ASIS, technique) or through the use of europium shift reagents. When the ASIS technique is applied to a sulfoxide such as **1-4**, the protons *trans* to the S—O bond are the more strongly shielded, an effect that is rationalized on the basis of a specific interaction of benzene with the positive end of the $S^+—O^-$ dipole (cf. **1-4c**; Ledaal, 1968; Fraser *et al.*, 1971; Barbarella *et al.*, 1974; and references therein). Using a europium shift

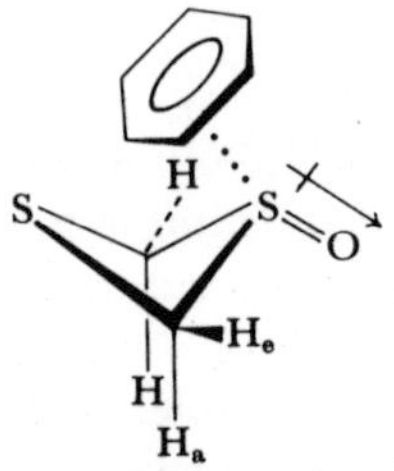

1-4c

reagent, the *cis*(axial) proton in **1-4** is found to suffer a greater downfield shift than the *trans* (equatorial) proton. This effect, which is general, is believed to rise from complexation of the sulfinyl oxygen with the europium atom causing greater paramagnetic shifts for those protons which are closer to oxygen (*cis* in a conformationally restricted system) (Fraser, 1971; Barbarella *et al.*, 1975; and references therein). The ability to distinguish between the α-sulfinylmethylene protons of a sulfoxide using NMR techniques is particularly useful in studying the differential kinetic acidity of these same protons (cf. Section 2.6).

The infrared spectra of 1,3-dithietanes **1-3–1-9** have also been examined. The most intense lines in both the infrared and Raman spectra of the parent 1,3-dithietane **1-3** appear in the region 600–780 cm^{-1}, characteristic for C—S stretching modes (see Appendix B, Table B-5; for a current treatment of Raman spectroscopy of sulfur compounds, see Freeman, 1974). The S-oxides **1-4–1-9** show characteristic sylfoxide and/or sulfone lines (cf. Table B-4). Symmetrical bissulfoxide **1-5** shows *two* S═O lines, perhaps as a result of coupled vibration of the two SO groups (Block *et al.*, 1976c).

1.8 Penicillin and Cephalosporin Chemistry: A Microcosm of Organosulfur Chemistry

The invaluable antibiotic penicillin and its close relative cephalosporin have played a preeminent role in the development of organosulfur chemistry. A brief description of aspects of the chemistry of these two compounds and their derivatives provides a fitting backdrop for the chapters that follow (for a more thorough discussion of the chemistry of these antibiotics, see Cooper *et al.*, 1973). Discovered in 1929 by Sir Alexander Fleming and first synthesized in the laboratory by John Sheehan and co-workers at MIT, penicillin is still obtained principally through fermentation processes.

Penicillin is believed to inhibit the growth of microorganisms by interfering with the synthesis of the cell wall (animal cells have cell membranes rather than cell walls). Penicillin acts on a particular stage of cell wall synthesis in which a glycinyl and D-alanyl unit are joined by a transpeptidase enzyme. The configuration as well as the steric features of the carbon atom in penicillin bearing the free carboxyl group mimic sufficiently well the analogous features of the D-alanyl–D-alanyl C-terminal end of the bacterial peptide† to inactivate (bind) the transpeptidase and thus block cell wall synthesis (Eq. 1-6; Rando, 1975). Penicillin is effective against Gram-positive bacteria but not Gram-

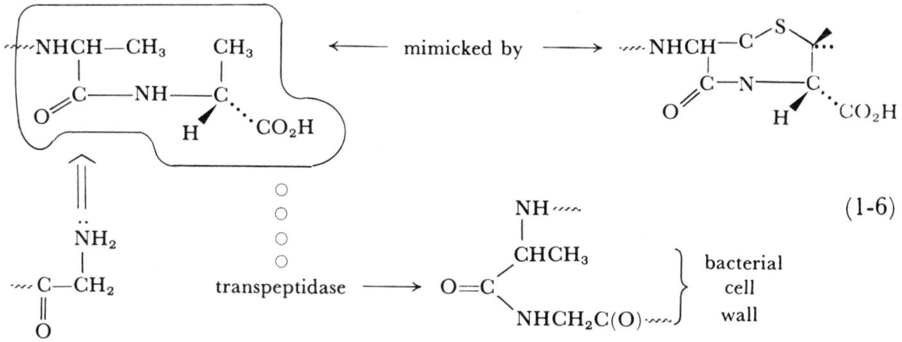

(1-6)

negative bacteria; the latter often contain an enzyme *penicillinase* which can hydrolyze and thereby deactivate pencillin. By a combination of enzymatic and chemical reactions, penicillin can be hydrolyzed to the amino acid penicillamine (Eq. 1-7). We shall return again in Chapter 2 to the use of

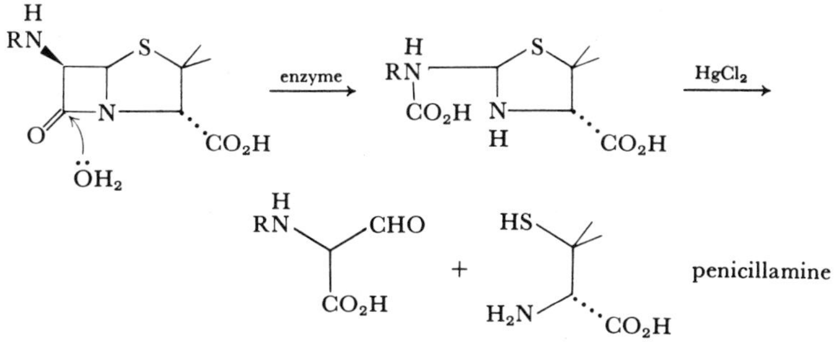

(1-7)

† It has been suggested (Rando, 1975) that the "isosteric" relationship of penicillin and the bacterial dipeptide terminus is enhanced by a "flattening out" of the penicillin ring system at the active site of the enzyme.

Chart 1-4. Laboratory Conversion of Penicillin to Cephalosporin

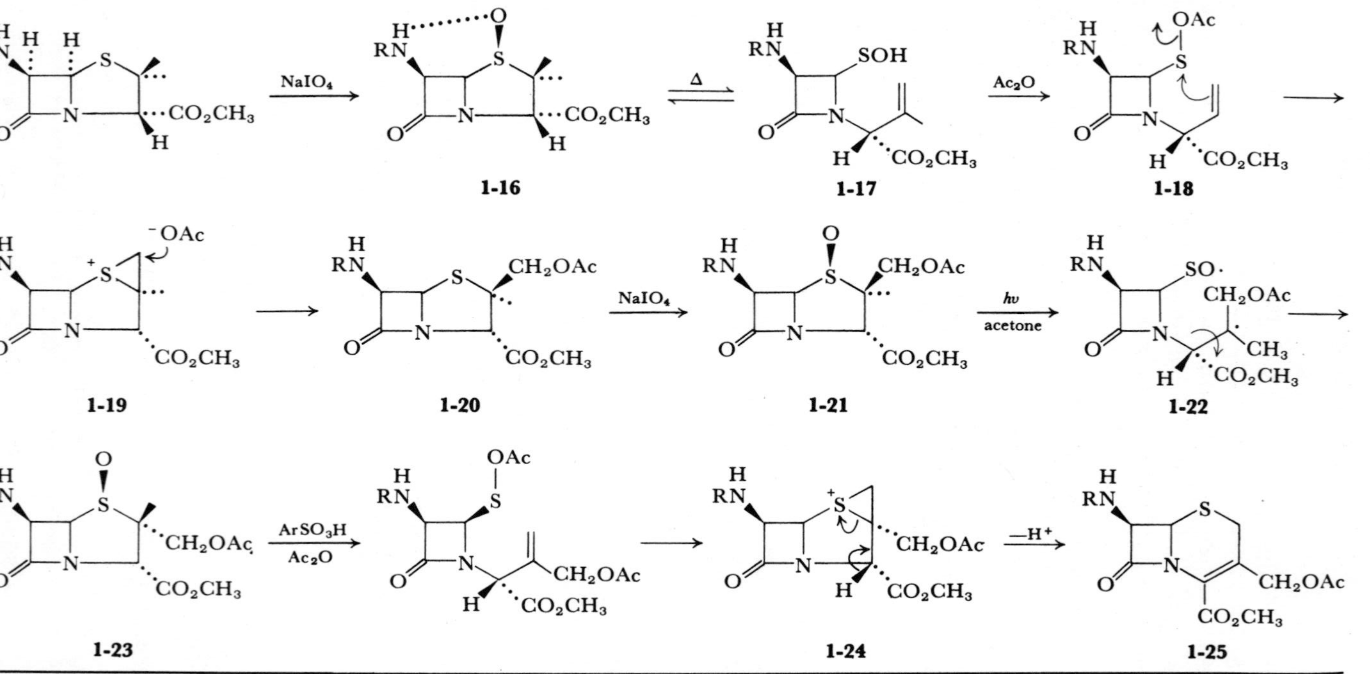

mercury(II) salts and other Lewis acids in the hydrolysis of sulfur compounds to give carbonyl compounds because of the importance of this reaction in many applications of sulfur compounds in organic synthesis.

One of the most remarkable chemical transformations of penicillin is its conversion into cephalosporin. The sequence, pioneered by researchers at the Eli Lilly Company, involves the microbiologically inactive penicillin β-sulfoxide and represents a microcosm of organic sulfur reactions and intermediates. An abbreviated outline of the reactions and postulated mechanisms involved appears in Chart 1-4 (Cooper *et al.*, 1973). The oxidation of penicillin to its S-oxide can be effected with a variety of oxidants previously considered in Table 1-2 including peracids, ozone, iodobenzene dichloride, and sodium metaperiodate. Two diastereomeric sulfoxides can result; with metaperiodate the more stable N—H hydrogen bonded β-sulfoxide **1-16** is formed. Pyrolysis of this sulfoxide leads to the formation of sulfenic acid **1-17** by a reaction analogous to the Cope elimination of amine oxides (this type of cycloelimination reaction is considered further in Section 7.5). Acid **1-17** has recently been isolated in crystalline form (Chou, 1974); on heating it reverts to compound **1-16**. The sulfenic acid can also react with a variety of external reagents as shown in Eq. 1-8 (Allan *et al.*, 1974; Cooper *et al.*, 1973;

RSOSi(CH$_3$)$_3$ ←[(CH$_3$)$_3$SiCl]— RSOH ≡ **1-17** —[D$_2$O]→ RSOD

RSOH —[(1) R$_2$NLi, (2) CH$_3$SO$_3$F]→ RSOCH$_3$

RSS-t-C$_4$H$_9$ ←[t-C$_4$H$_9$SH]— RSOH ≡ **1-17** (1-8)

RSOH —[norbornadiene]→ norbornenyl–S(O)R

RSSO$_2$Ar ←[ArSO$_2$H]— RSOH —[SO$_2$Cl$_2$]→ RS(O)Cl

Koppel and Kukolja, 1975). Acetylation converts the nucleophilic sulfur of acid **1-17** into the electrophilic sulfur of sulfenyl acetate **1-18** which presumably undergoes intramolecular electrophilic addition via episulfonium ion **1-19** affording the stable acetate **1-20** (see Sections 4.1 and 4.4 for further discussion of these cationic reactions). Compound **1-20** on oxidation to **1-21** followed by photoepimerization via diradical **1-22** (see Section 5.12 for details on this type of radical process) affords sulfoxide **1-23**. Finally, treatment of this sulfoxide under ring-opening conditions similar to those used in the sequence **1-16** → **1-20** but using p-toluenesulfonic acid instead of acetic anhydride gives cephalosporin **1-25**. Apparently with the poorly nucleophilic sulfonate anion, episulfonium ion **1-24** undergoes elimination rather than nucleophilic ring opening.

A second novel transformation of penicillin sulfoxide **1-16** to a cephalosporin derivative (Kukolja *et al.*, 1976) involves penicillin sulfinyl chloride **1-26** prepared by treatment of **1-16** with *N*-chlorosuccinimide. Compound **1-26** (and several of its derivatives as the sulfinite esters, amides, imides, etc.) on treatment with a Lewis acid gives the 3-methylenecepham sulfoxide **1-28** via an intramolecular ene reaction involving sulfinium cation **1-27** (see

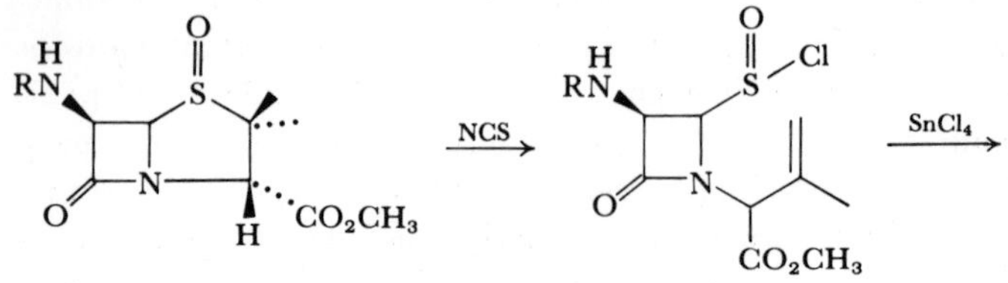

NCS → SnCl₄ →

1-16 **1-26**

→ PBr₃ DMF →

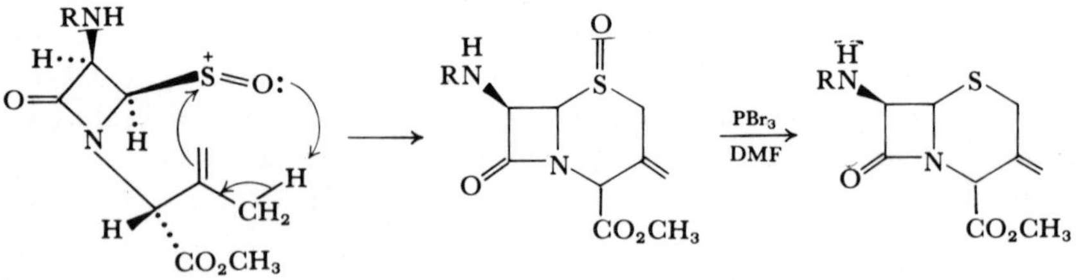

1-27 **1-28** **1-29**

Section 7.5 for a closely related ene reaction). Sulfoxide **1-28** can be conveniently reduced with phosphorus tribromide in dimethylformamide to the 3-methylenecepham **1-29**, said to be a versatile intermediate for the preparation of commercially significant cephalosporins (Kukolja *et al.*, 1976).

To bring this discussion full circle, it can be noted that a novel one-step procedure has been discovered for the transformation of cephalosporins into

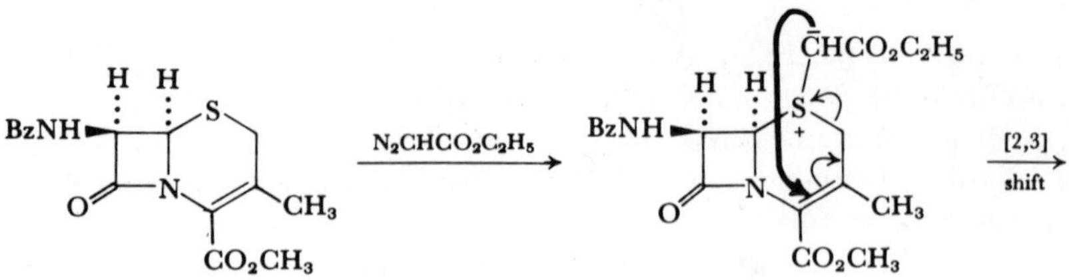

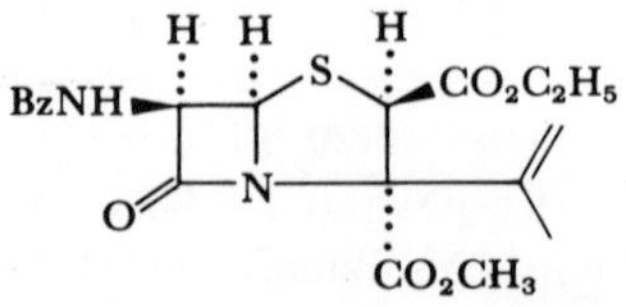

$N_2CHCO_2C_2H_5$ → [2,3] shift →

(1-9)

(53%)

penicillins (Eq. 1-9; Yoshimoto *et al.*, 1972). This remarkable reaction involves formation of a sulfonium ylide which spontaneously undergoes a [2,3] sigmatropic rearrangement (see Sections 3.5 and 7.2 for detailed discussions of this reaction type).

References

Allan, R. D., Barton, D. H. R., Girijavallabhan, M., and Sammes, P. G. (1974). *J. Chem. Soc. Perkin Trans.* **1**, 1456.

Alper, H., and Keung, E. C. H. (1970), *Tetrahedron Lett.* 53.

Anastassiou, A. G., Wetzel, J. C., and Chao, B. Y-H. (1975). *J. Am. Chem. Soc.* **97**, 1124.

Andersen, K. K., and Bernstein, D. T. (1975). *J. Chem. Ecol.* **1**, 493.

Andersen, K. K., Yeager, S. A., and Peynircioglu, N. B. (1970). *Tetrahedron Lett.* 2485.

Arduengo, A. J., and Burgess, E. M. (1977). *J. Am. Chem. Soc.* **99**, 2376.

Arnett, E. M., and Small, L. E. (1977). *J. Am. Chem. Soc.* **99**, 808.

Barbarella, G., Garbesi, A., and Fava, A. (1975). *J. Am. Chem. Soc.* **97**, 5883.

Barbieri, G., Cinquini, M., Colonna, S., and Montanari, F. (1968). *J. Chem. Soc. C* 659.

Beynon, J. H., Saunders, R. A., and Williams, A. E. (1968). "The Mass Spectra of Organic Molecules." Elsevier, Amsterdam.

Block, E., and O'Connor, J. (1974). *J. Am. Chem. Soc.* **96**, 3921, 3929.

Block, E., Penn, R. E., Olsen, R. J., and Sherwin, P. F. (1976a). *J. Am. Chem. Soc.* **98**, 1264.

Block, E., Bock, H., Mohmand, S., Rosmus, P., and Solouki, B. (1976b). *Angew. Chem. Int. Ed. English* **15**, 383.

Block, E., Corey, E. R., Penn, R. E., Renken, T. L., and Sherwin, P. F. (1976c). *J. Am. Chem. Soc.* **98**, 5715.

Boelens, M., de Valois, P. J., Wobben, H. J., and van der Gen, A. (1971). *J. Agric. Food Chem.* **19**, 984.

Bordia, A., and Bansal, H. C. (1973). *Lancet* 1491.

Bordignon, E., Cattalini, L., Natile, G., and Scatturin, A. (1973). *Chem. Commun.* 878.

Brandsma, L., and Arens, J. F. (1967). *In* "The Chemistry of the Ether Linkage" (S. Patai, ed.), Chapter 13. Wiley (Interscience), New York.

Brill, T. A. (1973). *J. Chem. Educ.* **50**, 392.

Brown, H. C., and Ravindran, N. (1973). *Synthesis* 42.

Calo, V., Ciminale, F., Lopez, G., and Todesco, P. E. (1971). *Int. J. Sulfur Chem. A* **1**, 130.

Capozzi, G., and Modena, G. (1974). *In* "The Chemistry of the Thiol Group" (S. Patai, ed.), Part 2, Chapter 17. Wiley, New York.

Casnati, G., Ricca, A., and Pavan, M. (1967). *Chem. Ind.* (*Milan*) **49**, 57.

Chalkley, G. R., Snodin, D. J., Stevens, G., and Whiting, M. C. (1970). *J. Chem. Soc.* (*C*) 682.

Chan, T. H., Melnyk, A., and Harpp, D. N. (1969). *Tetrahedron Lett.* 201.

Chaser, D. W. (1971). *J. Org. Chem.* **36**, 613.

Chemical Abstract Service (1972). "Index Guide," Chemical Abstracts, 76, Section IV.

Chen, M. M. L., and Hoffmann, R. (1976). *J. Am. Chem. Soc.* **98**, 1647.

Chou, T. S., Burgtorf, J. R., Ellis, A. L., Lammert, S. R., and Kukolja, S. P. (1974). *J. Am. Chem. Soc.* **96**, 1609.

Cilento, G. (1960). *Chem. Rev.* **60**, 146 (1960).

Coffen, D. L. (1969). *Rec. Chem. Progr.* **30**, 275.

Collins, J. B., Schleyer, P. von R., Binkley, J. S., and Pople, J. A. (1976). *J. Chem. Phys.* **64**, 5142.

Collins, M. P. S., and Duke, B. J. (1976). *Chem. Phys. Lett.* **42**, 364.

Connor, R. (1943). *In* "Organic Chemistry" (H. Gilman, ed.), 2nd ed., Vol. 1. Wiley, New York.

Cooper, R. D. G., Hatfield, L. D., and Spry, D. O. (1973). *Accounts Chem. Res.* **6**, 32.

Coulson, C. A. (1969). *Nature (London)* **221**, 1106

Coulson, C. A. (1973). *Proc. Robert A Welch Foundation Conf. Chem. Res.* (*Theoret. Chem.*) **16** 61–117.

Cox, B. G., and Gibson, A. (1973). *J. Chem. Soc. Perkin Trans.* **2**, 1355.

Craig, D. P., Maccoll, A., Nyholm, R. S., Orgel, L. E., and Sutton, L. E. (1954). *J. Chem. Soc.* 332.

Cram, D. J. (1965). "Fundamentals of Carbanion Chemistry." Academic Press, New York.

Cram, D. J., Day, J., Rayner, D. R., Von Schriltz, D. M., Duchame, D. J., and Garwood, D. C. (1970). *J. Am. Chem. Soc.* **92**, 7369.

Curci, R., Di Furia, F., Testi, R., and Modena, G. (1974). *J. Chem. Soc. Perkin Trans.* **2**, 752.

Dance, I. G., Conrad, R. C., and Cline, J. E. (1974). *Chem. Commun.* 13.

Danehy, J. P. (1971). *In* "Sulfur in Organic and Inorganic Chemistry" (A. Senning, ed.), Vol. I, Chapter 10. Dekker, New York.

DeRosa, M., Gambacorts, A., and Minale, L. (1975). *Chem. Commun.* 392.

Dreux, M., Lerous, Y., and Savignac, P. (1974). *Synthesis* 506.

Durst, H. D., Zubrick, J. W., and Kieczykowski, G. R. (1974). *Tetrahedron Lett.* 1777.

Durst, T., and Tin, K. C. (1971). *Can. J. Chem.* **49**, 2374.

Dyatkina, M. E., and Klimenko, N. M. (1973). *Zh. Struct. Khim.* **14**, 173.

Fabian, J. (1975). *In* "Organic Compounds of Sulphur, Selenium and Tellurium" (D. H. Reid, ed.), Vol. 3 (Specialist Periodical Report). Chemical Society, London.

Field, L. (1977). *In* "Organic Chemistry of Sulfur" (S. Oae, ed.). Plenum Press, New York.

Field, L., and Khim, Y. H. (1972). *J. Org. Chem.* **37**, 2710.

Filby, W. G., Gunther, K., and Penzhorn, R. D. (1973). *J. Org. Chem.* **38**, 4070.

Fletcher, J. H., Dermer, O. C., and Fox, R. B. (1974). "Nomenclature of Organic Compounds." Adv. in Chem. Ser. 126, Am. Chem. Soc., Washington, D.C.

Flood, E., and Boggs, J. E. (1976). *J. Mol. Struct.* **34**, 147; also see van Lierop, J., Van der Avoird, A., and Zwanenburg, B. *Tetrahedron* **33**, 539 (1977).

Florey, J. B., and Cusachs, L. C. (1972). *J. Am. Chem. Soc.* **94**, 3040.

Foote, C. S., and Peters, J. W. (1971). *J. Am. Chem. Soc.* **93**, 3795.

Fraser, R. R., Durst, T., McClary. M. R., Viau, R., and Wigfield, Y. Y. (1971). *Int. J. Sulfur Chem. Part A* **1**, 133.

Freeman, S. K. (1974). "Applications of Laser Raman Spectroscopy." Wiley, New York.

Fromm, E., and Baumann, E. (1889). *Chem. Ber.* **22**, 1035.

Gardner, J. N., Kaiser, S., Krubiner, A., and Lucas, H. (1973). *Can. J. Chem.* **51**, 1419.

Gorecki, M., and Patchornik, A. (1973). *Biochim. Biophys. Acta* **303**, 63.

Gurria, G. M., and Posner, G. H. (1973). *J. Org. Chem.* **38**, 2419.

Gutsche, C. D., and Pasto, D. J. (1975). "Fundamentals of Organic Chemistry." Prentice-Hall, Englewood Cliffs, New Jersey.

Haake, M. (1976). *In* "Topics in Sulfur Chemistry" (A. Senning, ed.), Vol. 1. Georg Thieme, Stuttgart.

Hardy, F. E., Speakman, P. R. H., and Robson, R. (1969). *J. Chem. Soc.* (*C*) 2334.

Harpp, D. N., and Back, T. G. (1973). *J. Org. Chem.* **38**, 4328.

Henbest, H. B., and Trocha-Grimshaw, J. (1974). *J. Chem. Soc. Perkin Trans.* **1**, 607.

Ho, T. L., and Wong, C. M. (1972). *Synthesis* 561.
Ho, T. L., and Wong, C. M. (1973a). *Synth. Commun.* **3**, 37.
Ho, T. L., and Wong, C. M. (1973b). *Synthesis* 206.
Hoffmann, R., Howell, J. M., and Muetterties, E. L. (1972). *J. Am. Chem. Soc.* **94**, 3047.
Hojo, M., and Masuda, R. (1976). *Tetrahedron Lett.* 613.
Isenberg, N., and Grdinic, M. (1973). *Int. J. Sulfur Chem.* **8**, 307.
IUPAC (1971). "Nomenclature of Organic Chemistry, Sections A, B, and C," 3rd ed. Butterworth, London.
Jain, R. C., Vyas, C. R., and Mahatma, O. P. (1973). *Lancet* 1491.
Janssen, M. J. (1969). *In* "The Chemistry of Carboxylic Acids and Esters" (S. Patai, ed.), Chapter 15. Wiley (Interscience), New York.
Johnson, C. R. (1973). *Accounts Chem. Res.* **6**, 341.
Johnson, C. R., and McCants, D. Jr. (1965). *J. Am. Chem. Soc.* **87**, 1109.
Johnson, C. R., Bacon, C. C., and Rigau, J. J. (1972). *J. Org. Chem.* **37**, 919.
Johnson, C. R., Kirchhoff, R. A., Jonsson, E. U., and Saukaitis, J. C. (1975). *In* "Organic Sulphur Chemistry" (C. J. M. Stirling, ed.). Butterworth, London.
Johnson, D. R., Powell, F. X., and Kirchhoff, W. H. (1971). *J. Mol. Spectrosc.* **39**, 136.
Kabalka, G. W., Baker, J. D., Jr., and Neal, G. W. (1977). *J. Org. Chem.* **42**, 512.
Kaiser, E. T. (1977). *In* "Organic Chemistry of Sulfur" (S. Oae, ed.). Plenum Press, New York.
Kamiyama, K., Minato, H., and Kobayashi, M. (1973). *Bull. Chem. Soc. Jpn.* **46**, 2255.
Keil, F., and Kutzelnigg, W. (1975). *J. Am. Chem. Soc.* **97**, 3623.
Kennewell, P. D., and Taylor, J. B. (1975). *Chem. Soc. Rev.* **4**, 189.
Kice, J. L. (1971). *In* "Sulfur in Organic and Inorganic Chemistry" (A. Senning, ed.), Vol. 1, Chapter 6. Dekker, New York.
King, J. (1975). *Accounts Chem. Res.* **8**, 10.
Kjaer, A. (1977). *Pure Appl. Chem.* **49**, 137.
Koppel, G. A., and Kukolja, S. (1975). *Chem. Commun.* 57.
Koutecky, V. B., and Musher, J. I. (1974). *Theoret. Chim. Acta* **33**, 227.
Kresze, G. (1975). *In* "Organic Sulphur Chemistry" (C. J. M. Stirling, ed.). Butterworth, London.
Kuhle, E. (1973). "The Chemistry of the Sulfenic Acids." Georg Thieme, Stuttgart.
Kukolja, S., Lammert, S. R., Gleissner, M. R. B., and Ellis, A. I. (1976). *J. Am. Chem. Soc.* **98**, 5040.
Kwart, H., and King, K. (1977). "The Question of d-Orbital Involvement in the Chemistry of Silicon, Phosphorus and Sulfur." Verlag Chemie, Weinheim, in press.
Ledaal, T. (1968). *Tetrahedron Lett.* 1683.
Liebig, J. (1834). *Justus Liebigs Ann. Chem.* **11**, 14.
Loening, K. L. (1972). *In* "Sulfur in Organic and Inorganic Chemistry" (A. Senning, ed.), Vol. 3, Chapter 28. Dekker, New York.
Louw, Ro, Vermeeren, H. P. W., Van Asten, J. J. A., and Ultee, W. J. (1976). *Chem. Commun.* 496.
Magee, P. S. (1971). *In* "Sulfur in Organic and Inorganic Chemistry." (A. Senning, ed.), Vol. 1, Chapter 9. Dekker, New York.
Marino, J. P. (1976). *In* "Topics in Sulfur Chemistry" (A. Senning, ed.), Vol. 1. Georg Thieme, Stuttgart.
Martin, J. C., and Perozzi, E. F. (1974). *J. Amer. Chem. Soc.* **96**, 3155.
Meienhofer, J., Czombos, J., and Maeda, H. (1971). *J. Am. Chem. Soc.* **93**, 3080.
Mitchell, K. A. R. (1969). *Chem. Rev.* **69**, 157.
Moore, R. E. (1971). *Chem. Commun.* 1168.

Murray, R. W., Smetana, R. D., and Block, E. (1971). *Tetrahedron Lett.* 299.

Musher, J. I. (1969). *Angew Chem. Int. Ed. English* **8**, 54.

Musher, J. I. (1972). *J. Amer. Chem. Soc.* **94**, 1370.

Naimie, H., Samek, Z., Doleis, L., and Reltacek, Z. (1972). *Coll. Czech. Chem. Commun.* **37**, 1166.

Nakaya, T., Arabori, H., and Imoto, M. (1970). *Bull. Chem. Soc. Jpn.* **43**, 1888.

Naumann, K., Zon, G., and Mislow, K. (1969). *J. Am. Chem. Soc.* **91**, 2788.

Numata, T., and Oae, S. (1973). *Chem. Ind. (London)* 277.

Nuzzo, R. G., Simon, H. J., and Filippo, J. S., Jr. (1977). *J. Org. Chem.* **42**, 568.

Oae, S. (1977). *In* "Organic Chemistry of Sulfur" (S. Oae, ed.). Plenum Press, New York.

Oae, S., and Kunieda, N. (1977). *In* "Organic Chemistry of Sulfur" (S. Oae, ed.). Plenum Press, New York.

Oae, S., Nakanishi, A., and Kozuka, S. (1972a). *Tetrahedron* **28**, 549.

Oae, S., Nakanishi, A., and Tsujimoto, N. (1972b). *Tetrahedron* **28**, 2981.

Oae, S., Yagihara, T., and Okabe, T. (1972c). *Tetrahedron* **28**, 3203.

Ohloff, G. (1969). *Fortschr. Chem. Forsch.* **12**, 185.

Ohno, A. (1977). *In* "Organic Chemistry of Sulfur" (S. Oae, ed.). Plenum Press, New York.

Ohno, A., and Oae, S. (1977). *In* "Organic Chemistry of Sulfur" (S. Oae, ed.). Plenum Press, New York.

Olmstead, W. N., and Brauman, J. I. (1977). *J. Am. Chem. Soc.* **99**, 4219.

Oppenheimer, M., and Dalgarno, A. (1974). *Astrophys. J.* **187**, 231.

Overman, L. E., Smoot, J., and Overman, J. D. (1974). *Synthesis* 59.

Patai, S. (ed.) (1974). "The Chemistry of the Thiol Group," Parts 1 and 2. Wiley (Interscience), New York.

Pauling, L. (1939). "The Nature of the Chemical Bond," 1st ed. Cornell Univ. Press, Ithaca, New York.

Pearson, R. G. (1973). "Hard and Soft Acids and Bases." Dowden, Hutchinson and Ross, Stroudsburg, Pennsylvania.

Penn, R. E., and Olsen, R. J. (1976). *J. Mol. Spectros.* **61**, 21.

Perozzi, E. F., and Martin, J. C. (1972). *J. Am. Chem. Soc.* **94**, 5519.

Peterson, D. J. (1972). *Organomet. Chem. Rev. Sect. A* **7**, 295.

Pews, R. G., and Corson, F. P. (1969). *Chem. Commun.* 1187.

Polzhofer, K. P., and Ney, K. H. (1971). *Tetrahedron* **27**, 1997.

Price, C. C., and Oae, S. (1962). "Sulfur Bonding." Ronald Press, New York.

Rae, I. D., and Pojer, P. M. (1971). *Tetrahedron Lett.* 3077.

Rando, R. R. (1975). *Accounts Chem. Res.* **8**, 281.

Rauk, A., Allen, L. C., and Mislow, K. (1972). *J. Am. Chem. Soc.* **94**, 3035.

Reid, E. E. (1966). "Organic Chemistry of Bivalent Sulfur," Vol. VI. Chemical Publ., New York.

Richmond, D. V. (1973). *In* "Phytochemistry" (L. P. Miller, ed.), Vol. III, Chapter 2. Van Nostrand–Reinhold, Princeton, New Jersey.

Rigau, J. J., Bacon, C. C., and Johnson, C. R. (1970). *J. Org. Chem.* **35**, 3655.

Rivett, D. E. A. (1974). *Tetrahedron Lett.* 1253.

Roesky, H. W. (1971). *In* "Sulfur in Organic and Inorganic Chemistry" (A. Senning, ed.), Vol. 1, Chapter 9. Dekker, New York.

Rösch, N., Smith, V. H., Jr., and Whangbo, M. H. (1976). *Inorg. Chem.* **15**, 1768.

Rundle, R. E. (1963). *Surv. Progr. Chem.* **1**, 81.

Russ, C. R., and Douglass, I. B. (1971). *In* "Sulfur in Organic and Inorganic Chemistry" (A. Senning, ed.), Chapter 8. Dekker, New York.

Salmond, W. G. (1968). *Quart. Rev.* **22**, 253.
Sandler, S. R., and Karo, W. (1968). "Organic Functional Group Preparations." Academic Press, New York.
Sato, Y., Kunieda, N., and Kinoshita, M. (1972). *Chem. Lett. Jpn.* 1023.
Schildknecht, H., Wilz, J., Enzmann, F., Grund, N., and Ziegler, M. (1976). *Angew. Chem.* **88**, 228.
Schöllkopf, U., and Hilbert, P. (1973). *Justus Liebigs Ann. Chem.* 1061.
Schroek, C. W., and Johnson, C. R. (1971). *J. Am. Chem. Soc.* **93**, 5305 and following papers.
Schwenzer, G. M., and Schaefer, H. F. III (1975). *J. Am. Chem. Soc.* **97**, 1393.
Seebach, D. (1969). *Angew. Chem. Int. Ed. Engl.* **8**, 639.
Singer, A. G., Agosta, W. C., O'Connell, R. J., Pfaffmann, C., Bowen, D. V., and Field, F. H. (1976). *Science* **191**, 948.
Siegl, W. O., and Johnson, C. R. (1970). *J. Org. Chem.* **35**, 3657.
Siegl, W. O., and Johnson, C. R. (1971). *Tetrahedron* **27**, 341.
Simons, G., Zandler, M. E., and Talaty, E. R. (1976). *J. Am. Chem. Soc.* **98**, 7869.
Stirling, C. J. M. (1977). *In* "Organic Chemistry of Sulfur" (S. Oae, ed.). Plenum Press, New York.
Stout, E. I., Shasha, B. S., and Doane, W. M. (1974). *J. Org. Chem.* **39**, 562.
Tagaki, W. (1977). *In* "Organic Chemistry of Sulfur" (S. Oae, ed.). Plenum Press, New York.
Thompson, Q. E. (1965). *J. Org. Chem.* **30**, 2703.
Truce, W. E., Klinger, T. C., and Brand, W. W. (1977). *In* "Organic Chemistry of Sulfur" (S. Oae, ed.). Plenum Press, New York.
Ungemach, S. R., and Shaefer, H. F. III (1976). *J. Am. Chem. Soc.* **98**, 1658.
Walter, W., and Voss, J. (1970). *In* "The Chemistry of Amides" (J. Zabicky, ed.), Chapter 8. Wiley (Interscience), New York.
Walti, M., and Hope, D. B. (1971). *J. Chem. Soc.* (*C*) 2326.
Wheeler, J. W., von Endt, D. W., and Wemmer, C. (1975). *J. Am. Chem. Soc.* **97**, 441.
Wilson, R. M., Buchanan, D. N., and Davis, J. E. (1971). *Tetrahedron Lett.* 3919.
Wratten, S. J., and Faulkner, D. J. (1976). *J. Org. Chem.* **41**, 2465.
Yoshimoto, M., Ishihara, S., Nakayama, E., and Soma, N. (1972). *Tetrahedron Lett.* 2923.
Yoshimura, T., Omata, T., Furukawa, N., and Oae, S. (1976). *J. Org. Chem.* **41**, 1728.
Zeise, W. C. (1834). *Justus Liebigs Ann. Chem.* **11**, 1.

Chapter 2 | Sulfur-Containing Carbanions

2.1 Introduction

Carbanions are most simply described as conjugate bases of carbon acids (Cram, 1965), for example, as defined by the equilibrium indicated in Eq. 2-1.

$$\gt C{-}H \; + \; B: \underset{k_{-1}}{\overset{k_1}{\rightleftharpoons}} \; \gt C:^- \; + \; BH^+ \tag{2-1}$$

$$K_a = k_1/k_{-1}$$

The stability of a carbanion may be equated with its basicity and therefore with the acidity of the conjugate acid in the base system used, as indicated by the equilibrium constant K_a or more commonly by the pK_a value. Another widely used means of assessing carbanion stability involves determination of the *kinetic acidity* of a carbon acid through measurement of the ionization rate k_1.

Kinetic and thermodynamic acidity are related through the Brønsted equation,

$$\log k = \alpha \log K_a + C$$

In "well-behaved" systems the values of α range between 0.4 and 0.6 so that a change of one pK_a unit roughly corresponds to a change of about 100 in the rate constant for proton removal (Cram, 1968). We shall deal below with problems associated with the interpretation of kinetic acidity data.

The geometry of a carbanion may be pyramidal with the lone pair occupying an sp^3-hybridized orbital, or planar (sp^2 hybridized) with the lone pair in a p orbital. Since s electrons are, on the average, closer to the carbon nucleus than p electrons, electrons in an sp^3-hybrid orbital would be more stable than in a p orbital. However sp^2 hybridization is favored in cases involving π-conjugation, as in the benzyl anion (Buncel, 1975).

The generation and study of carbanions derived from "weak" carbon acids (p$K_a \sim$ 16–30) requires the use of nonaqueous solvents. In these solvents carbanions show a gradation of ionic character which is a function of the solvent polarity, the carbanion stability and the electropositive character of the metal. A useful description of the stages of charge separation is shown:

$\geq\overset{\delta^-}{C}$—$\overset{\delta^+}{M}$	$\geq C:^-\ M^+$	$\geq C:^-\mid solv \mid M^+$	$\geq C:^-\quad M^+$
Polar covalent	Intimate (contact) ion pair	Solvent-separated ion pair	Free ions

Metals such as sodium, potassium, and cesium tend to favor the ionic state while less electropositive metals such as lithium, Group II, Group III, and transition metals increasingly favor the covalent state. In solvents such as ether, tetrahydrofuran, benzene, ammonia, and cyclohexylamine, carbanions generally exist as contact ion pairs (of variable "tightness") even with cesium as cation, while ion pairing is thought to be absent in dimethyl sulfoxide and related highly polar aprotic solvents. The formation in organic solvents of free anions from ion pairs involving alkali metals can be greatly facilitated through the use of various cyclic or polycyclic polyethers (crown ethers such as 18-crown-6 (**2-1**) or cryptates such as [2.2.2]cryptate (**2-2**)) which efficiently complex the alkali–metal cations. As we shall see, ion pairing greatly

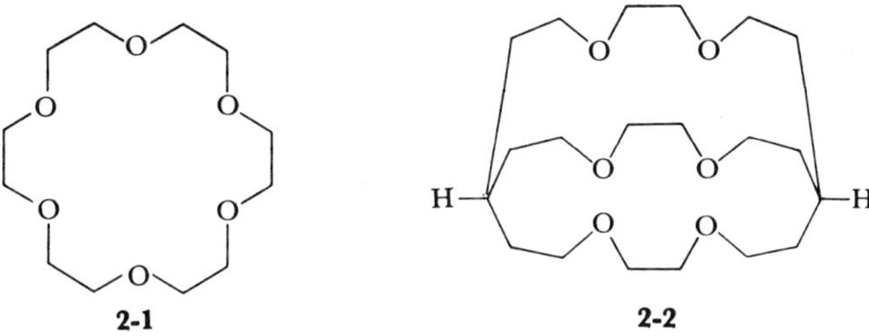

2-1 **2-2**

complicates the determination of kinetic and equilibrium acidities.

The capacity of sulfur in its various oxidation states to enhance significantly the acidity of adjacent C—H bonds has been described as early as 1889 (Fromm, 1889). While the precise mechanism of stabilization of carbanions by adjacent groups RS, RS(O), RSO_2, R_2S^+, $R_2S^+(O)$, etc., is a subject of some controversy, the great importance of this effect in synthetic organic chemistry is incontrovertible. We shall postpone discussion of carbanions adjacent to a heteroatom bearing a formal positive charge (*ylides*, e.g., $(CH_3)_2S^+$—CH_2^-) until Chapter 3.

2.2 Preparation

General

Methods for the generation of sulfur-containing carbanions include proton abstraction (metalation), decarboxylation, nucleophilic addition, insertion of a metal into a carbon–halogen bond, transmetalation and related metal–heteroatom exchange processes, nucleophilic trapping of thiocarbenes, and

Scheme 2-1. Generation of Sulfur-Containing Carbanions by Proton Abstraction (Metalation)

1.	$PhSCH_3 + n\text{-BuLi} \xrightarrow[\text{THF, 0°}]{\text{DABCO}} PhSCH_2Li$	Corey and Seebach (1966)
2.	$CH_2{=}CHSC_2H_5 + sec\text{-BuLi} \xrightarrow[-78°]{\text{THF–HMPA}} CH_2{=}C(Li)SC_2H_5$	Oshima *et al.* (1973)
3.	$(CH_3)_2C{=}CHCH_2SPh + n\text{-BuLi} \longrightarrow$ (allyl anion)–SPh	Biellmann and Ducep (1968)
4.	cyclopropyl(H)–SPh + n-BuLi $\longrightarrow$ cyclopropyl(Li)–SPh	Trost *et al.* (1973; 1977)
5.	1,3-dithiane (H, H) + n-BuLi $\longrightarrow$ 2-lithio-1,3-dithiane (Li, H)	Corey and Seebach (1965)
6.	$CH_3S(O)CH_3 + NaH$ or $KH \xrightarrow{\Delta} CH_3S(O)CH_2Na + H_2$	Corey and Chaykovsky (1965); Brown (1974)
7.	$CH_3S(O)$–N(morpholine)O $\xrightarrow[\text{THF, }-78°]{t\text{-BuLi}}$ $LiCH_2S(O)$N(morpholine)O	Corey and Durst (1966)
8.	H–(cyclic SO_2)–H $\xrightarrow{t\text{-BuLi}}$ H–(cyclic SO_2)–Li	Corey and Block (1969)
9.	$(C_2H_5SO_2)_3CH \xrightarrow{H_2O} (C_2H_5SO_2)_3C{:}^-\ H_3O^+$	Doering and Levy (1955)
10.	$CH_3SO_2{-}OCH_3 \xrightarrow[-78°]{n\text{-BuLi/THF}} LiCH_2SO_2{-}OCH_3$	Corey and Durst (1966)
11.	$CH_3SO_2N(CH_3)_2 \xrightarrow[\text{THF, 0°}]{n\text{-BuLi}} LiCH_2SO_2N(CH_3)_2$	Corey and Chaykovsky (1965)
12.	$PhS(O)(=NTs)CH_3 \xrightarrow{NaH} PhS(O)(=NTs)CH_2Na$	Johnson and Katekar (1970)
13.	$C_2H_5S(O)CH_2SC_2H_5 \xrightarrow[\text{THF}]{n\text{-BuLi}} C_2H_5S(O)CHLiSC_2H_5$	Herrmann *et al.* (1973)
14.	$PhSCH_2OCH_3 \xrightarrow[\text{THF, }-30°]{n\text{-BuLi}} PhSCH(Li)OCH_3$	Trost and Miller (1975)

Scheme 2-2. Other Methods for Preparation of Sulfur-Containing Carbanions

1. Decarboxylation

$PhSO_2C(CH_3)(C_6H_{13})CO_2^- \xrightarrow{150°} PhSO_2\bar{C}(CH_3)C_6H_{13}$ Corey *et al.* (1962)

2. Nucleophilic addition

a to C=C

2-Methylene-1,3-dithiane $\xrightarrow{n\text{-BuLi}}$ 2-lithio-2-(C_5H_{11}-*n*)-1,3-dithiane — Seebach *et al.* (1973)

2-(2-Butenylidene)-1,3-dithiane $\xrightarrow{t\text{-BuLi}}$ 2-lithio-2-alkenyl-1,3-dithiane — Seebach *et al.* (1973)

b to C=S

$Ph_2C{=}S \xrightarrow{PhLi} Ph_2C(Li)SPh$ Beak *et al.* (1975)

3. Insertion of metal into carbon–halogen bond

$PhCH_2SCH_2Cl + Mg \xrightarrow{THF} PhCH_2SCH_2MgCl$ Normant and Castro (1964)

4. Transmetalation and metal-heteroatom exchange routes

$CH_3SCH_2SnBu_3 \xrightarrow{n\text{-BuLi}} CH_3SCH_2Li + Bu_4Sn$ Peterson (1972)

$PhSC(CH_3)_2SePh \xrightarrow{n\text{-BuLi}} PhSC(CH_3)_2Li + BuSePh$ Anciaux *et al.* (1975); Seebach and Beck (1974)

$(PhS)_4C \xrightarrow{n\text{-BuLi}} (PhS)_3CLi + BuSPh$ Bos *et al.* (1967)

5. Nucleophilic trapping of thiocarbene

$(PhS)_2C{:} + PhLi \longrightarrow (PhS)_2C(Li)Ph$ Seebach *et al.* (1972)

6. Retroaldol-type reaction

3-Hydroxy-2-phenyl-2-methyltetrahydrothiophene 1,1-dioxide (OH, Ph, CH_3, S, O_2) $\xrightarrow{OH^-}$ alkoxide (O^-, Ph, CH_3, S, O_2) $\longrightarrow$ open-chain carbanion (CHO, Ph, CH_3, S, O_2) — Corey and Lowry (1965)

retroaldol-type reactions. Proton abstraction is the most significant of these routes and a variety of examples are given in Scheme 2-1. Scheme 2-2 provides examples of the other methods.

Preparative Aspects of Carbanion Generation by Proton Abstraction

By far the most important and general method for preparing sulfur-containing carbanions involves proton abstraction by base. In these reactions the practical matter of maximizing the yield of carbanion (or generating any carbanion at all!) often depends critically on the choice of base, solvent, temperature, and on structural features of the organosulfur compound. All reactions are best run under scrupulously anhydrous conditions in a nitrogen (or argon) atmosphere. Additions to, and withdrawals from, the reaction vessel are most conveniently made by syringe techniques. Figure 2-1 illustrates a typical experimental setup for carbanion generation and reaction. Three examples, reactions 1–3 from Scheme 2-1, will be discussed in more detail to illustrate the problems sometimes encountered in carbanion generation.

In diethyl ether, the yield of $PhSCH_2Li$ from reaction of butyllithium with thioanisole is only 35%; in tetrahydrofuran (THF), 15 hr at room

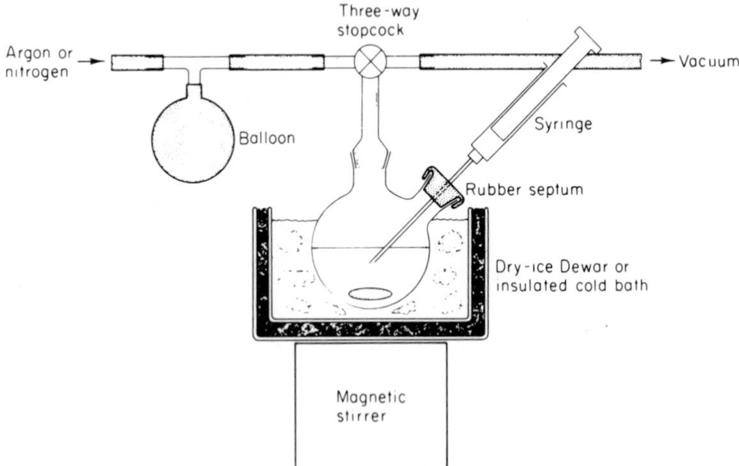

Figure 2-1. Apparatus for generation and reactions of sulfur-containing carbanions. A simple insulated cold bath may be made by sandwiching glass wool between two crystallizing dishes. Generally the system is repeatedly evacuated and flushed with train purified argon or nitrogen. Argon is often preferred since it is significantly heavier than air. A dry ice acetone or methanol slush provides a bath temperture of $-80°$; methanol or methylcyclohexane partially solidified with liquid nitrogen gives a bath temperature of $-100°$ or $-130°$, respectively.

temperature are required to obtain $PhSCH_2Li$ in 90% yield. However, on addition of an equivalent of 1,4-diazabicyclo[2.2.2]octane (**2-3**, "DABCO") a 97% yield of $PhSCH_2Li$ is realized after 45 min at 0° (Corey and Seebach, 1966)! The DABCO (and other tertiary amines) is believed to enhance the basicity of organolithium reagents at the expense of their nucleophilicity (Seebach, 1969; Ebel, 1969). Peterson has reported the analogous use of

2-3

N,N,N′,N′-tetramethylethylenediamine (TMEDA)–*n*-butyllithium complexes to generate methylthiomethyllithium in 84% yield (Eq. 2-2; Peterson, 1967).

$$CH_3SCH_3 + n\text{-}BuLi \cdot (N,N) \xrightarrow[4\ hr]{20^\circ} CH_3SCH_2Li \cdot (N,N) \qquad (2\text{-}2)$$

The yield of 1-(ethylthio)vinyllithium (see Eq. 2, Scheme 2-1), prepared by treatment of ethyl vinyl sulfide with an alkyllithium reagent, is critically dependent on the reaction parameters as demonstrated by the following data (yield is given in parentheses): *n*-BuLi/THF (2%), *n*-BuLi/TMEDA–THF (10%), *n*-BuLi/HMPA (68%), *sec*-BuLi/THF (72%), *sec*-BuLi/10% HMPA–THF (90%) (Oshima *et al.*, 1973b). The use of *sec*-butyllithium in the solvent pair 10% hexamethylphosphoric amide (HMPA, $[(CH_3)_2N]_3P{=}O$)–THF is apparently particularly effective in suppressing significant side reactions such as Michael-type addition (Eq. 2-3) (Parham and Matter, 1959). The

$$n\text{-}BuLi + RSCH{=}CH_2 \longrightarrow RSCH(Li)CH_2Bu\text{-}n \qquad (2\text{-}3)$$

highly polar solvent HMPA probably enhances the degree of dissociation of the alkyllithium and therefore, its basicity.†

Some novel alkylthio compounds have been prepared that contain within the molecule metal-coordinating groups (Eq. 2-4, Hirai *et al.*, 1971; Eqs. 2-5, 2-6, Narasaka *et al.*, 1972, and Evans and Andrews, 1974). The presence of these groups improves both the stability of the organolithium derivative as

† HMPA should be used with caution as recent studies suggest that it may be carcinogenic (*Chemical and Engineering News*, Feb. 2, 1976, p. 3). It should also be noted here that under certain circumstances HMPA can function as a source of the imine $CH_3N{=}CH_2$ in reactions with alkyllithiums (Abatjoglou and Eliel, 1974).

well as its nucleophilicity. In the case of allylic thiocarbanions (Eqs. 2-5 and 2-6), the coordinating group also plays an important role in controlling the regiospecificity of ensuing reactions, as will be discussed later in this chapter.

CH_3S–(2-thiazolinyl) $\xrightarrow[\text{THF–HMPA}]{n\text{-BuLi}/-30°}$ CH_2(Li····N)–S–(2-thiazolinyl) (2-4)

allyl–S–(2-pyridyl) $\xrightarrow[-70°]{n\text{-BuLi}}$ $CH_2{=}CH{-}CH$(Li····N)–S–(2-pyridyl) (2-5)

allyl–S–(1-methyl-2-imidazolyl) $\xrightarrow[\text{THF, } -30°]{n\text{-BuLi}}$ $CH_2{=}CH{-}CH$(Li····N)–S–(1-methyl-2-imidazolyl) (2-6)

2.3 Kinetic and Thermodynamic Assessment of Stability of Sulfur-Containing Carbanions

There is substantial evidence of both a qualitative and quantitative nature supporting the view that α-hydrogens in sulfides (as well as higher valent organosulfur compounds) are more acidic than α-hydrogens in ethers or other compounds involving only first-row elements (for a review of the early literature, see Price and Oae, 1962). For example, in 1940 Gilman reported that while thioanisole could be metalated at the methyl group, anisole underwent only nuclear metalation (Eq. 2-7; Gilman and Webb, 1940). Also of interest are the more quantitative data of Shatenshtein on the

$$PhSCH_3 \xrightarrow{n\text{-BuLi}} PhSCH_2Li \quad \text{but} \quad PhOCH_3 \xrightarrow{n\text{-BuLi}} o\text{-}(OCH_3)C_6H_4Li \qquad (2\text{-}7)$$

relative rates of H/D exchange with the base system KND_2/ND_3 for a series of substituted benzenes (given is the compound followed by the relative exchange rate of a methyl proton; Shatenshtein and Gvozdeva, 1969): $C_6H_5N(CH_3)_2$, 1; $C_6H_5OCH_3$, 40; $C_6H_5CH_3$, 10^4; $C_6H_5P(CH_3)_2$, 3×10^4; $C_6H_5SeCH_3$, 2×10^7; $C_6H_5SCH_3$, 2×10^8. On this same scale, the relative exchange rate of a methyl proton of cyclohexylmethyl sulfide is 9×10^4 (Shatenshtein and Gvozdeva, 1969). Additional quantitative exchange data are included in Table 2.1 (Oae *et al.*, 1964).

There are problems associated with the interpretation of the above data. Thus, a careful study of the metalation of thioanisole as a function of time

Table 2-1. Relative Rates of D/H or T/H Exchange for Acetals, Thioacetals, Orthoformates and Trithioorthoformates with KO-*t*-Bu in *t*-BuOH[a,b]

Compound	Relative rate (per H)	Compound	Relative rate (per H)
$C_2H_5CH(SC_2H_5)_2$	1.00	 H, S, H, S [structure]	531
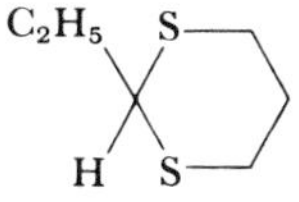 C_2H_5, S, H, S [structure]	1.43	$HC(SC_2H_5)_3$	1.5×10^4
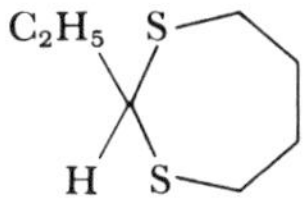 C_2H_5, S, H, S [structure]	5.37	$PhCH(SC_2H_5)_2$	8.7×10^4
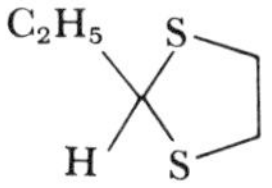 C_2H_5, S, H, S [structure]	19.1	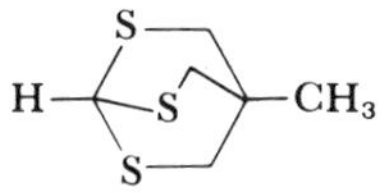 H, S, S, S, CH_3 [structure]	1.1×10^7
$H_2C(SC_2H_5)_2$	38.3	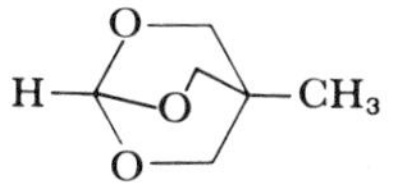 H, O, O, O, CH_3 [structure]	no exchange
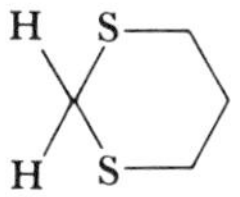 H, S, H, S [structure]	212	$HC(OC_2H_5)_3$	no exchange

[a] Oae *et al.* (1964).
[b] See also Coffen *et al.* (1971).

(Table 2.2) indicated that substantial nuclear metalation did occur although methyl metalation was thermodynamically favored (Shirley and Reeves, 1969). Furthermore, studies by Shatenshtein reveal that the exchange rate of thioanisole relative to toluene shows dramatic variation with base system varying from 10^4 in KNH_2/NH_3 to 40 in $NaCH_2S(O)CH_3$/DMSO to 0.2 in KO-*t*-Bu/DMSO (Shatenshtein and Gvozdeva, 1969). In general, *kinetic acidity data* are difficult to interpret because of *internal return* and *ion pairing*. In the former case the rate-limiting step is the rate of exchange of solvent molecules at the carbanion site rather than carbanion formation, while in the latter case relative rates vary markedly with cation, anion, and solvent (as we have seen above) particularly in solvents of only moderate polarity such as ethers, alcohols, amines, and ammonia (Matthews *et al.*, 1975). Bordwell and others have argued that kinetic acidities may indeed misrepresent carbanion stabilities; steric, polar, or conjugative effects of substituents near the acidic site may in some cases affect kinetic acidities in such a way as to indicate an order of carbanion stability the inverse of that actually present (Matthews *et al.*, 1975)!

Difficulty may also be encountered in *equilibrium acidity measurements*

Table 2-2. Product Distribution in Metalation of Thioanisole by Butyllithium as a Function of Time

Product	Distribution (%)		
	5 min	60 min	900 min
$C_6H_5SCH_2Li$	63	90	96
o-$LiC_6H_4SCH_3$ and m-$LiC_6H_4SCH_3$	37	9	4
p-$LiC_6H_4SCH_3$	Trace	1.3	Trace

(determination of relative or absolute pK_a values) because of ion-pairing effects. Thus phenylacetylene has an apparent pK_a of ~17.8 in ether, ~22.8 in cyclohexylamine, and 28.8 in dimethylsulfoxide (Matthews *et al.*, 1975). In dimethyl sulfoxide, a solvent of high dielectric constant, potassium salts of carbanions have been shown by Bordwell to exist as fully dissociated ions; in cyclohexylamine and ether, ion pairing is thought to occur (Matthews *et al.*, 1975). In addition to dielectric constant, other solvent properties that are thought to determine the equilibrium position in acidity measurements are (a) hydrogen bonding, (b) basicity, (c) dipole moment, (d) polarizability, and (e) structural organization. All things considered, DMSO is thought to be the most satisfactory solvent for equilibrium acidity measurements over a wide range of pK (about 30 pK units) with apparently little or no interference from ion association effects. Bordwell has accordingly established an absolute scale of acidities in DMSO solution, a portion of which is given as Table 2.3 (Matthews *et al.*, 1975; Bordwell *et al.*, 1975, 1976).

2.4 The Nature of Carbanion Stabilization by Sulfur

The much higher acidity of C—H bonds α to sulfur as compared to C—H bonds in hydrocarbons or C—H bonds α to oxygen has usually been ascribed to $d\pi$–$p\pi$ back-bonding of the carbanion lone pair into the vacant sulfur 3d orbital. (See for example, Cilento, 1960; Price and Oae, 1962; Cram, 1965;

Salmond, 1968; Coffen, 1969; Seebach, 1969; Mitchell, 1969; Peterson, 1972; Bordwell *et al.*, 1977, and references therein.)

Several representations of dπ–pπ bonding involving the orbitals C_{2p_z}–$S_{3d_{xz}}$ or C_{2p_y}–$S_{3d_{yz}}$ (the C—S bond is taken as the *x*-axis) are shown in Eq. 2-8. Early theoretical work (Kimball, 1940; Koch and Moffitt, 1951)

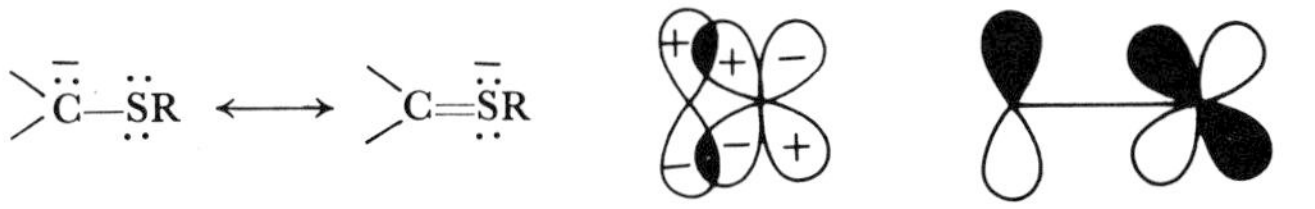

(2-8)

indicated that there is little angular requirement for d–p π-overlap, in contrast to p–p π-overlap. The observations that bicyclic structures **2-4**–**2-6** showed evidence of C—H activation (or carbanion stability) which was quite comparable with that of their respective acyclic analogs **2-7**–**2-9** (Doering and Levy, 1955; Doering and Hoffmann, 1955; Oae *et al.*, 1974), in striking contrast to the greatly diminished bridgehead acidities of **2-10** and **2-11** compared to their respective acyclic analogs (acetylacetone and triphenylmethane have respective pK_a values of 9 and 31.5) (Bartlett and Woods, 1940; Bartlett and Lewis, 1950) have been frequently cited as compelling evidence for d–p π-bonding by sulfur.

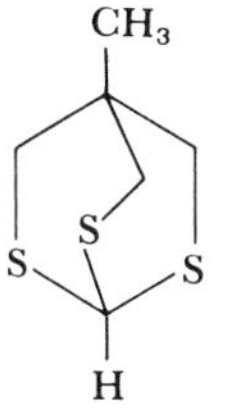

2-4 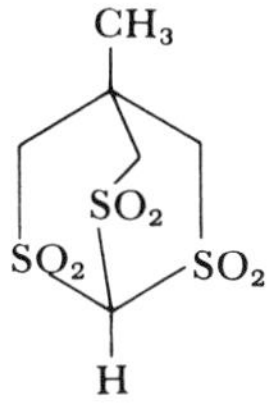 **2-5** 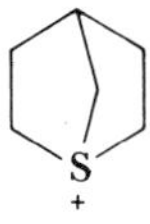 **2-6** $(C_2H_5S)_3CH$ **2-7** $(C_2H_5SO_2)_3CH$ **2-8**

$(CH_3)_3S^+$ **2-9** **2-10** 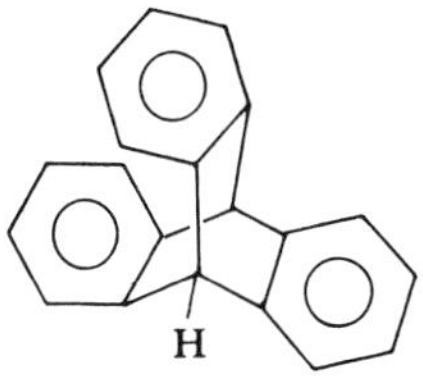 **2-11**

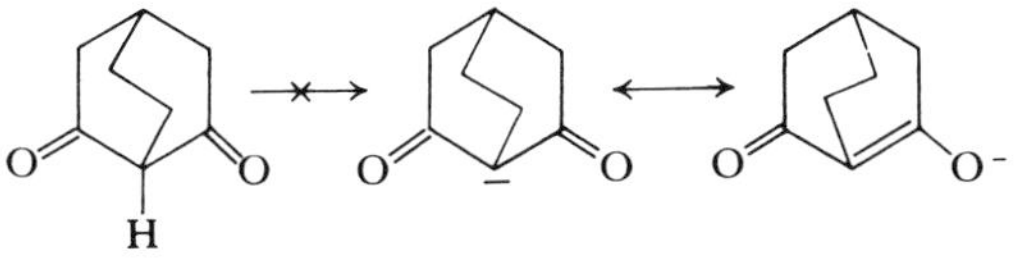

Inductive effects by the strongly electronegative sulfonyl and sulfonium groups could account for the comparable kinetic acidities of **2-5** and **2-8** and of **2-6** and **2-9** since it has been reported that 1 *H*-undecafluorobicyclo[2.2.1]-heptane (**2-12**) undergoes H/D exchange five times faster than acyclic analog **2-13** (Streitweiser and Holtz 1967). However, the enhanced acidities of sulfides **2-4** and **2-7** compared to ethers **2-14** and **2-15**, or of $PhSO_2CH_2SPh$

Table 2-3. Absolute Equilibrium Acidities of Carbon Acids in Dimethyl Sulfoxide[a]

Entry	Carbon acid	p*K*
1	$PhSO_2CH_2NO_2$	7.1
2	$H_2C(CN)_2$	11.1
3	$PhSO_2CH_2COPh$	11.4
4	$PhSO_2CH_2CN$	12.0
5	$PhSO_2CH_2SO_2Ph$	12.2
6	$PhSO_2CH_2COCH_3$	12.5
7	$C_2H_5SO_2CH_2SO_2C_2H_5$	14.4
8	CH_3NO_2	17.2
9	$CF_3SO_2CH_3$	18.8
10	$PhSO_2CH_2N^+(CH_3)_3$	19.4
11	$PhSO_2CH_2PPh_2$	20.2
12	$PhSO_2CH_2SPh$	20.3
13	$CF_3SO_2CH_2CH_3$	20.4
14	$CF_3SO_2CH(CH_3)_2$	21.9
15	$(PhS)_3CH$	22.5
16	$PhCH(SPh)_2$	23.0
17	$PhSO_2CH_2Ph$	23.4
18	$PhSO_2CH_2SCH_3$	23.4
19	$Ph_3P^+CH_3$	24.0
20	$PhCH_2SO_2CH_2Ph$	24.0
21	$PhCOCH_3$	24.7
22	$(PhCH(CH_3))_2SO_2$	25.7
23	CH_3COCH_3	26.5
24	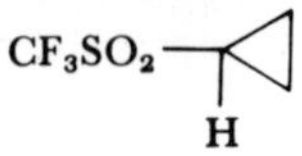 CF_3SO_2— (cyclopropane ring) H	26.6
25	Ph_2CHSPh	26.7
26	$PhSO_2CH_2OPh$	27.9
27	$PhC{\equiv}CH$	28.8
28	$PhCH_2S(O)CH_3$	29.0
29	$PhSO_2CH_3$	29.0
30	$(n\text{-}C_3H_7S)_2CHPh$	29.2
31	$(Ph_2P)_2CH_2$	29.9
32	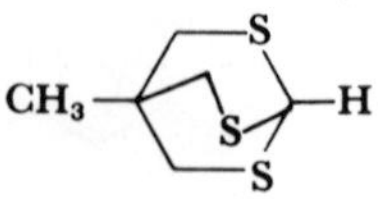 CH_3— (bicyclic structure: S, S, S) —H	30.5
33	Ph_3CH	30.6
34	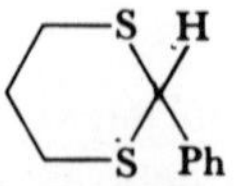 (ring structure: S, H, S, Ph)	30.7
35	$PhSO_2CH_2OCH_3$	30.7
36	$PhCH_2SPh$	30.8
37	$(PhS)_2CH_2$	30.8
38	$PhSO_2CH_2CH_3$	31.0

39	$CH_3SO_2CH_3$	31.1
40	$PhSO_2CH_2C(CH_3)_3$	31.2
41	CH_3CN	31.3
42	$(n\text{-}C_3H_7S)_3CH$	31.3
43	Ph_2CH_2	32.3
44	$CH_3S(O)CH_3$	35.1
45	$PhCH_3$	44 (estimated)
46	$PhSCH_3$	48 (estimated)
47	CH_4	65 (estimated)

[a] Matthews *et al.* (1975); Bordwell *et al.* (1975, 1976, 1977).

and $PhSO_2CH_2SCH_3$ compared to oxygen analogs $PhSO_2CH_2OPh$ and $PhSO_2CH_2OCH_3$ (from Table 2.3, pK_a values 20.3, 23.4, 27.9, and 30.7, respectively) cannot be rationalized on the basis of inductive effects since oxygen is more electronegative than divalent sulfur. Indeed, the fact that $PhSO_2CH_2OCH_3$ and $PhSO_2CH_2CH_3$ have virtually identical acidities (from Table 2.3, pK_a values 30.7 and 31.0, respectively) suggests that there is a destabilizing effect associated with an ether function which cancels out

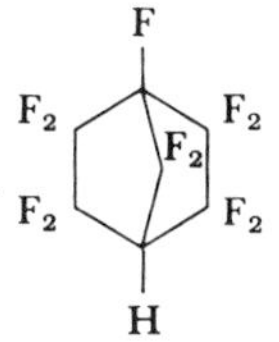

2-12 $(CF_3)_3C—H$ **2-13** **2-14** $(C_2H_5O)_3C—H$ **2-15**

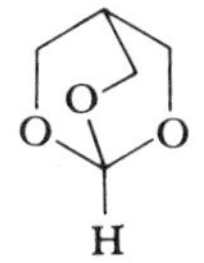

any carbanion stabilization through polar effect (Bordwell *et al.*, 1976). The ability of α-alkoxy and α-fluoro substituents to destabilize planar (sp^2) carbanions had been previously noted by Hine and Dalsin (1972) and by Streitwieser and Mares (1968) and has recently been examined by molecular orbital calculations (Lehn and Wipff, 1976). Polar effects for the trimethylammonio group $(CH_3)_3N^+$ should be larger than for any uncharged group so that the large carbanion stabilizing effect seen for the NO_2, COPh, CN, and $PhSO_2$ groups compared to the $(CH_3)_3N^+$ group† must be attributed to additional resonance (conjugative) effects (Bordwell *et al.*, 1976). Consequently, Bordwell concludes that the phenylsulfonyl ($PhSO_2$) function "is capable of a strong conjugative interaction with an α-carbanion, comparable in size with that of the cyano function but somewhat smaller than that of carbonyl or nitro functions" (Bordwell *et al.*, 1976).

The relevance of d orbital conjugation to ground state properties of organosulfur compounds has been challenged on the basis of molecular orbital

† If $PhSO_2CH_2C(CH_3)_3$ is used as a reference point, the acidifying effects from Table 2.3 expressed as ΔpK for $(CH_3)_3N^+$, $PhSO_2$, CN, COPh, and NO_2 are 11.8, 19, 19.2, 19.8, and 24.1, respectively.

calculations by Wolfe and co-workers (Wolfe *et al.*, 1967, 1969, 1970; Rauk *et al.*, 1965, 1969; Bernardi *et al.*, 1975; Epiotis *et al.*, 1976), Coulson (1969), Florey and Cusachs (1972), Musher (1972), Streitweiser and Williams (1975), and Lehn and Wipff (1976). For example, *ab initio* SCF–MO calculations on $RSCH_2^-$, $ROCH_2^-$, and $RCH_2CH_2^-$ (R = H or CH_3) predict the order of gas phase carbanion stabilization as S > O > C whether or not sulfur 3d orbitals are used in the calculations. Indeed in the calculations of proton affinities of these anions (inversely proportional to carbanion stability) the inclusion of d orbitals has essentially no effect on the numbers obtained. Table 2.4 summarizes proton affinity data as calculated by three different research groups (Bernardi *et al.*, 1975; Streitwieser and Williams, 1975; Lehn and Wipff, 1976). Other conclusions from these *ab initio* calculations are

1. The C—S bond in $^-CH_2SH$ is longer than in CH_3SH which argues against d–p π-bonding in the anion (bond shortening should occur if π-bonding is involved).

2. The preferred conformation of $^-CH_2SH$ is predicted to be **2-16** whether or not 3d orbitals are included in the basis set (Bernardi *et al.*, 1975). This latter point finds support in the observation that the equatorial 2*H* of 1,3-dithiane has a greater kinetic acidity than the axial 2*H* (Eliel *et al.*, 1974) (note that the equatorial 2-carbanion **2-17** corresponds to conformer **2-16**).

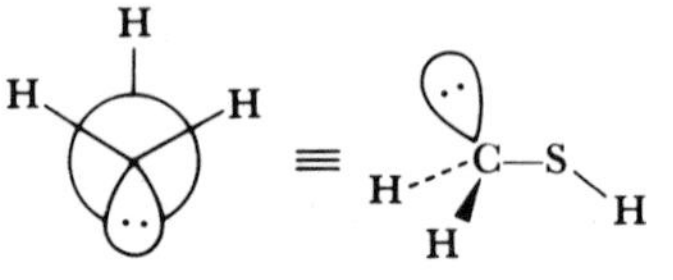

2-16 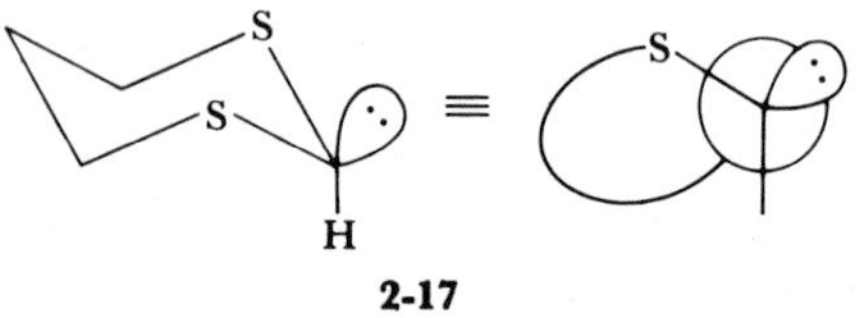 **2-17**

3. An important mechanism of stabilization of carbanions by adjacent sulfur is by *polarization* of the electron distribution, i.e., dispersal of the charge over the molecule.† A perturbational molecular orbital (PMO) approach indicates that in addition to the polarization effects, n_C-σ^*_{SR} charge transfer interactions of the carbon lone pair with the antibonding σ orbital of the antiperiplanar SR group contribute strongly to carbanion stabilization (Lehn and Wipff, 1976; Epiotis *et al.*, 1976). The PMO approach indicates that the $^-CH_2$—SR bond should be strengthened while the $^-CH_2S$—R bond should be lengthened and weakened. It is also argued that since σ^*(S—R) is

† Sulfur is much more polarizable than either carbon or oxygen as indicated by the polarizability values for the atoms (S, 3.45 Å^3; C, 1.75 Å^3; O, 0.73 Å^3) and bonds to carbon (C—S, 1.9 Å^3; C—C, 1.0 Å^3; C—O, 0.8 Å^3) (Lehn *et al.*, 1976). Polarizability α can also be calculated from an equation $\alpha = 2/3\langle r^2\rangle/\Delta$ which relates polarizability directly to the size of the atom (r^2), and inversely to the average energy difference (Δ) between the ground and exicted state (Bernardi *et al.*, 1975).

expected to lie below σ^*(O—R), interaction of the carbanion with the former should be stronger and more stabilizing than with the latter, in agreement with the greater stability of α-thiocarbanions than α-oxycarbanions (Lehn and Wipff, 1976; Epiotis *et al.*, 1976).

Table 2-4. Calculated Proton Affinities for $HSCH_2^-$, $HOCH_2^-$, $CH_3SCH_2^-$, and RCH_2^- (kcal/mole)

Basis set	$HSCH_2^-$	$HOCH_2^-$	$CH_3SCH_2^-$	RCH_2^-		Reference
sp	−423.3	−438.3	—	—		Bernardi *et al.* (1975)
spd	−421.4	−443.5	—	—		Bernardi *et al.* (1975)
sp	−423.5	—	—	−453.0	R = CH_3	Streitweiser and Williams (1975)
spd	−423.2	—	—	−452.7	R = CH_3	Streitweiser and Williams (1975)
sp	—	—	−428.0	−457.0	R = C_2H_5	Lehn and Wipff (1976)
spd	—	—	−428.0	−457.0	R = C_2H_5	Lehn and Wipff (1976)

Bicyclic compound **2-4**, in which the bridgehead carbanion is constrained to the W conformational relationship (analogous to **2-16**) with respect to the adjacent C—S bonds, has been suggested to be a particularly favorable case for strong n_C-σ^*_{SR} charge transfer interaction (Epiotis *et al.*, 1976; Lehn and Wipff, 1976). Oae had previously noted that the kinetic acidity of the bridgehead proton of **2-4** is greater by a factor of 10^3 than that of the tertiary proton of the open chain analog tris(ethylthio)orthoformate **2-7** (Oae *et al.*, 1964). Such a rate difference could translate into a difference in equilibrium acidities between **2-4** and **2-7** as large as 6 pK units (assuming a "normal" Brønsted coefficient of 0.5, as discussed in Section 2.1). However Bordwell has recently found that **2-4** is only 0.8 pK units more acidic than **2-7** (n-propyl groups instead of ethyl groups; see Table 2.3) (Bordwell *et al.*, 1977). Bordwell therefore concludes that the differences in acidities between **2-4** and **2-7** can be explained by differences in solvation of the anions rather than by stereoelectronic effects as argued by Epiotis and Lehn (Bordwell *et al.*, 1977).

As experimental evidence for the localized nature of the 2-dithiane cesium salt, Streitweiser cites the unusually large pK_a changes observed in going from 1,3-dithiane (pK_a, 31.1) to 2-methyl 1,3-dithiane (pK_a, 38.3) and to 2-benzyl 1,3-dithiane (pK_a, 34.0) suggesting that delocalization or rehybridization effects which could ameliorate the carbanion-destabilizing effect of a methyl group are absent (Streitweiser and Ewing 1975).

An additional point that is relevant to the discussion of d orbital overlap is that dicoordinate selenium, the d shells of which have a higher energy level and should be more diffuse than those of sulfur, also effectively stabilize carbanions (Seebach, 1969; Seebach and Peleties, 1972; Shatenshtein and Gvozdeva, 1969; Reich, 1975). Indeed, it is known that selenophen is metalated in the α-position by organolithium reagents five to six times *faster* than thiophene which in turn is metalated appreciably faster than furan and 1-methylpyrrole (Gronowitz, 1975). Even tellurophen is readily metalated. Bordwell reports pK_a values of 17.1 and 18.6 for $PhSCH_2COPh$ and $PhSeCH_2COPh$, respectively (Bordwell *et al.*, 1976). When corrections are made for polar effects, the sulfide is determined to be 0.8 pK_a units more acidic than the selenide.

It has been predicted that increasing the electron-withdrawing power of the group attached to sulfur (e.g., converting sulfides to sulfoxides, sulfones, and sulfonium salts or increasing the electronegativity of attached carbon groups) will contract the sulfur 3d orbitals and make them more available for conjugative overlap (Craig and Thirunamachandian, 1965). Earlier calculations on α-sulfinyl and α-sulfonyl carbanions by Wolfe and co-workers led to the conclusion that postulation of d-orbital conjugation is unnecessary in these systems (Wolfe *et al.*, 1967, 1969, 1970; Wolfe, 1972; Rauk *et al.*, 1969). However, since it now appears that the sulfoxide bond *cannot* be made without d orbitals on sulfur, the proper assessment of the role of $p\pi$–$d\pi$ conjugation in the stabilization of α-sulfinyl and α-sulfonyl carbanions requires considerable further work (Wolfe, 1976). Further aspects of bonding in α-sulfinyl and α-sulfonyl carbanions will be discussed in the next section.

2.5 Stereochemical Aspects of α-Sulfonyl Carbanions

The remarkable observation that α-sulfonyl carbanions could be generated from optically active precursors in polar, protic solvents and protonated to give optically active products with retention of configuration (Eq. 2-9; Cram, 1965; Corey and Lowry, 1965a, b and references therein) stimulated

$$\begin{array}{l}
PhSO_2{*}C(CH_3)(C_6H_{13}\text{-}n){-}D \xrightarrow{B^-} \\
PhSO_2{*}C(CH_3)(C_6H_{13}\text{-}n){-}CO_2^- \xrightarrow{-CO_2} \left[PhSO_2\overset{CH_3}{\underset{C_6H_{13}\text{-}n}{C}}{:}^- \right] \xrightarrow{HB} PhSO_2{*}\overset{CH_3}{\underset{C_6H_{13}\text{-}n}{C}}{-}H \\
PhSO_2{*}C(CH_3)(C_6H_{13}\text{-}n){-}C(O^-)(CH_3)_2 \xrightarrow{-(CH_3)_2CO}
\end{array} \tag{2-9}$$

considerable research and debate on the stereolectronic basis for this effect (Gresser, 1969; Roitman and Cram, 1971; Henderson, 1973). In this system the barrier to racemization is the barrier to C—S bond rotation if the carbanion center is planar as in Eq. 2-10 or the barrier to carbanion inversion with C—S bond rotation if the carbanion is nonplanar as in Eq. 2-11 (Henderson, 1973). If the carbanion center is planar, the involvement of structure B, Eq. 2-10, can be excluded since it is achiral and would lead to racemization. Several experiments were designed to distinguish between a

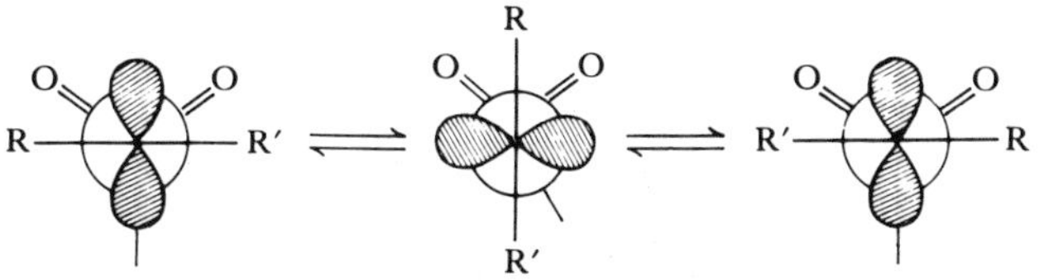

A ⇌ B ⇌ A′ (2-10)

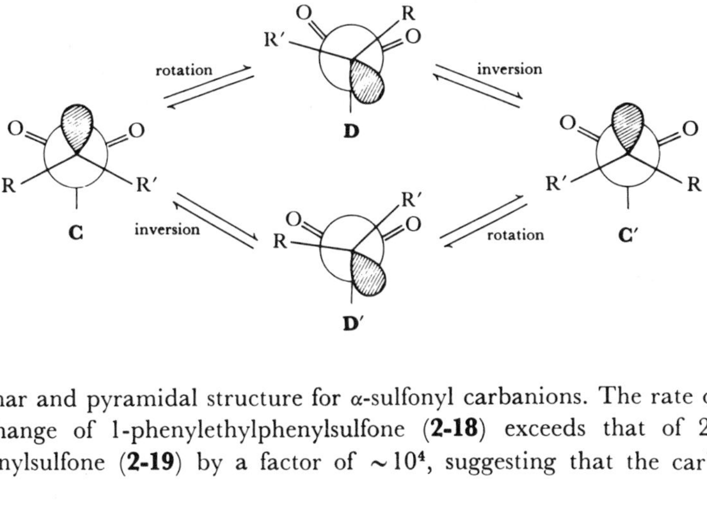

C ⇌ D (rotation); D ⇌ C′ (inversion); C ⇌ D′ (inversion); D′ ⇌ C′ (rotation) (2-11)

planar and pyramidal structure for α-sulfonyl carbanions. The rate of H/D exchange of 1-phenylethylphenylsulfone (**2-18**) exceeds that of 2-octylphenylsulfone (**2-19**) by a factor of $\sim 10^4$, suggesting that the carbanion

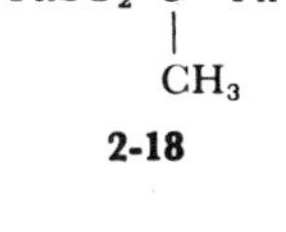

$PhSO_2{}^*CH(CH_3)—Ph$ **2-18**

$PhSO_2{}^*CH(CH_3)—C_6H_{13}\text{-}n$ **2-19**

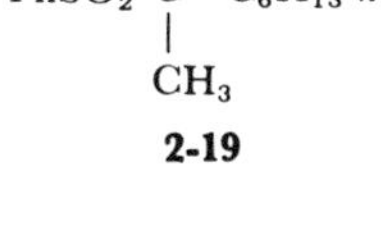

center of **2-18** is planar (sp^2 hybridized) to facilitate π-conjugation. However, in spite of this *increase* in planarity, the barrier to racemization is actually slightly higher in the carbanion from **2-18** than in the carbanion from **2-19**

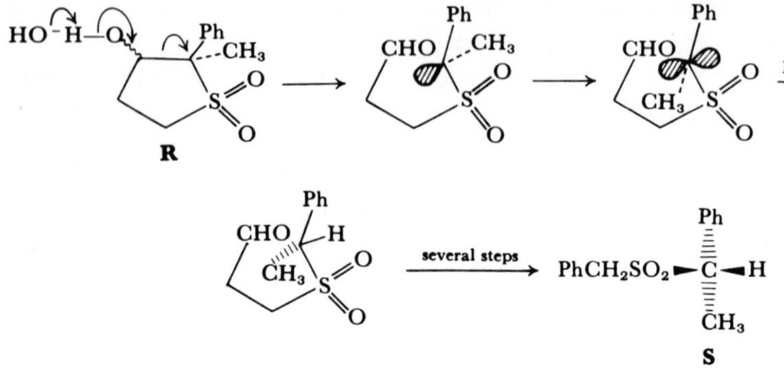

(2-12)

(Corey and Lowry, 1965b)! Corey and Lowry have demonstrated the stereochemical course of protonation of an asymmetric α-sulfonyl carbanion (Eq. 2-12; Corey and Lowry, 1965a) which proceeds with inversion of configuration. This datum requires that the barrier to rotation about C_α—S be the significant factor in maintaining asymmetry of α-sulfonyl carbanions (i.e., sequences **C** ⇌ **D** and **C′** ⇌ **D′**, Eq. 2-11, can be excluded). Another result which indicates that pyramidal α-sulfonyl carbanions can readily invert is the facile Ramberg–Bäcklund reaction of 1-bromo-9-thiabicyclo[3.3.1]nonane 9,9-dioxide (Eq. 2-13; Corey and Block, 1969).

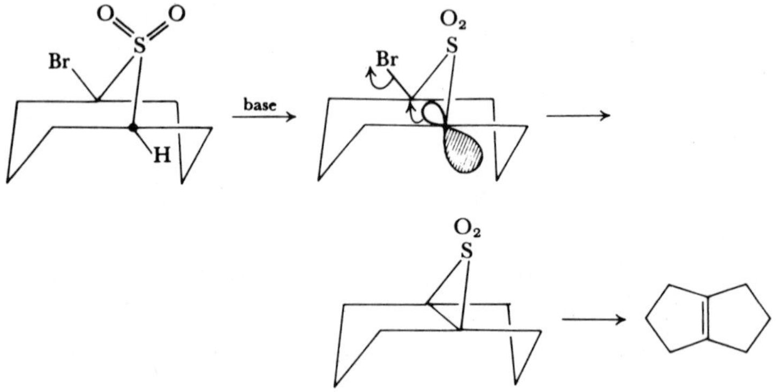

(2-13)

Ab initio calculations by Wolfe on the hypothetical α-sulfonyl carbanion $HSO_2CH_2^-$ indicate that the preferred conformation of the carbanion places the lone pair between the sulfone oxygens (*syn*). The geometry is suggested to be intermediate between forms **A** and **C** (Eqs. 2-10 and 2-11, respectively) with an HCH angle (RCR′ in C, Eq. 2-11) of 115° (Wolfe *et al.*, 1969). A number of studies of carbanion generation and H/D exchange in cyclic sulfones provide support for the theoretically predicted preferred conformation of α-sulfonyl carbanions which places the lone pair on the internal bisector of the O—S—O angle (Durst, 1971b; Barbarella *et al.*, 1975; Kattenberg *et al.*, 1975).

A recent study by Bordwell and Hendrickson provides evidence on the geometry of α-sulfonyl carbanions (Bordwell *et al.*, 1975). As indicated in

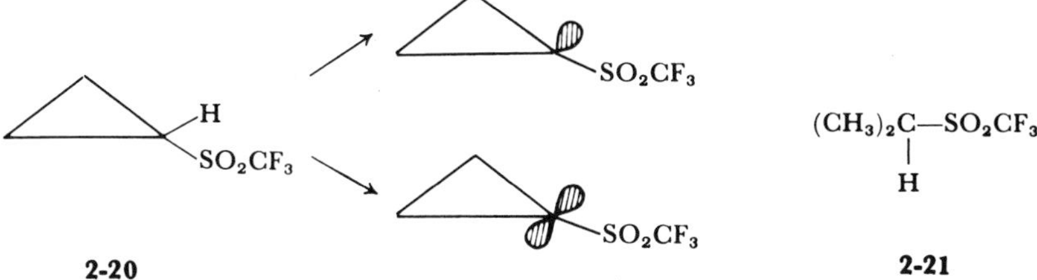

Table 2.3, trifluoromethylcyclopropyl sulfone (**2-20**) is substantially less acidic than trifluoromethyl isopropyl sulfone (**2-21**) ($\Delta pK \sim 5$) contrary to the intrinsically greater acidity of cyclopropyl hydrogens. The implication is that α-carbanion stabilization of CF_3SO_2 is maximum with a planar carbanion and that planarity is far more difficult to achieve because of ring strain in the anion derived from **2-20** than in the anion derived from **2-21**. Similar trends are seen on comparing the acidities of isopropyl and cyclopropyl nitro and carbonyl compounds in which the anions are known to be stabilized by π-conjugation (Bordwell *et al.*, 1975). From a ^{13}C NMR study of $PhCHLiSO_2CH_3$ and *t*-$BuSO_2CH_2Li$, it is suggested that the former carbanion is primarily sp^2 hybridized while the latter carbanion is intermediate between sp^2 and sp^3 hybridization (Lett *et al.*, 1976).

2.6 Stereochemical Aspects of α-Sulfinyl Carbanions

It has been known for decades that suitably substituted sulfoxides **2-22** are dissymmetric, configurationally stable under ordinary conditions, and resolvable into optically active enantiomeric forms (Mislow, 1967). Enantiomers may interconvert by a number of processes, one of which is pyramidal

2-22

inversion, corresponding to a $t_{1/2}$ of about 5000 yr at room temperature ($\Delta H^{\ddagger} = 35$–42 kcal/mol and $\Delta S^{\ddagger} = -8$ to $+4$ e.u. (Rayner, *et al.*, 1968)). Because the sulfoxide group is pyramidal, a diastereotopic relationship exists for α-methylene protons of sulfoxides. The diamagnetic anisotropy of the sulfoxide group results in the nonequivalence of these α-methylene protons in NMR spectra; peak separation can generally be improved by addition of shift reagents or benzene (Barbarella *et al.*, 1975 and references therein). The first evidence for the differential acidity of diastereotopic α-methylene protons such as H_A and H_B in **2-23** was provided in 1965 by Rauk and co-workers who demonstrated that one of the protons in **2-23** (later shown to be the

$$PhC^{-}H_AS(O)CH_3 \xleftarrow{k_{rel}\,=\,1} PhCH_AH_BS(O)CH_3 \xrightarrow{k_{rel}\,=\,16} PhC^{-}H_BS(O)CH_3$$

2-23

pro-R hydrogen H_A) undergoes exchange in D_2O/NaOD at a rate approximately 16 times faster than the other (H_B or pro-S hydrogen) (Rauk *et al.*, 1965; also see Baldwin *et al.*, 1969 and Durst and Viau, 1973). The basis for proton exchange in sulfoxides with retention and inversion of configuration was examined by molecular orbital calculations on the hypothetical anion $^{-}CH_2S(O)H$ (Rauk *et al.*, 1969). These studies indicated

1. a preference for a nonplanar geometry at the carbanionic center,
2. the relative insignificance of 3d-orbital conjugation effects, and
3. the order of carbanion stability as **2-24 A** > **B** > **C**.

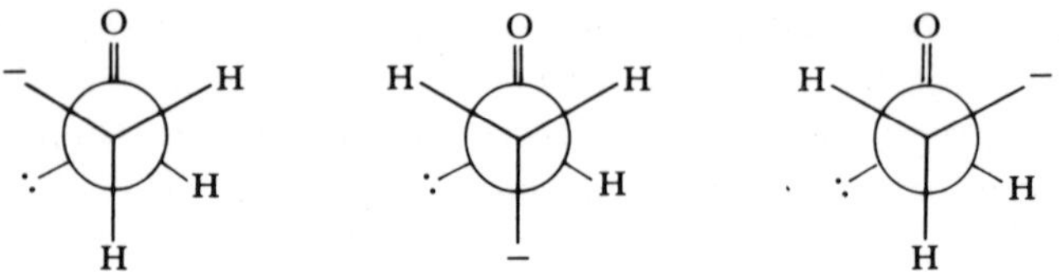

A **B** **C**

2-24

The particular stability predicted for **2-24 A**, wherein the lone-pair on carbon lies in the bisector of the oxygen–sulfur: lone-pair angle, is attributed to the "gauche effect," i.e., the tendency by a molecule or reactive intermediate "to adopt that structure which has the maximum number of gauche interactions between the adjacent electron pairs and/or polar bonds" (Wolfe, 1972).

A large number of studies have been conducted to determine the conformational preferences of α-sulfinyl carbanions in solution (Durst and Viau, 1973; Henderson, 1973; Bory and Marquet, 1973 and references therein). The fact that agreement with the prediction **2-24 A** > **B** > **C** is rare has been attributed to the overriding importance of solvation and ion-pairing effects, which are neglected in the theoretical calculations (Montanari, 1975; Barbarella *et al.*, 1975). For example, in a study of the stereoselectivity of reactions of α-lithiosulfoxides, Biellmann finds product stereochemistry to vary with solvent, temperature, and with the presence or absence of lithium chelating agents such as cryptate **2-2**. (Biellmann and Vicens, 1974). It is argued that cation–oxygen chelation **2-25** may control the conformation. The ratio of diastereomers produced by quenching α-lithiobenzyl methyl sulfoxide **2-23** has

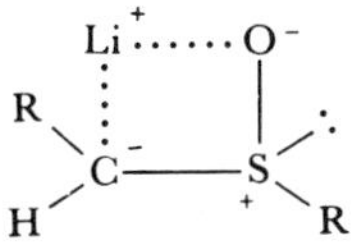

2-25

been shown by Durst to be strongly affected by the presence of other tetrahydrofuran-soluble lithium salts, certain of which (LiBr, LiI) may be present in commercial preparations of organolithium reagents (Durst and Molin 1975). Fraser has shown that internal return must be considered when H–D exchange data on sulfoxides are used to assess relative stabilities of the transition states leading to anion formation (Fraser and Ng, 1976). Even the widely accepted assumption of pyramidal sp^3 hybridized α-sulfinyl carbanions has been challenged. On the basis of a ^{13}C NMR study of $PhCHLiS(O)CH_3$ and $t\text{-}BuS(O)CH_2Li$, Marquet argues that the former carbanion is primarily

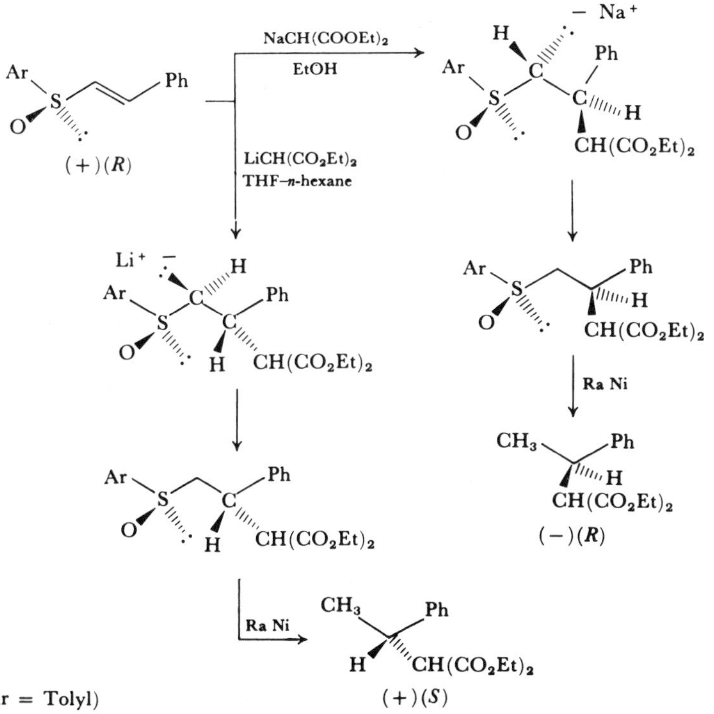

$NaCH(COOEt)_2$, EtOH

$LiCH(CO_2Et)_2$, THF–*n*-hexane

(+)(*R*)

Ra Ni

(−)(*R*)

(2-14)

(+)(*S*)

(Ar = Tolyl)

sp^2 hybridized while the latter is intermediate between sp^2 and sp^3 hybridized (Lett *et al.*, 1976). It is hoped that some of the controversy associated with the question of the preferred stereochemistry of α-sulfinyl carbanions will be resolved by more sophisticated calculations that take solvation effects into consideration and by gas phase acidity measurements (e.g., see McMahon and Kebarle, 1976).

Exploitation of α-sulfinyl carbanion ion-pair effects permits the creation of both (*R*)- and (*S*)-asymmetric centers from a single chiral origin as shown in Eq. 2-14 (Tsuchihashi *et al.*, 1976).

2.7 Reactions of Sulfur-Containing Carbanions

General

Organosulfur carbanions derived from saturated and unsaturated sulfides, sulfoxides and sulfones, from dithioacetals, hemithioacetals, and trithioformates (and their S-oxides), from sulfinate and sulfonate esters, as well as from other sulfur-containing systems may undergo the entire gamut of reactions characteristic of organometallic compounds. A partial list of such reactions would include protonation (proton-abstraction), carbonation, inter- and intramolecular alkylation, hydroxyalkylation with epoxides (nucleophilic ring opening) and with aldehydes and ketones (carbonyl addition), acylation with carboxylic acid derivatives, addition to carbon–carbon and carbon–nitrogen multiple bonds (including conjugate addition), elimination and fragmentation, oxidative coupling, and rearrangement. In addition to their ease of preparation and versatile chemistry, organosulfur carbanions have been widely and artfully used in organic synthesis because of the diverse possibilities for further functionalization of the primary products as well as removal of sulfur through elimination, reduction, or hydrolysis.

This latter point bears elaboration since it forms the basis of a widely used synthetic strategem. A variety of organosulfur compounds are known that have the same level of oxidation as aldehydes and ketones (e.g., **2-26–2-29** or as carboxylic acids (e.g., **2-30**) and which are readily converted into the carbonyl equivalents by hydrolysis. Thioacetals and thioketals have, of course, long been used as protecting groups for the carbonyl function.

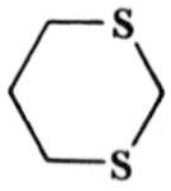

(1,3-dithiane) $\equiv CH_2{=}O$ **2-26**

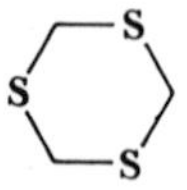

(1,3,5-trithiane) $\equiv CH_2{=}O$ **2-27**

$CH_2{=}C(SR)H \equiv CH_3CHO$ **2-28**

$RSCH_2S(O)R \equiv CH_2{=}O$ **2-29**

$(RS)_3C{-}H \equiv HCO_2H$ **2-30**

While the carbanion-forming capacity of certain of these organosulfur carbonyl equivalents had been recognized by a number of researchers (e.g., Arens *et al.*, 1959; Truce and Roberts, 1963), it remained for Corey and Seebach to demonstrate with the 1,3-dithiane carbanion the great synthetic utility of *nucleophilic acylation* (Corey and Seebach, 1965; Seebach and Kolb, 1974; Lever, 1976). By means of the 1,3-dithianyl group, the normal direction of polarity of the carbonyl group has in effect been temporarily reversed, transforming the normally electrophilic carbonyl carbon into a nucleophilic center (termed *umpolung* or *dipole inversion*) as represented in Eq. 2-15 (Seebach

$$\mathrm{HSCH_2CH_2CH_2SH} + \mathrm{R(H)}\overset{\delta+}{\mathrm{C}}{=}\overset{\delta-}{\mathrm{O}} \longrightarrow \text{2-R-2-H-1,3-dithiane} \xrightarrow{\mathrm{BuLi}} \text{2-R-2-Li-1,3-dithiane} \equiv \mathrm{R}\underline{\mathrm{C}}{=}\mathrm{O} \qquad (2\text{-}15)$$

and Beck, 1974; Gröbel and Seebach, 1977). This procedure is of obvious great synthetic importance since the whole range of reactions characteristic of carbanions can now be effected with these nucleophilic carbonyl equivalents. Following the original publications by Corey and Seebach, reports of other useful organosulfur (and sulfur-free) nucleophilic acylating agents have appeared frequently in the chemical literature. Certain of these newer methods are claimed to offer advantages over the dithiane method because of greater ease of hydrolysis to carbonyl compounds.†

The number of useful applications of organosulfur carbanions in organic synthesis is legion. The survey of reactions of sulfur-containing carbanions that follows will point up the variety of reactions possible and will emphasize the value of combining sulfur carbanion reactions with sulfur removal reactions in the synthesis of natural products. More complete coverage of the literature is to be found in the chapters by Block (1977) and Johnson *et al.* (1975).

Alkylation

The controlled formation of C—C bonds is often a critical step in organic syntheses. Organosulfur carbanions have proven to be extremely useful in this process.

† A survey of methods for the hydrolysis of thioacetals, thioketals, and vinyl sulfides may be found in Tang *et al.* (1975), Corey and Erickson (1971), Mura *et al.* (1975), Vedejs and Fuchs (1971), Gröbel and Seebach (1977).

Carbon–carbon bond formation by displacement on an alkyl halide by a thiocarbanion such as a lithio dithiane may be viewed as involving initial complex formation as in Eq. 2-16 (Eliel *et al.*, 1974). Selected examples of

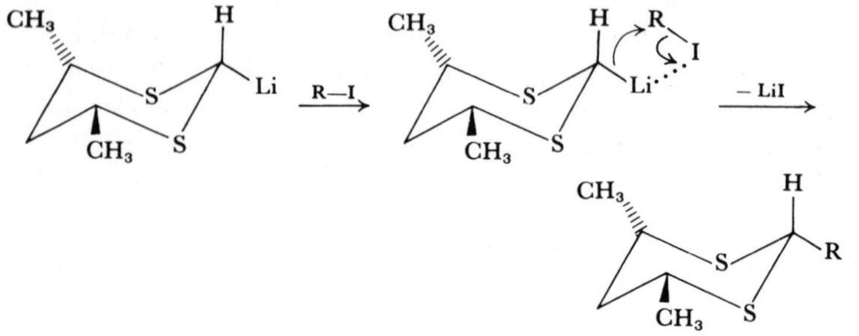

(2-16)

C—C bond formation involving displacement by the dithiane anion on alkyl halides and epoxides include the synthesis of a bark beetle sex attractant (Eq. 2-17; Reece *et al.*, 1968), cyclobutanone (Eq. 2-18; Seebach and Beck, 1971), cyclophanes (Eq. 2-19; Boekelheide *et al.*, 1974), and alnusone dimethyl ether (Eq. 2-20; Semmelhack and Ryono, 1975). In the total synthesis of the antibiotic (±)-gliotoxin, Kishi ingeniously protects the sensitive disulfide bridge as the thioacetal of anisaldehyde. The reduced acidity of the thioacetal proton permits double alkylation at the bridgehead position (Eq. 2-21; Fukuyama and Kishi, 1976; also see Kishi *et al.*, 1973).

A synthetically valuable method of coupling allylic groupings via allylic thiocarbanions was discovered by J. F. Biellmann and is termed *Biellmann alkylation*. Two significant applications of this procedure are the syntheses of squalene (Eq. 2-22) by Biellmann and Ducep (1971) and of the cecropia juvenile hormone (Eq. 2-23) by Kondo *et al.* (1972) and independently by

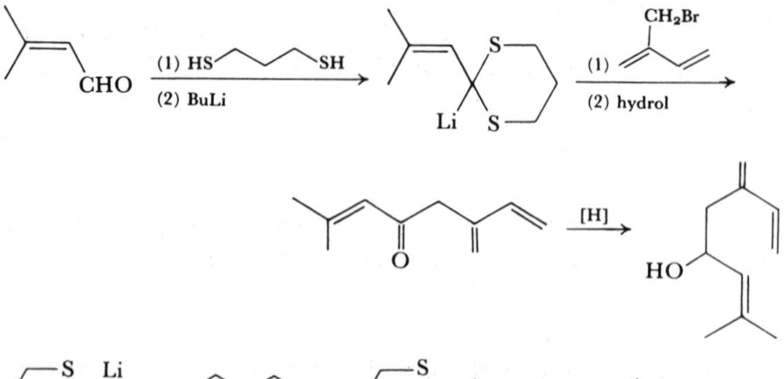

(2-17)

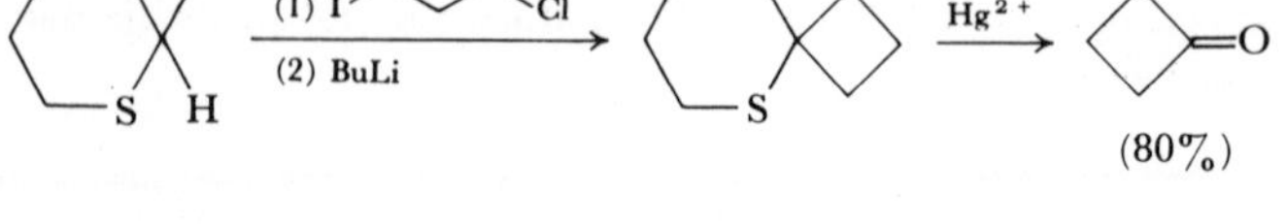

(2-18)

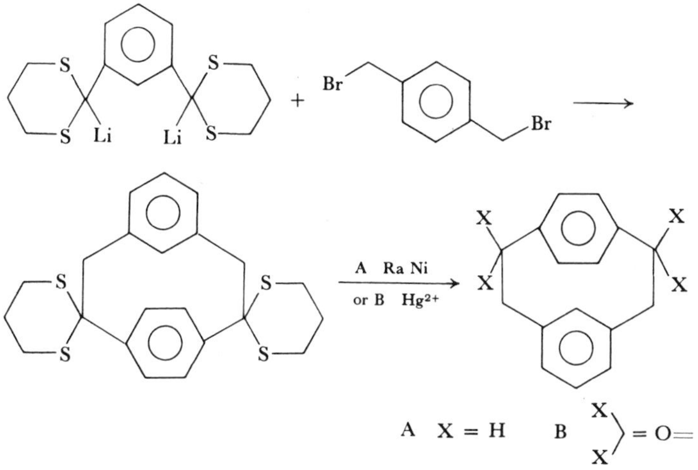

A Ra Ni
or B Hg^{2+}

(2-19)

A X = H B X, X = O=

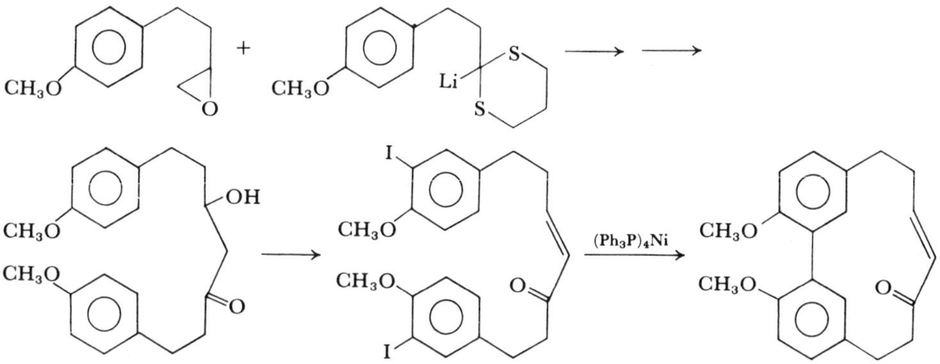

$(Ph_3P)_4Ni$

(2-20)

PhLi, THF
$PhCH_2OCH_2Cl$
−78°

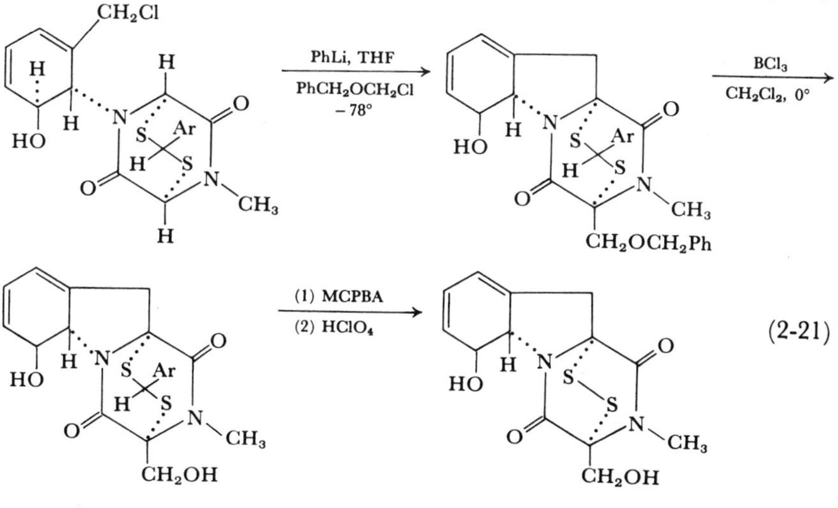

BCl_3
CH_2Cl_2, 0°

(1) MCPBA
(2) $HClO_4$

(2-21)

Br

Farnesyl bromide

(1) PhS^-
(2) BuLi/DABCO

Li SPh

Farnesyl bromide

PhS

$Li/C_2H_5NH_2$

(2-22)

S

BuLi
DABCO

Li

Br

(1) BuLi/DABCO
(2) Cl OTHP

OTHP

(1) H_3O^+
(2) $Li/EtNH_2$

SH SH

OH

Ra– Ni

OH

[O]

O

CO_2CH_3

(2-23)

Stotter and Hornish (1973) and by van Tamelen *et al.* (1971). Notable features of both of these syntheses are the retention of double bond positions and stereochemistry during both the alkylation and the C—S bond cleavage steps. While studies by Biellmann indicate that allylic thiocarbanions do not readily undergo *cis–trans* interconversion (Eq. 2-24; Biellmann and Ducep, 1969), other studies indicate that alkylation with exclusive retention of the double bond position (e.g., α-alkylation) occurs only if the terminal olefinic carbon is dialkyl substituted (see below; Biellmann and Ducep, 1968). In the

SPh SPh

(2-24)

execution of biosynthetic type syntheses of macrocyclic diterpenes and sesquiterpenes via anion-induced cyclization, Ito encountered serious problems with double bond migration and isomerization during *both* the alkylation and desulfurization steps (Eq. 2-25; Kodama *et al.*, 1976; see also van Tamelen *et al.*, 1972). As a result of certain of the difficulties associated with the original Biellmann alkylation procedure employing the thiophenyl group in stabilizing the anion and lithium–ethylamine in removing the thiophenyl group following coupling, several modifications of the original procedure were developed. Table 2.5 summarizes the effect of the substituent on sulfur on the regiospecificity (α versus γ) of alkylation. The remarkable effects realized with the heterocyclic sulfur substituents in favoring α over γ alkylation are thought to be a consequence of the metal chelating effect of the nitrogen atom (as discussed earlier in this chapter). A novel reagent recommended for

PhS— (1) [O] (2) BuLi DABCO THF → PhS— Li O →

HO PhS— Li $C_2H_5NH_2$ $-78°$ → HO (20%) (30%) HO ↓

Cembrene A

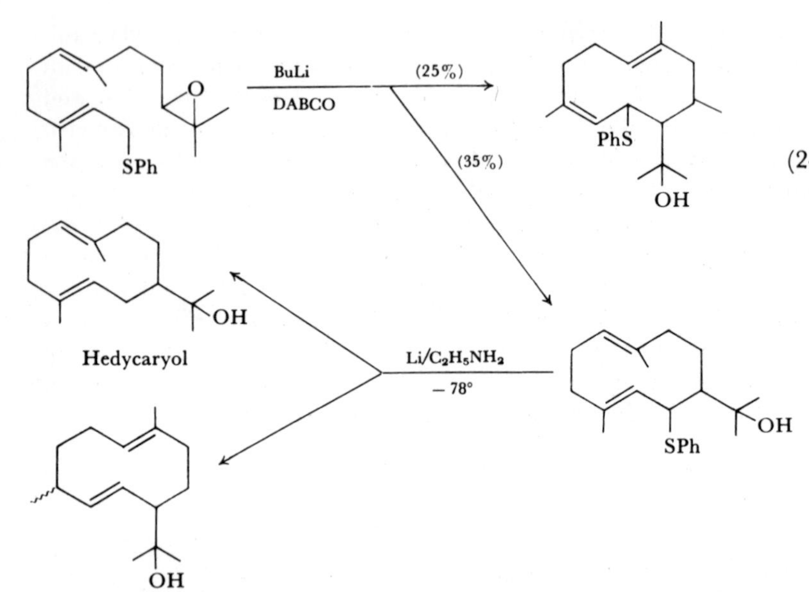

SPh

BuLi

DABCO

(25%)

(35%)

PhS

OH

(2-25)

OH

Hedycaryol

$Li/C_2H_5NH_2$

− 78°

OH

SPh

OH

the replacement of the 2-pyridylthiyl function by hydrogen is $LiAlH_4/CuCl_2$ (Mukaiyama *et al.*, 1971).

It is notable that a favorable ratio of α- to γ-alkylation (>95:5) can be achieved with allylphenyl sulfide itself if *sec*-butyllithium is used in place of

Table 2-5. Regiospecificity in the Alkylation of Various Thioallylic Carbanions (α:γ Ratio)

R	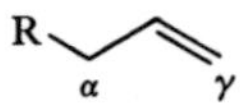 Substrate: R α γ	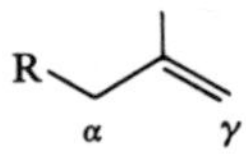 R α γ	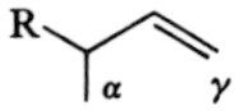 R α γ	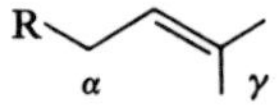 R α γ	Reference
PhS—	62:29	—	—	~100	Biellmann and Ducep (1968)
	75:25	75:25	—	—	Evans and Andrews (1974)
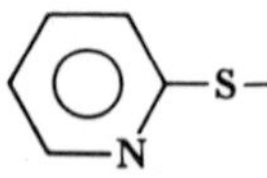 2-pyridyl-S—	>99:1	99:10	80:20	—	Evans and Andrews (1974)
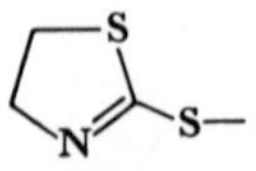 2-thiazolinyl-S—	>99:1	—	—	—	Evans and Andrews (1974)
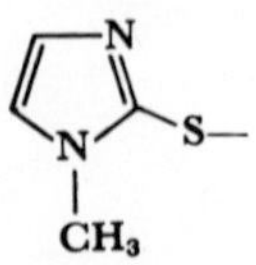 1-methyl-2-imidazolyl-S— (CH_3)	>99:1	92:8	90:10	—	Evans and Andrews (1974)

n-butyllithium as base (Oshima *et al.*, 1973a). *sec*-Butyllithium is also the base of choice for the preparation of synthetically useful vinylic thiocarbanions (Eq. 2-26; Oshima *et al.*, 1973b).

$$C_2H_5SCH{=}CH_2 \xrightarrow[\text{THF/HMPA}]{sec\text{-BuLi}} C_2H_5SCLi{=}CH_2 \xrightarrow{Br(CH_2)_4Br}$$

$$[CH_2{=}C(SC_2H_5)CH_2CH_2\text{-}]_2 \xrightarrow[\Delta]{HgCl_2} CH_3CO(CH_2)_4COCH_3 \quad (60\%) \qquad (2\text{-}26)$$

A dramatic reversal in the position of alkylation occurs on replacement of the lithium in allyl thiocarbanions with copper. Thus isopropylthioallylcopper undergoes alkylation exclusively γ to sulfur upon reaction with allylic halides. Halide displacement is said to be exclusively S_N2' (Eq. 2-27; Oshima *et al.*, 1973c).

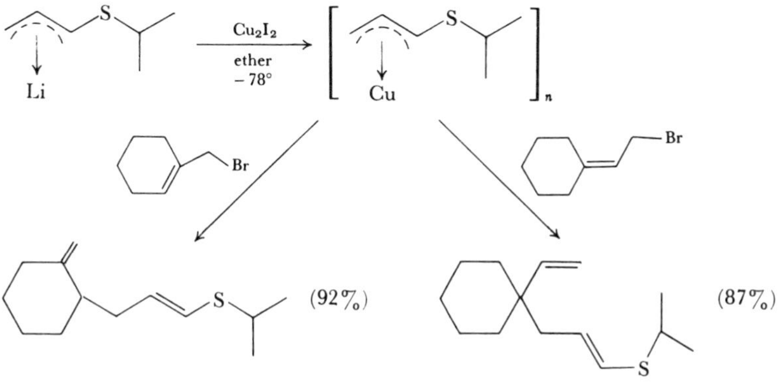

Li ... S — $\xrightarrow[\text{ether, }-78°]{Cu_2I_2}$ [Cu ... S]$_n$

Br (92%) S — Br (87%) S (2-27)

We have already described a variety of reagents that may be considered synthetic equivalents ("synthons," using the terminology of Corey *et al.*, 1967) for acyl anions, $RC{=}O^-$. The 1,3-bis(methylthio)allyl anion has been shown by Corey to be a β-acylvinyl anion synthon (see Eq. 2-28) and has been used in the synthesis of a key prostaglandin precursor (Eq. 2-28; Corey *et al.*, 1971). A different β-acylvinyl anion synthon involving an α-sulfonyl carbanion has been employed by Kondo in the synthesis of the terpene ($\pm$)-nuciferal (Eq. 2-29; Kondo and Tunemoto, 1975). Table 2-6 brings together a selection of sulfur-containing synthons which are described in this and subsequent chapters.

Alkylations of anions of sulfonates, sultones, and sulfones have been used to advantage as in Durst's sultone synthesis (Eq. 2-30; Durst and Tin, 1970), and in syntheses of the bicyclic sesquiterpenes α-santalene (Julia and Arnould, 1973), β-santalene (Eq. 2-31; Wolinsky *et al.*, 1972), and sesquifenchene (Eq. 2-32; Grieco and Masaki, 1975). In these reactions the sulfonyl group is replaced with hydrogen using reagents such as lithium in ethylamine, aluminium hydride or sodium amalgam/Na_2HPO_4 (Trost *et al.*, 1976). A

Table 2-6. Selected Organosulfur Carbanionic Synthons

Synthon	Sulfur-free equivalent	Examples
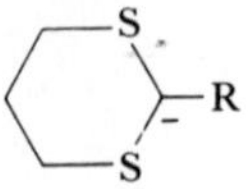 [ring structure: S, S, R, −]	$R\bar{C}{=}O$	2-17
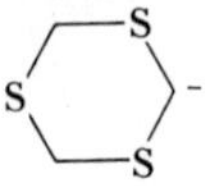 [ring structure: S, S, S, −]	$H\bar{C}{=}O$	—
$RS\bar{C}{=}CH_2$	$CH_3\bar{C}{=}O$	2-26
$RS\bar{C}HS(O)R$	$H\bar{C}{=}O$	2-47
$(RS)_3C^-$	$^-CO_2H$	2-48
$CH_3SCH\cdots\bar{C}H\cdots CHSCH_3$	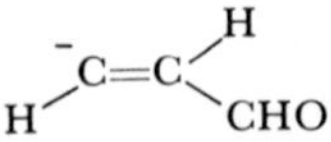 $H\bar{C}{=}C(H)CHO$	2-28
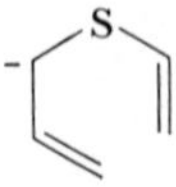 [structure: −, S]	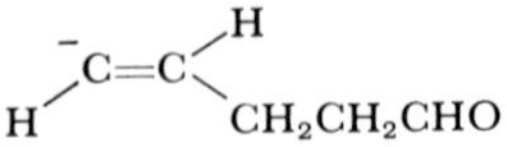 $H\bar{C}{=}C(H)CH_2CH_2CHO$	7-16
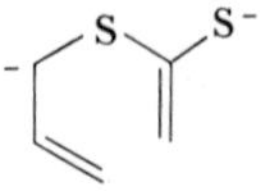 [structure: −, S, S−]	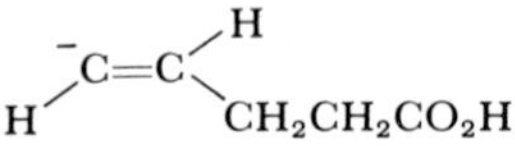 $H\bar{C}{=}C(H)CH_2CH_2CO_2H$	7-24
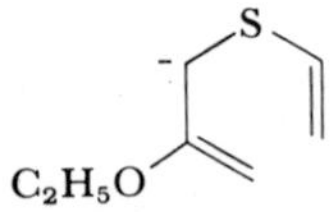 [structure: −, S, C_2H_5O]	$^-CH_2C(O)CH_2CH_2C(O)H$	7-17
$PhS(O)\bar{C}HCH{=}CH_2$	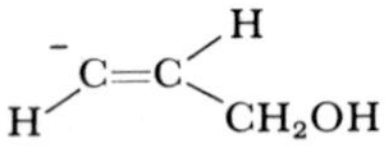 $H\bar{C}{=}C(H)CH_2OH$	7-41, 7-42
$PhS\bar{C}HCH{=}CHR$	$^-CH_2CH\overset{t}{=}CHR$	2-22
$ArSO_2\bar{C}HCH{=}CHR$	$^-CH_2CH\overset{t}{=}CHR$	2-32
$CH_2{=}CH\bar{C}HSC(S)N(CH_3)_2$	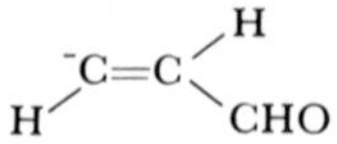 $H\bar{C}{=}C(H)CHO$	7-22
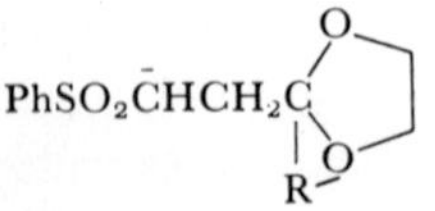 $PhSO_2\bar{C}HCH_2C(R)$ [dioxolane ring: O, O]	$^-CH_2CH_2C(O)R$ or 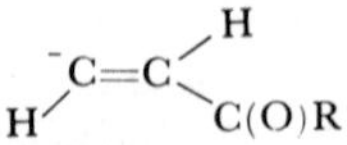 $H\bar{C}{=}C(H)C(O)R$	2-29

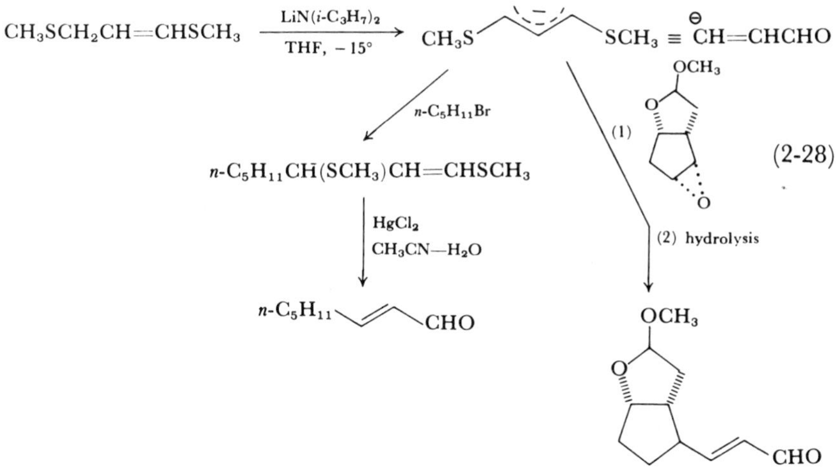

(2-28)

second useful approach involves base catalyzed β-elimination of sulfinate anion as used in the synthesis of vitamin A ester (Julia and Arnould, 1973; Manchand *et al.*, 1976), 2,2′-dinorcarotenoids (Kienzle and Minder, 1976), apocarotenoids and β-carotene (Fischli and Mayer, 1975) and the sesquiterpene deoxytrisporone (Uneyama and Torii, 1976) as illustrated in Eq. 2-33. Cyclofragmentation of a γ-hydroxysulfone is a key step in a three-atom ring expansion approach to muscenone **2-31** (Fischli *et al.*, 1976). The synthesis of

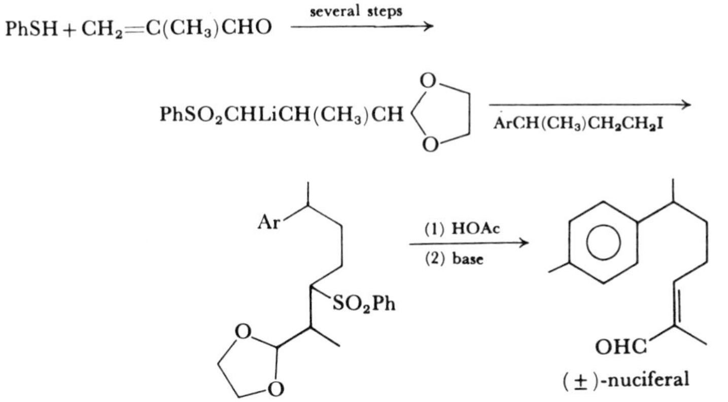

(2-29)

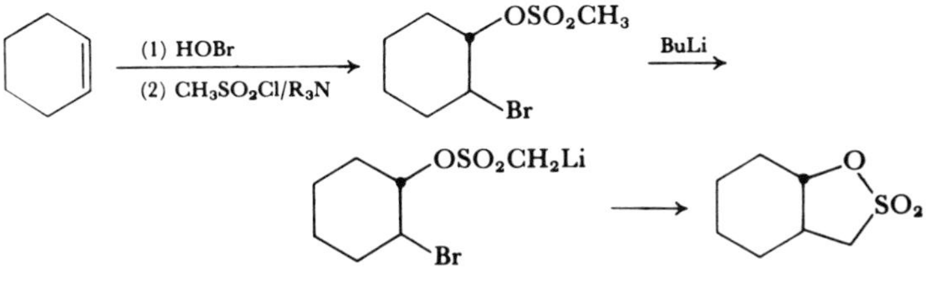

(2-30)

Li
Br
CH_2
SO_2
AlH_3
OH
$POCl_3$
(56% overall) (2-31)
β-santalene

I
several steps
SO_2Ar
Li
O
Br
SO_2Ar
(1) $Li/C_2H_5NH_2$
(2) HOAc
(3) $Ph_3P{=}CH_2$
(2-32)
sesquifenchene

OH
2RLi
SO_2Ph
OLi
Li
Br
CO_2CH_3
$^-OCH_3$
H
MeONa
(2-33)
O
several steps
OH
CO_2CH_3

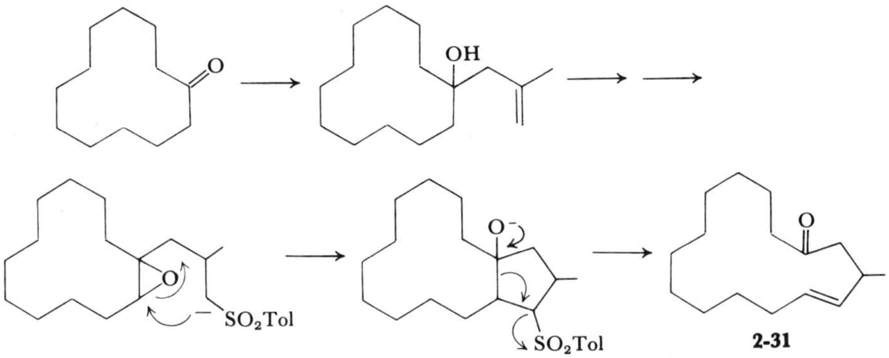
O
OH
O⁻
SO₂Tol
SO₂Tol
O

2-31

cyclopropanes via γ-elimination of sulfinate anions will be discussed in the section on conjugate addition.

A total synthesis of the growth factor biotin utilizes the stereospecific alkylation of an α-sulfinyl carbanion (Eq. 2-34; Bory *et al.*, 1975). This route should be useful for the preparation of biotin analogs. A number of examples involving alkylation of allylic α-sulfinylcarbanions have been described (Evans and Andrews, 1974); some of these will be discussed in Chapter 7.

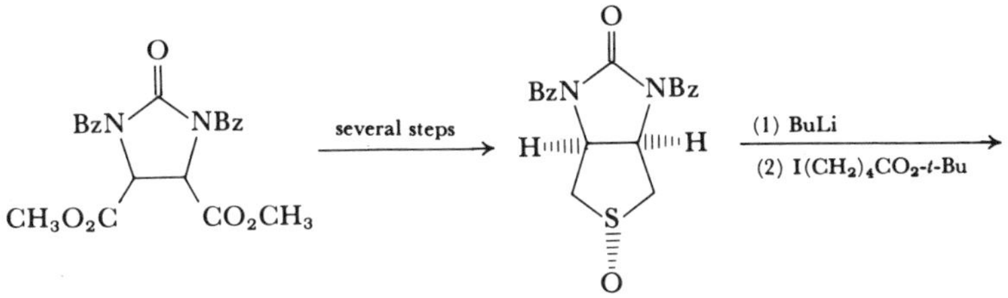
O
BzN
NBz
CH₃O₂C
CO₂CH₃
several steps
O
BzN
NBz
H
H
S
O
(1) BuLi
(2) I(CH₂)₄CO₂-t-Bu

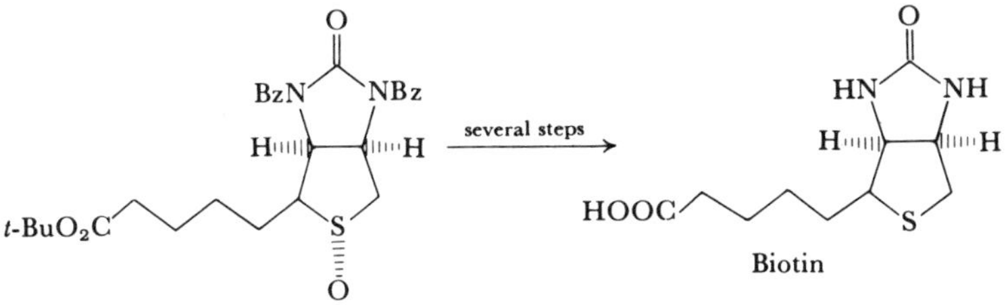
O
BzN
NBz
H
H
t-BuO₂C
S
O
several steps
O
HN
NH
H
H
HOOC
S
Biotin

(2-34)

Carbonyl Addition

The addition of a variety of types of sulfur-containing carbanions to carbonyl groups has formed the basis for a number of artful syntheses. Illustrative are Trost's annelation procedures (Eq. 2-35; Trost, 1973), Coates' ketone methylenation method (Eq. 2-36; Sowerby and Coates, 1972), syntheses of α-methylene-δ-lactones (Eq. 2-37; Trost and Miller, 1975), Meyer's and Johnson's episulfide syntheses (Eq. 2-38; Meyers and Ford, 1976; Johnson,

(2-35)

(2-36)

1975), preparation of optically active epoxides (Eq. 2-39; Durst *et al.*, 1971) using optically active α-sulfinylcarbanions, conversion of esters to ketones (Eq. 2-40; Russell and Weiner, 1966; Corey and Chaykovsky, 1965), construction of bridged ring systems (Eq. 2-41; House and Larson, 1968), and syntheses of the sesterterpene (±)-diumycinol (Eq. 2-42; Grieco *et al.*, 1975) and the monosaccharide L-streptose (Eq. 2-43; Paulsen *et al.*, 1972).

The regioselectivity of carbonyl addition of various thioallyl anions has been examined and compared with alkylation, particularly with regard to

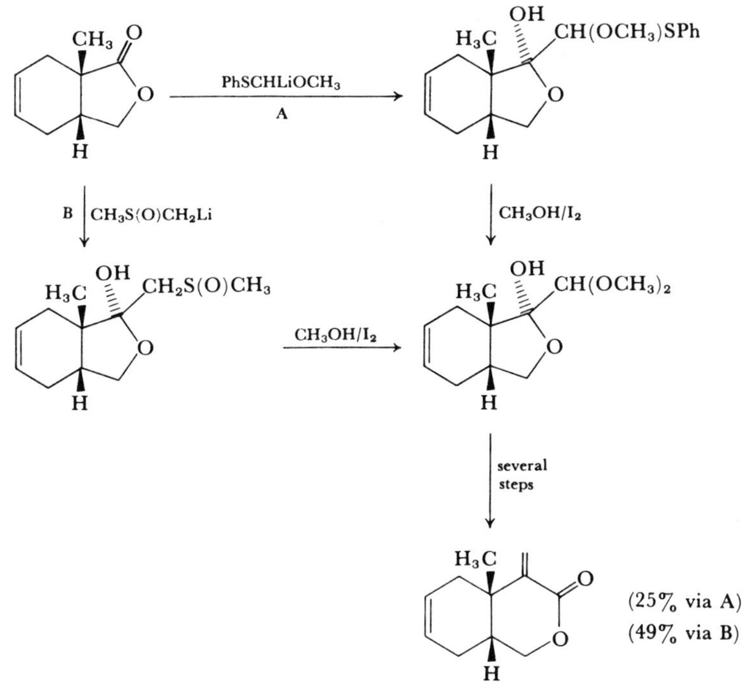
PhSCHLiOCH3
A
B
CH3S(O)CH2Li
CH3OH/I2
CH3OH/I2
several steps
(25% via A)
(49% via B)

(2-37)

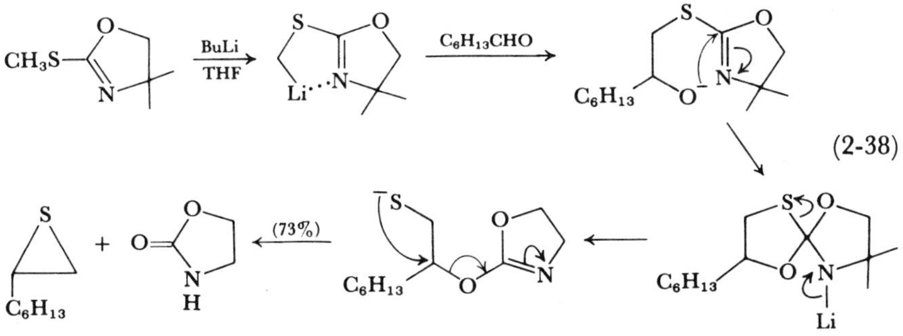
BuLi
THF
C6H13CHO
(73%)

(2-38)

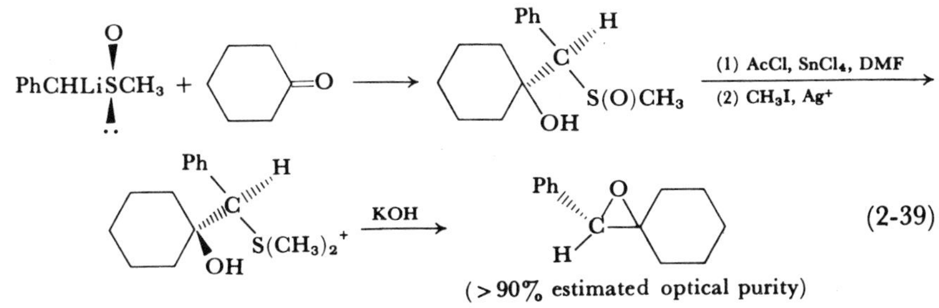
(1) AcCl, SnCl4, DMF
(2) CH3I, Ag+
KOH
(>90% estimated optical purity)

(2-39)

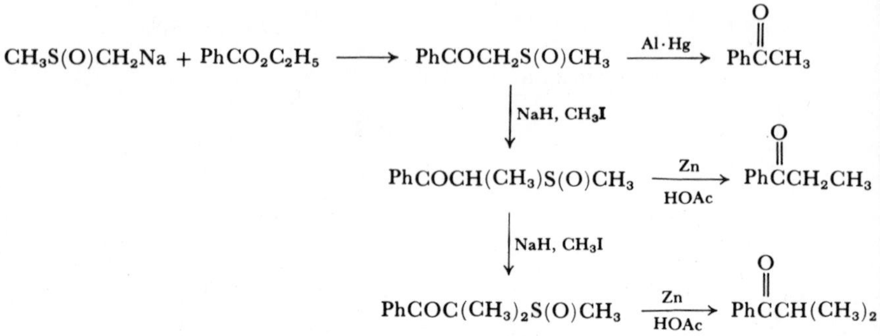

$$CH_3S(O)CH_2Na + PhCO_2C_2H_5 \longrightarrow PhCOCH_2S(O)CH_3 \xrightarrow{Al\cdot Hg} PhCCH_3 \ (C{=}O)$$

$$PhCOCH_2S(O)CH_3 \xrightarrow{NaH,\ CH_3I} PhCOCH(CH_3)S(O)CH_3 \xrightarrow[HOAc]{Zn} PhCCH_2CH_3 \ (C{=}O)$$

$$PhCOCH(CH_3)S(O)CH_3 \xrightarrow{NaH,\ CH_3I} PhCOC(CH_3)_2S(O)CH_3 \xrightarrow[HOAc]{Zn} PhCCH(CH_3)_2 \ (C{=}O) \qquad (2\text{-}40)$$

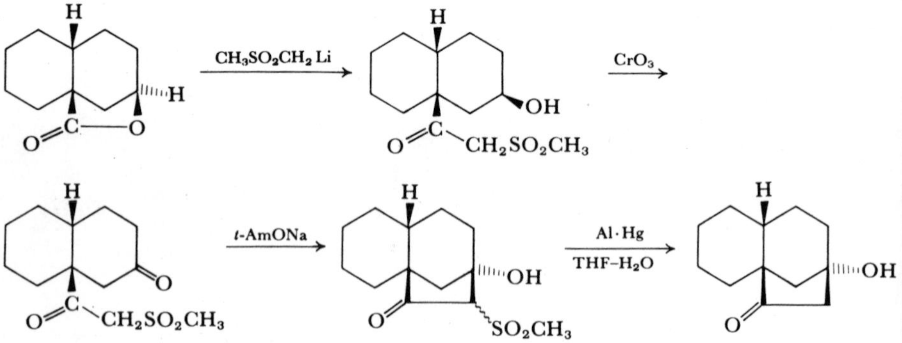

H, H, O=C—O $\xrightarrow{CH_3SO_2CH_2Li}$ H, OH, O=C—$CH_2SO_2CH_3$ $\xrightarrow{CrO_3}$

H, O, O=C—$CH_2SO_2CH_3$ $\xrightarrow{t\text{-AmONa}}$ H, OH, O, SO_2CH_3 $\xrightarrow[THF\text{–}H_2O]{Al\cdot Hg}$ H, OH, O

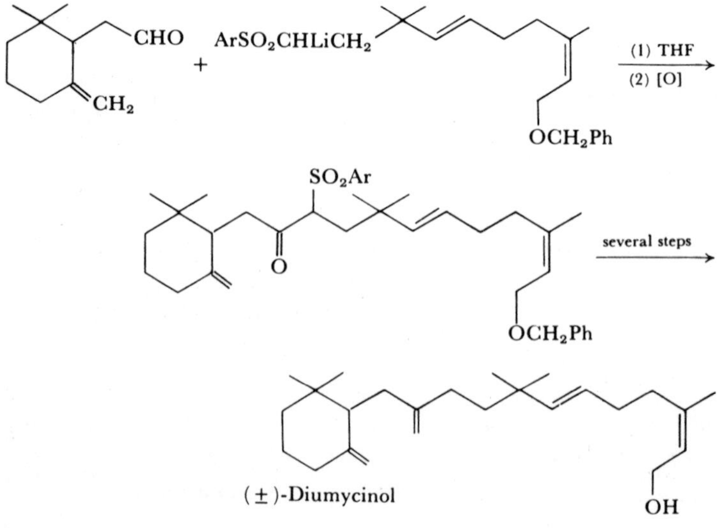

(2-41)

CHO, CH_2 + $ArSO_2CHLiCH_2$, OCH_2Ph $\xrightarrow[(2)\ [O]]{(1)\ THF}$

SO_2Ar, O, OCH_2Ph $\xrightarrow{\text{several steps}}$

(±)-Diumycinol, OH (2-42)

(1) HgO/BF_3 (2) H_3O^+

CHO OH OH OH (52%)

L-streptose (2-43)

chelation and ion-pairing effects. Thus Biellmann and co-workers (Atlanti *et al.*, 1974) find that thioallyllithium **2-32** in the presence of [2.2.2]cryptate **2-2** adds to acetone exclusively from the α-position of **2-32** while in the presence of DABCO or TMEDA addition occurs exclusively from the γ-position. Biellmann argues that cryptate favors formation of a free anion (by complexing lithium), which, due to a combination of steric and charge concentration factors, adds at the α-position. Where steric factors are less important, as with methyl iodide, a 60/40 mixture of α/γ alkylated products results when cryptate is present. Under conditions favoring intimate or solvated ion pairs (DABCO, TMEDA, or no added reagent) complex formation involving acetone (see **2-33**) is postulated to explain the formation of product derived from bonding at the γ-position (Atlanti *et al.*, 1974).

PhS, HO (γ) ← $(CH_3)_2CO$ / DABCO or TMEDA — PhS, Li **2-32** — $(CH_3)_2CO$ / [2.2.2.]cryptate → PhS, OH (α)

PhS, Li, C, O **2-33** → PhS, HO

2-34 (S, S, OH) — $NaOCH_3$ / CH_3OH / 91% → (S, S, O) — (1) CH_3OH (2) CH_3I CH_3CN/H_2O → CHO, O

The reversal of thiocarbanion–carbonyl addition, more properly categorized as a fragmentation reaction, is also known and is of some synthetic value as illustrated with **2-34** (Trost *et al.*, 1975a; also see Marshall and Seitz, 1974).

Addition to Carbon-Nitrogen Multiple Bonds

Representative examples of the addition of α-sulfinyl carbanions to C—N multiple bonds include a synthesis of optically active amines (Eq. 2-44) (Tsuchihashi *et al.*, 1973) and a useful procedure for ring methylation of polynuclear heteroaromatics such as isoquinoline (Eq. 2-45) (Russell and Weiner, 1966). This latter procedure can also be applied to polynuclear aromatic hydrocarbons.

$$PhCH{=}NPh + ArS(O)CH_2Li \longrightarrow Ph{-}\overset{*}{C}H(CH_2S(O)Ar){-}NHPh \xrightarrow[\text{Ra Ni}]{} Ph{-}CH(CH_3){-}NHPh$$

>90% optical purity

(2-44)

$$CH_3S(O)CH_2Na + \text{isoquinoline} \longrightarrow \text{1-H-1-}(CH_2S(O)CH_3)\text{-1,2-dihydroisoquinolin-2-ide } (N^-) \longrightarrow$$

$$\text{1-}(CH_2S(O)CH_3)\text{-1,2-dihydroisoquinoline (NH)} \xrightarrow{-CH_3SO^-} \text{1-methylene-1,2-dihydroisoquinoline (NH, =}CH_2) \xrightarrow{\sim H} \text{1-methylisoquinoline } (CH_3)$$

(2-45)

Conjugate (Michael) Addition

Addition of thiocarbanions to isolated double bonds has been observed only under the influence of transition metal–olefin complexes such as **2-35** (Julia and Saussine, 1974). Even with enones, thiocarbanions generally add in a 1,2 rather than 1,4 manner. Conjugate addition does occur if the carbanion is converted to a cuprate salt, as in Mukaiyama's synthesis of dihydrojasmone (Eq. 2-46; Mukaiyama *et al.*, 1972), or if less reactive thiocarbanions

$$PhSO_2CH_2{-}Pd(Cl)(\text{cyclooctadiene}) + CH_2{=}CHC_4H_9 \longrightarrow PhSO_2CH_2CH_2CH(Pd(Cl)COD)C_4H_9$$

2-35

are employed, such as that derived from a thioacetal *S*-oxide (Eq. 2-47; Herrmann *et al.*, 1973), from tris(phenylthio)methane (Eq. 2-48; Manas and Smith, 1975), or from bis(methylthio)(silyl)- and bis(methylthio)(stannyl)-methanes (Seebach and Burstinghaus, 1975). Michael addition of sulfonyl carbanions to α,β-unsaturated esters followed by 1,3-elimination of sulfinate

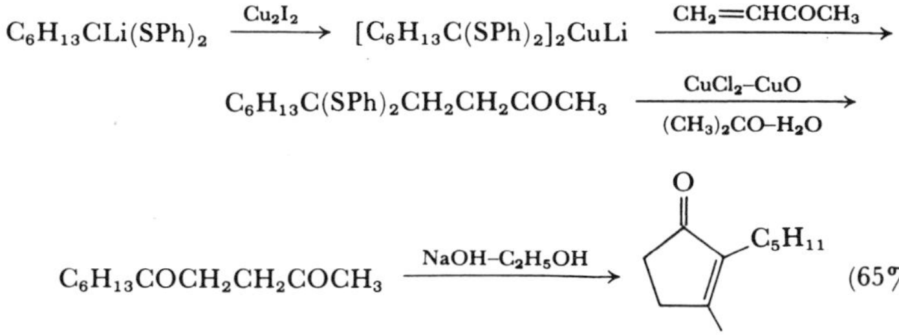

$$C_6H_{13}CLi(SPh)_2 \xrightarrow{Cu_2I_2} [C_6H_{13}C(SPh)_2]_2CuLi \xrightarrow{CH_2{=}CHCOCH_3}$$

$$C_6H_{13}C(SPh)_2CH_2CH_2COCH_3 \xrightarrow[(CH_3)_2CO\text{–}H_2O]{CuCl_2\text{–}CuO}$$

$$C_6H_{13}COCH_2CH_2COCH_3 \xrightarrow{NaOH\text{–}C_2H_5OH} \text{[2-pentyl-3-methylcyclopent-2-enone, } C_5H_{11}\text{]} \quad (65\%) \qquad (2\text{-}46)$$

Dihydrojasmone

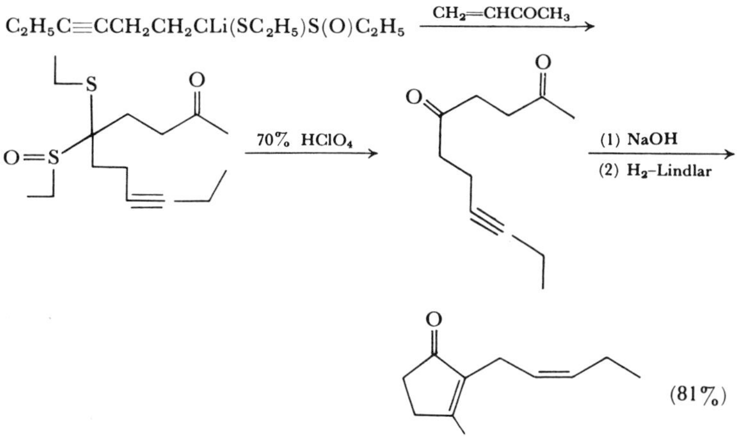

$$C_2H_5C{\equiv}CCH_2CH_2CLi(SC_2H_5)S(O)C_2H_5 \xrightarrow{CH_2{=}CHCOCH_3}$$

[structure] $\xrightarrow{70\%\ HClO_4}$ [structure] $\xrightarrow[(2)\ H_2\text{–Lindlar}]{(1)\ NaOH}$ [structure] (81%) (2-47)

cis-Jasmone

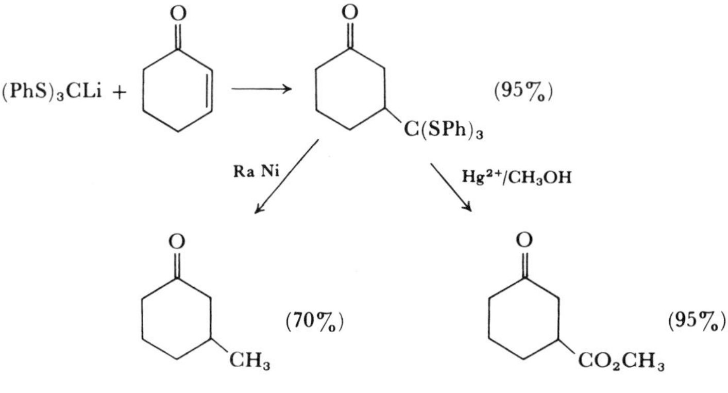

$$(PhS)_3CLi + \text{[cyclohexenone]} \longrightarrow \text{[3-substituted cyclohexanone, } C(SPh)_3\text{]} \quad (95\%) \qquad (2\text{-}48)$$

Ra Ni → [3-methylcyclohexanone, CH_3] (70%)

Hg^{2+}/CH_3OH → [3-substituted cyclohexanone, CO_2CH_3] (95%)

anion provides a route to cyclopropanes utilized in the synthesis of chrysanthemic acid (**2-36**) (Julia and Guy-Roualt, 1967; Martell and Huynh, 1967) and more recently in the total synthesis of presqualene and prephytoene alcohols (Campbell *et al.*, 1975).

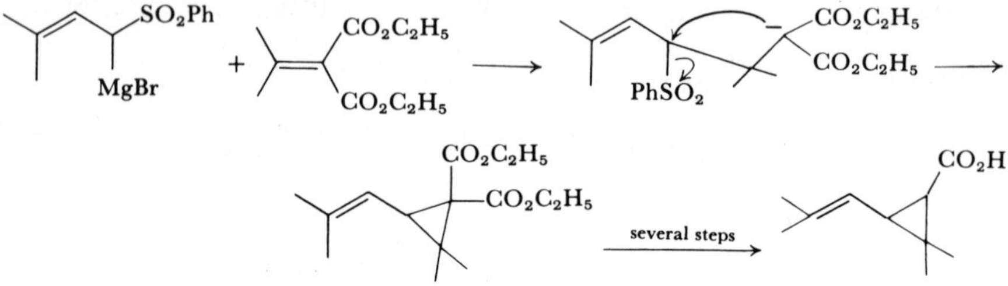

SO_2Ph MgBr + $CO_2C_2H_5$ $CO_2C_2H_5$ → $PhSO_2$ $CO_2C_2H_5$ $CO_2C_2H_5$ →

$CO_2C_2H_5$ $CO_2C_2H_5$ —several steps→ CO_2H

2-36

Protonation

While the stereochemical aspects of protonation of sulfur-containing carbanions have already been discussed, it is worth noting the synthetic utility of this reaction for the synthesis of deuterated (or tritiated) compounds, for example as illustrated in Eq. 2-49 (Seebach *et al.*, 1966; Mutterer and

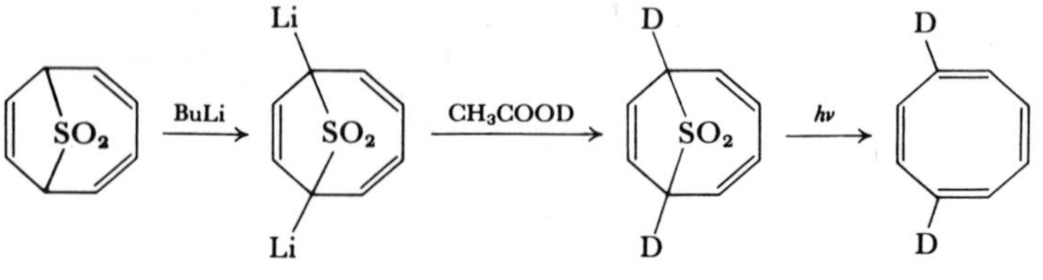

SO_2 —BuLi→ Li SO_2 Li —CH_3COOD→ D SO_2 D —$h\nu$→ D D

2-37

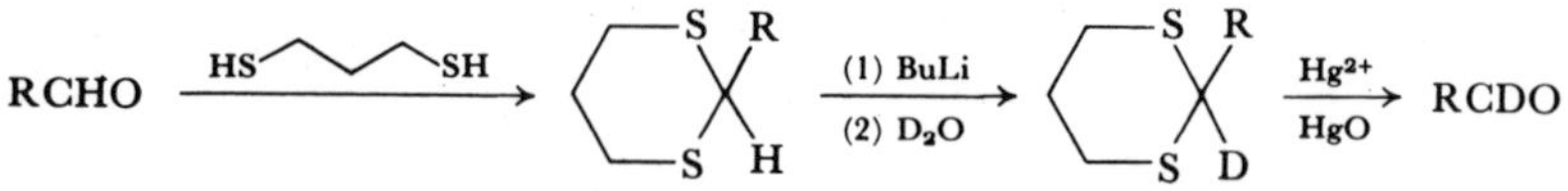

RCHO —HS⁀⁀SH→ S R / S H —(1) BuLi (2) D_2O→ S R / S D —Hg^{2+}, HgO→ RCDO

R = Ph 80% yield
R = *t*-Bu 50% yield

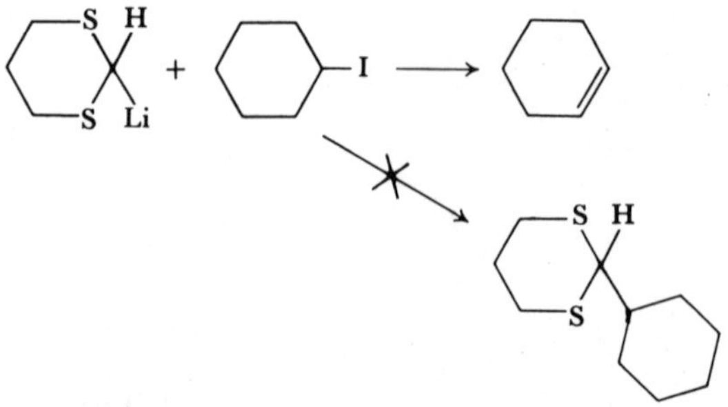

(2-49)

S H / S Li + cyclohexyl–I → cyclohexene

✕→ S H / S cyclohexyl

(2-50)

Fluery, 1974), and by the preparation of 1,4-dideuteriocyclooctatetraene **2-37** (Paquette *et al.*, 1974). Sometimes the protonation of thiocarbanions can be an undesired process, as in the attempted alkylation of cyclohexyl iodide (Eq. 2-50; Corey and Seebach, 1965).

Elimination: The Ramberg–Bäcklund Reaction

Elimination reactions may be variously categorized according to the timing of carbanion generation (or deprotonation) and leaving group ejection (E1cb versus E2, etc.) or according to the relationship of the departing groups (e.g., α- or 1,1-elimination, β- or 1,2-elimination, γ- or 1,3-elimination). Special classes of elimination reactions include concerted cycloeliminations (covered in Chapter 7) and those in which ejection of the leaving group triggers or accompanies additional elimination processes (*fragmentation reactions*). This section will concern itself primarily with elimination reactions involving sulfur-containing carbanions.

A large number of α-eliminations involving thiocarbanions are known (e.g., Eq. 2-51; Schöllkopf *et al.*, 1966) but discussion of these will be deferred until Chapter 6. On rare occassions sulfur-containing carbanions will undergo α-elimination with loss of the sulfur group (Eq. 2-52; Schaumann and Walter, 1974; see also the early work by Hine and Porter, 1960). Carbanions α to sulfur may undergo β-elimination in two directions:

1. toward sulfur forming a carbon–sulfur double bond as in thiocarbonyl compounds (Eq. 2-53) (Shirley and Reeves, 1969), sulfines (Eq. 2-54) (Corey and Durst, 1966b) or sulfenes (Eq. 2-55) (King and Beatson, 1975), or

2. away from sulfur as in Eq. 2-56. The 2-(*p*-tritylphenyl)sulfonylethyl group (p-$Ph_3C{-}C_6H_4SO_2CH_2CH_2{-}$) has been ingeniously employed by Khorana (Agarwal *et al.*, 1976) as a highly lipophilic protecting group for the 5′-phosphate group of nucleotides in polynucleotide synthesis. Mild alkaline

$$PhSCHCl_2 \xrightarrow{KO\text{-}t\text{-}Bu} PhSCKCl_2 \longrightarrow [PhSCCl:] \qquad (2\text{-}51)$$

$$PhSO_2CHF_2 \xrightarrow{NaOCH_3} PhSO_2CNaF_2 \longrightarrow F_2C: + PhSO_2Na \qquad (2\text{-}52)$$

$$PhSCHLiCH_3 \longrightarrow PhLi + CH_3CH{=}S \qquad (2\text{-}53)$$

$$t\text{-BuLi} + CH_3S(O)N\langle\text{morpholine}\rangle O \longrightarrow LiCH_2S(O)N\langle\text{morpholine}\rangle O \longrightarrow$$

$$LiN\langle\text{morpholine}\rangle O + CH_2{=}S{=}O \xrightarrow{t\text{-BuLi}} LiCH_2S(O)t\text{-Bu} \qquad (2\text{-}54)$$

$$PhCH_2SO_2OAr \underset{}{\overset{R_3N}{\rightleftharpoons}} Ph\bar{C}HSO_2OAr \xrightarrow{-ArO^-}$$

$$PhCH{=}SO_2 \xrightarrow[\text{(pyrrolidino)NCH=C(CH}_3)_2]{} \text{(pyrrolidino)N-cyclobutane(CH}_3)_2\text{-}SO_2 \quad (2\text{-}55)$$

Ar $= p\text{-}NO_2\text{-}C_6H_4$

$$RSO_2CH_2CH_2X \xrightarrow{\text{base}} RSO_2CH{=}CH_2 + X^- \quad (2\text{-}56)$$

treatment releases the phosphate group as in Eq. 2-56. A number of *fragmentation reactions* are known that are initiated by β-elimination toward or away from sulfur [Eq. 2-57 (Marshall and Belletire, 1971); Eq. 2-58 (Jones *et al.*, 1974); Eq. 2-59 (Schönberg *et al.*, 1931); Eq. 2-60 (Shank *et al.*, 1973)]. The first two of these reactions (Eqs. 2-57 and 2-58) are of possible synthetic utility (a mechanism related to that shown in Eq. 2-58 may be involved in the Corey–Winter olefin synthesis, discussed in Chapter 6) while reaction 2-59 explains why dithiolanes may not be used in place of dithianes in carbanionic reactions such as alkylation or carbonyl addition.

OTs, S, S, Li → S, S —hydrol→ CO_2H (2-57)

Ph Ph, O S, Ph H —R_2NLi→ Ph Ph, O S, Ph Li → Ph Ph (71%) + PhC(O)SLi (2-58)

Ph_2 Ph_2, S S, H H —PhLi→ Ph_2 Ph_2, S S, H Li → $Ph_2C{=}CPh_2$ + HC(S)SLi (2-59)

O, S, O_n —KO-*t*-Bu→ O, S ⊖, O_n → $CH_2O + CH_2{=}CHSO_nK$ (2-60)

As with β-eliminations involving sulfur-containing carbanions, γ-eliminations may also occur away from sulfur (e.g., Eq. 2-61; Cristol *et al.*, 1966) or toward (across) sulfur (e.g., Eq. 2-62; Ramberg and Bäcklund, 1940). The latter process known as the *Ramberg–Bäcklund reaction* is the sulfur analog of the Favorskii rearrangement. Because of the considerable synthetic utility of the Ramberg–Bäcklund reaction, it will be discussed in some detail (for reviews see Paquette, 1968; Bordwell, 1968, 1970). The Ramberg–Bäcklund

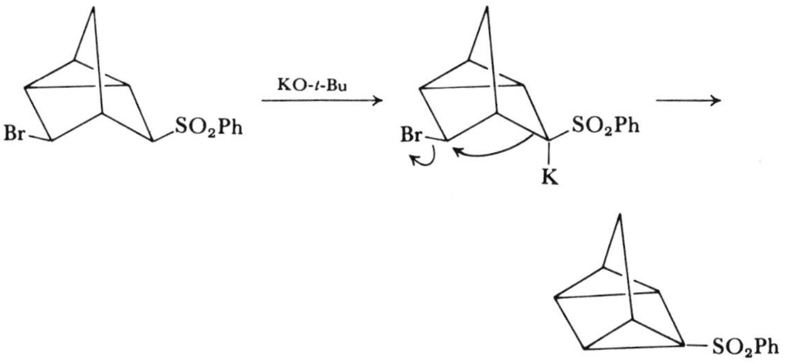

Br ... SO_2Ph $\xrightarrow{\text{KO-}t\text{-Bu}}$ Br ... SO_2Ph (K) $\longrightarrow$ ... SO_2Ph (2-61)

$$CH_3CH_2SO_2CHBrCH_3 \xrightarrow[a]{KOH} CH_3CHKSO_2CHBrCH_3 \xrightarrow[b]{}$$

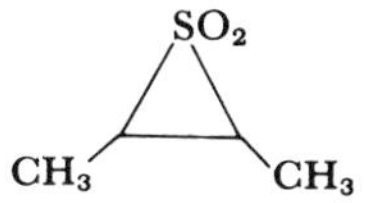

$$\text{(episulfone: } SO_2, CH_3, CH_3) \xrightarrow[c]{-SO_2} CH_3CH{=}CHCH_3 \qquad (2\text{-}62)$$

reaction is general for molecules containing the structural elements of a sulfonyl group, an α-halogen (or other suitable leaving group), and at least one α′-hydrogen atom and with few exceptions allows the clean replacement of a sulfonyl group by a double bond. The required α-halogen atom may be introduced by treatment of the corresponding α-sulfonyl carbanion with a source of X^+ (BrCN, I_2, and Cl_3CSO_2Cl are convenient sources of Br^+, I^+, and Cl^+, respectively (Corey and Block, 1969)). A particularly useful modification of the Ramberg–Bäcklund reaction has been developed by Meyers *et al.* (1969) whereby sulfones may be taken directly to olefin without the isolation of α-halosulfones. As illustrated in Eq. 2-63, carbon tetrachloride serves as the halogen source. A variety of synthetic applications of the

$$R_2C(H)SO_2CHR_2' \xrightarrow[t\text{-BuOH–}H_2O]{KOH,\ CCl_4} R_2\bar{C}SO_2CHR_2' \xrightarrow{CCl_4} Cl_3C^- + R_2C(Cl)SO_2CHR_2'$$

$$Cl_3C^- \longrightarrow Cl_2C\text{: (can be trapped)} \qquad R_2C(Cl)SO_2CHR_2' \longrightarrow R_2C{=}CR_2' \qquad (2\text{-}63)$$

Ramberg–Bäcklund reaction and the Meyers modification of this reaction are listed in Table 2-7. Several mechanistic features of the Ramberg–Bäcklund reaction as outlined in Eq. 2-62 are of interest. Step a in Eq. 2-62 is

Table 2-7. Some Synthetic Applications of the Ramberg–Bäcklund Reaction

Sulfone	Product	Yield (%)	Reference
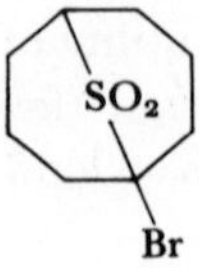	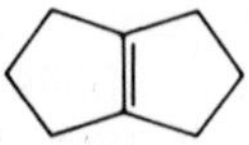	81	Corey and Block (1969)
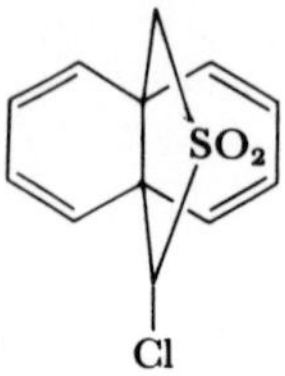	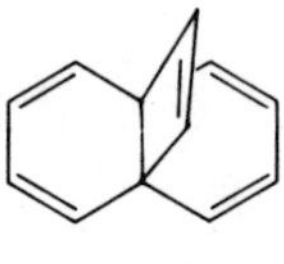	14	Paquette and Philips (1969)
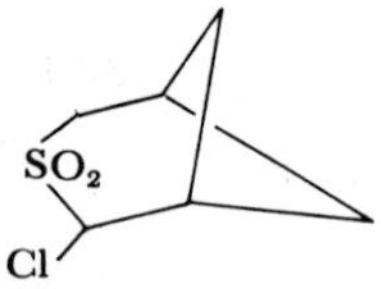	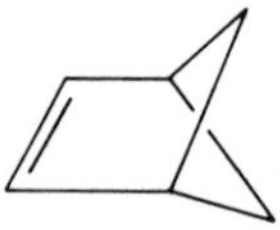	68	Carlson and May (1975)
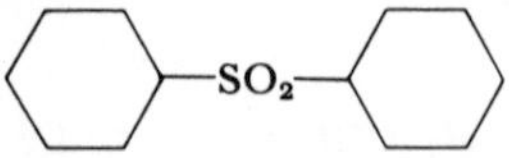	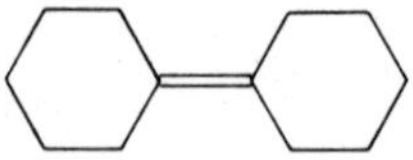	32	Meyers *et al.* (1969)
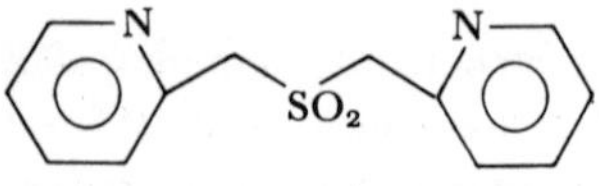	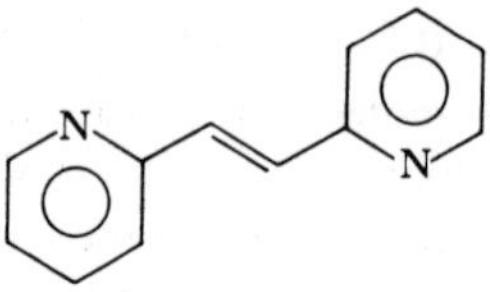	—	Martel and Rasmussen (1971)
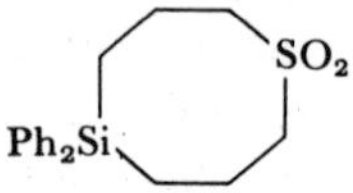	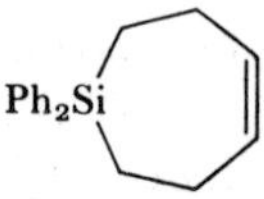	40	Koenig *et al.* (1974)
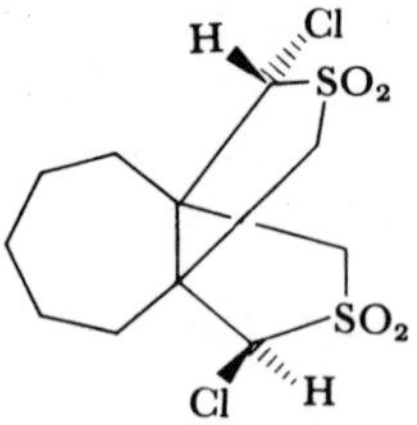	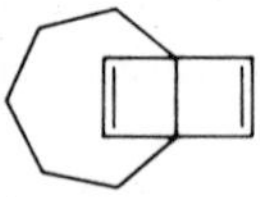	12	Weinges and Klessing (1976)
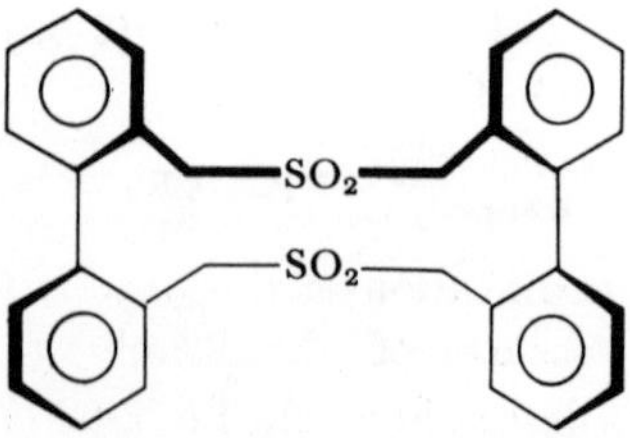	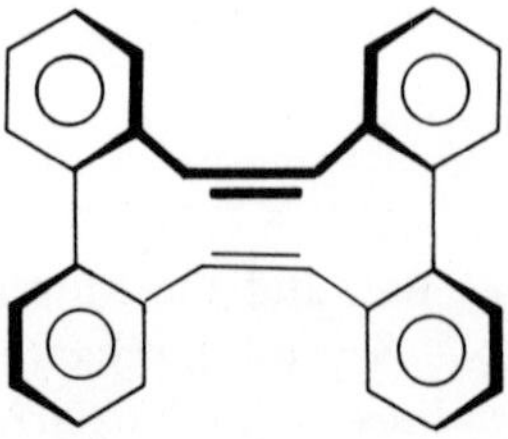	65	Bestmann, and Schaper (1975)

reversible as shown by deuterium exchange studies. Indeed by conducting the Ramberg–Bäcklund reaction in D_2O, good yields of deuterated olefins can be easily obtained (Eq. 2-64) (Neureiter, 1965). A double inversion displacement

$$ClCH_2SO_2CH_2CH(CH_3)C_2H_5 \xrightarrow[D_2O]{NaOD} CD_2{=}CDCH(CH_3)C_2H_5 \quad (82\%) \qquad (2\text{-}64)$$

mechanism (W geometry) seems likely for step b in Eq. 2-62, based on studies by Bordwell and Doomes (1974) as well as the finding by Corey and Block that a bridgehead α-halosulfone readily undergoes rearrangement (Eq. 2-65) (Corey and Block, 1969).

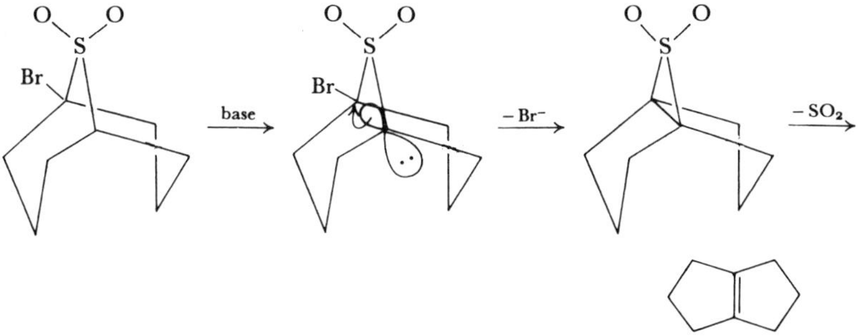

(2-65)

While episulfones cannot usually be isolated from Ramberg–Bäcklund reactions because of the facility with which they lose SO_2 (see Chapter 7), the thermally more stable thiirene dioxides have been isolated from the reaction of α,α′-dihalosulfones with base (Eq. 2-66) (Carpino *et al.*, 1971).

$$PhCHBrSO_2CHBrPh \xrightarrow[CH_2Cl_2]{(C_2H_5)_3N} \quad (2\text{-}66)$$

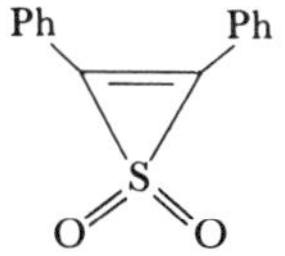

A novel "bishomoconjugative" analog of the Ramberg–Bäcklund reaction has been reported (Eq. 2-67; Paquette *et al.*, 1973).

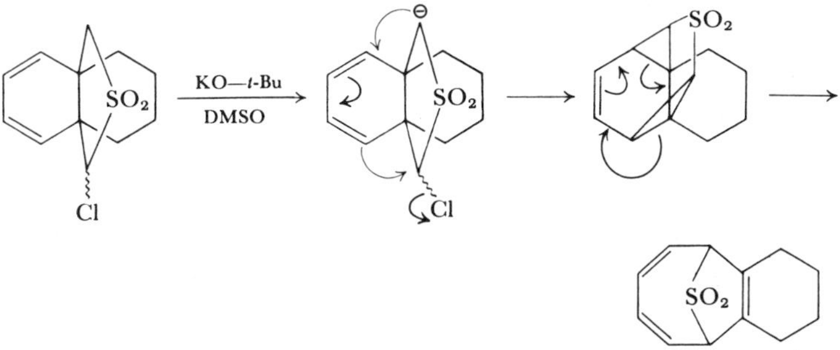

(2-67)

Rearrangement

Several rearrangement pathways are available to sulfur-containing carbanions. For example, dibenzyl sulfide has been shown to undergo *Sommelet rearrangement* (Eq. 2-68; Hauser *et al.*, 1953). In addition to this [2,3] sigmatropic rearrangement (which will be discussed in greater detail in Chapter 7), dibenzyl sulfide also undergoes *Wittig rearrangement*, a [1,2] sigmatropic process (Eq. 2-69) (Biellmann and Schmitt, 1973) of some synthetic utility (Eq. 2-70) (Mitchell *et al.*, 1975), (Eq. 2-71) (Wright and West, 1974). A rearrangement observed with benzyl dithiocarbamate on

(2-68)

$$PhCH(S^-)CH_2Ph \quad (2\text{-}69)$$

(99%) (2-70)

$$PhCH_2SSi(CH_3)_3 \xrightarrow[\text{THF, } -78^\circ]{t\text{-BuLi}} Ph\overset{\ominus}{C}HS{-}Si(CH_3)_3 \longrightarrow$$

$$PhCH(S^-)Si(CH_3)_3 \xrightarrow{CH_3I} PhCH(SCH_3)Si(CH_3)_3 \quad (93\%) \qquad (2\text{-}71)$$

treatment with base and claimed to be an example of a Wittig rearrangement (Eq. 2-72; Hayashi and Baba, 1975) may actually occur by an intramolecular addition–elimination route as depicted. In an important study of solvation

$$\mathrm{PhCH_2SC(S)N(CH_3)_2} \xrightarrow[\text{THF–HMPA}]{\mathrm{LiN}(i\text{-}\mathrm{C_3H_7})_2} \mathrm{PhCHLiSC(S)N(CH_3)_2}$$

$$\mathrm{PhCHLiSC(S)N(CH_3)_2} \xrightarrow{?} \mathrm{PhCH{-}C(S^-)(N(CH_3)_2){-}S} \text{ (three-membered ring)} \longrightarrow \mathrm{PhCH(SLi)C(S)N(CH_3)_2}$$

$$\mathrm{PhCHLiSC(S)N(CH_3)_2} \longrightarrow \mathrm{PhCH(SLi)C(S)N(CH_3)_2} \longrightarrow \mathrm{PhCH(SCH_3)C(S)N(CH_3)_2}\ (96\%) \qquad (2\text{-}72)$$

and ion-pairing effects on the rearrangement of the dibenzyl sulfide carbanion, Biellmann and Schmitt conclude that the more dissociated species leads to the *Sommelet rearrangement* (with delocalized negative charge in the transition state) while the less dissociated species (contact ion pair) leads only to *Wittig rearrangement.* It is argued that *Wittig rearrangement* involves homolytic processes such as dissociation and recombination of a radical pair generated by *homolysis of the thiocarbanion* (Eq. 2-73; Biellmann and Schmitt, 1973). Another case thought to involve thiocarbanion homolysis is shown in Eq. 2-74 (Dagonneau, 1973). Further discussion of the possible

$$\mathrm{PhCH(Li)S{-}CH_2Ph} \longrightarrow \mathrm{PhCH(Li)S\cdot} + \mathrm{PhCH_2\cdot} \longrightarrow$$

$$\mathrm{PhCH_2CH_2Ph} + \mathrm{PhCH(SLi){-}CH(SLi)Ph} + \mathrm{PhCH(SLi)CH_2Ph} \qquad (2\text{-}73)$$

$$t\text{-}\mathrm{BuC(S)Ph} + \mathrm{CH_3MgBr} \xrightarrow{\text{THF}} t\text{-}\mathrm{BuC(Ph)(SCH_3){-}MgBr} \rightleftharpoons$$

$$t\text{-}\mathrm{BuC(Ph)(SCH_3)\cdot} + \mathrm{MgBr\cdot} \xrightarrow{\mathrm{CH_3MgBr}} t\text{-}\mathrm{BuC(Ph)(SCH_3){-}CH_3}\ (100\%) \qquad (2\text{-}74)$$

involvement of free radicals in these processes will be reserved for Chapter 5. The formation of stilbene on base treatment of dibenzylsulfide presumably involves elimination of sulfur from the initial product of Wittig rearrangement (Eq. 2-75) (Wallace *et al.*, 1965).

$$\mathrm{PhCH_2SCH_2Ph} \xrightarrow[\text{DMF}]{\mathrm{KO}t\text{-}\mathrm{Bu}} [\mathrm{PhCH(S^-)CH_2Ph}] \longrightarrow \mathrm{PhCH{=}CHPh}\ (50\%) \qquad (2\text{-}75)$$

Another reaction that may involve a rearrangement–elimination sequence has been reported for sulfones by Paquette (Eqs. 2-76 and 2-77) (Photis and Paquette, 1974). Similar examples of carbanionic rearrangement of sulfones

$$\text{[bridged sulfone]} \xrightarrow[\text{(2) } CH_3I]{\text{(1) 2BuLi}} \text{[dimethyl bridged sulfone]} \xrightarrow[\text{(2) LiAlH}_4,\ \Delta]{\text{(1) } n\text{BuLi}} \text{[product with } H_3C,\ CH_3\text{]} \quad (2\text{-}76)$$

$$PhCH_2SO_2CH_2Ph \xrightarrow[\text{(2) LiAlH}_4]{\text{(1) BuLi}} PhCH{=}CHPh \quad (56\%) \qquad (2\text{-}77)$$

have been reported previously by Dodson (Eq. 2-78; Dodson *et al.*, 1971). It would seem that an additional driving force, such as complexation of an oxygen atom of the α-sulfonyl carbanion with an aluminium-containing species, is required to explain Paquette's results with compounds that should form relatively stable carbanions.

$$\text{[2,4-diphenylthietane } SO_2\text{]} \xrightarrow{C_2H_5MgBr} \text{[cyclopropane: H, Ph, Ph, } SO_2MgBr\text{]} \quad (77\%) \qquad (2\text{-}78)$$

Oxidation

One-electron oxidation of a carbanion gives a free radical that could couple with a second radical. Such a process becomes particularly favorable if the entire process is conducted within the coordination sphere of a metal such as copper, as illustrated by the inter- and intramolecular coupling reactions

$$\underset{\mathbf{R}}{\text{Ar}\overset{O}{S}CH_3} \xrightarrow[\text{(2) } CuCl_2,\ O_2]{\text{(1) } LiN(C_2H_5)_2} \underset{\mathbf{R},\ \mathbf{R};\ \geqslant 97\%\ \text{optical purity}}{\text{Ar}\overset{O}{S}CH_2CH_2\underset{O}{S}\text{Ar}} \quad (45\%) \qquad (2\text{-}79)$$

$$n\text{-}C_4H_9SO_2C_4H_9\text{-}n \xrightarrow[\text{BuLi}]{\text{excess}} n\text{-}C_3H_7CHLiSO_2CHLiC_3H_7\text{-}n \xrightarrow[CuCl_2]{}$$

$$\text{[2,3-dipropylthiirane } SO_2\text{]} \xrightarrow{-SO_2} C_3H_7CH{=}CHC_3H_7 \quad (45\%) \qquad (2\text{-}80)$$

of Eq. 2-79 (Maryanoff *et al.*, 1973) and Eq. 2-80 (Frossert *et al.*, 1974). Oxidation may also be effected by a halogen, although here the possibility exists in certain cases of a mechanism involving halogenation followed by alkylation. Examples include Eq. 2-81 (Seebach and Beck, 1972), Eq. 2-82 (Corey and Seebach, 1965; also see Baarschers and Loh, 1971) and Büchi's conversion of vitamin A to β-carotene (Eq. 2-83) (Büchi and Friedinger, 1974)

$$(PhS)_3CLi \xrightarrow{I_2} (PhS)_3C{-}C(SPh)_3 \quad (95\%) \qquad (2\text{-}81)$$

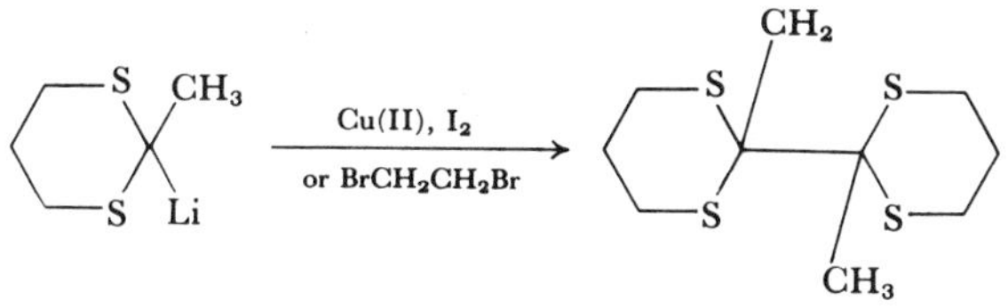

CH_3, Li; Cu(II), I_2 or $BrCH_2CH_2Br$; CH_3, CH_3 (2-82)

(Vitamin A) $\xrightarrow{SCl_2}$ … $\xrightarrow{\Delta}$ … $\xrightarrow{iPr_2NLi}$ … $\xrightarrow{I_2 \text{ or } -Br_2}$ … $\xrightarrow{-SO_2}$ … $\xrightarrow{\text{heat or } I_2}$ (2-83)

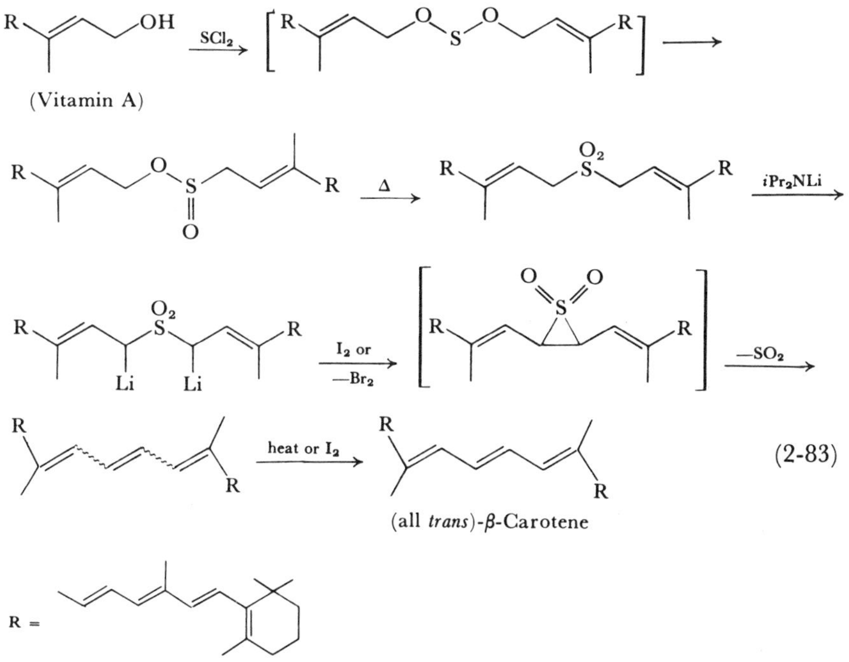

(all *trans*)-β-Carotene

R =

Reaction with Electrophiles

We have already considered the reaction of sulfur-containing carbanions with carbon electrophiles, with protons, and with sources of positive halogen. Examples of other electrophilic species which may be used are shown in

Eqs. 2-84 to 2-87 (Corey, 1967; Seebach *et al.*, 1972; Almog and Weissman, 1973; Yamamoto *et al.*, 1973, respectively). Reaction 2-87 is of interest since

2-Lithio-2-*t*-butyl-1,3-dithiane $\xrightarrow[\text{(2) hydrol}]{\text{(1) }(CH_3)_3SiCl}$ tBu—C(=O)—$Si(CH_3)_3$ (29%) (2-84)

2-Lithio-1,3-dithiane $\xrightarrow{CH_3SSCH_3}$ 2-SCH_3-1,3-dithiane $\xrightarrow[\text{(2) }CH_3SSCH_3]{\text{(1) 2-lithio-1,3-dithiane}}$ 2,2-bis(SCH_3)-1,3-dithiane (63%) (2-85)

$$CH_3S(O)CH_2Na \xrightarrow{ClP(O)(OC_2H_5)_2} CH_3S(O)CH_2P(O)(OC_2H_5)_2 \qquad (2\text{-}86)$$

n-$C_4H_9CLi(SPh)_2$ $\xrightarrow[(C_6H_{11})_3B]{-LiSPh}$ $(C_6H_{11})_2$B—C(C_4H_9-n)(SPh)—C_6H_{11} $\xrightarrow{H_2O_2}$ C_6H_{11}—C(O)C_4H_9-n (84%) (2-87)

$\longrightarrow \longrightarrow \xrightarrow{\text{9-BBN}}$

$\xrightarrow[\text{THF, }-40^\circ]{LiC(CH_3)(SPh)_2}$ $\xrightarrow{-PhS^-}$

(1) NaOH, H_2O_2
(2) $HgCl_2$, $CaCO_3$

Mg(Hg)–$TiCl_4$

(2-88)

it can be used in cases where attack on alkyl halides (e.g., cyclohexyl, halides) proceeded poorly if at all. In a modification of reaction 2-87, Corey has found that unsymmetrical trialkylboranes can be utilized in a selective manner as shown in Eq. 2-88 (Corey *et al.*, 1976).

General References

Buncel, E. (1975). "Carbanions: Mechanistic and Isotopic Aspects." Elsevier, Amsterdam.

Cram, D. J. (1965). "Fundamentals of Carbanion Chemistry." Academic Press, New York.

Cram, D. J. (1968). *Surv. Progr. Chem.* **4**, 45.

Ebel, H. F. (1969). *Fortschr. Chem. Forsch.* **12**, 387.

References

Abatjoglou, A. G., and Eliel, E. L. (1974). *J. Org. Chem.* **39**, 3043.

Agarwal, K. L., Berlin, Y. A., Fritz, H.-J., Gait, M. J., Kleid, D. G., Lees, R. G., Norris, K. E., Ramamoorthy, B., and Khorana, H. G. (1976). *J. Am. Chem. Soc.* **98**, 1065.

Almog, J., and Weissman, B. A. (1973). *Synthesis* 164; also see Corey, E. J., and Shulman, J. I., *J. Org. Chem.* **35**, 777 (1970).

Anciaux, A., Eman, A., Dumont, W., and Krief, A. (1975). *Tetrahedron Lett.* p. 1617.

Arens, J. F., Fröling, M., and Fröling, A. (1959). *Rec. Trav. Chim. Pays-Bas* **78**, 663; also see Froling, A. and Arens, J. F., *ibid.* **81**, 1009 (1962).

Atlanti, P. M., Biellmann, J. F., Dube, S., and Vincens, J. J. (1974). *Tetrahedron Lett.* 2665.

Baarschers, W. H., and Loh, T. L. (1971). *Tetrahedron Lett.* 3483.

Baldwin, J. E., Hackler, R. E., and Scott, R. M. (1969). *Chem. Commun.* 1415.

Barbarella, G., Garbesi, A., and Fava, A. (1975). *J. Am. Chem. Soc.* **97**, 5883.

Bartlett, P. D., and Lewis, E. S. (1950). *J. Am. Chem. Soc.* **72**, 1005; also see Streitwieser, A., Jr., Caldwell, R. A., and Granger, M. R., *J. Am. Chem. Soc.* **86**, 3578 (1964).

Bartlett, P. D., and Woods, G. F. (1940). *J. Am. Chem. Soc.* **62**, 2933.

Beak, P., Yamamoto, J., and Upton, C. J. (1975). *J. Org. Chem.* **40**, 3053; also see Beak, P., and Worley, J. W., *J. Am. Chem. Soc.* **94**, 597 (1972); also see Inagaki, S., Fujimoto, H., and Fukui, K., *J. Am. Chem. Soc.* **98**, 4054 (1976).

Bernardi, F., Csizmadia, I. G., Mangini, A., Schlegel, H. B., Whangbo, M-H., and Wolfe, S. (1975). *J. Am. Chem. Soc.* **97**, 2209.

Bestmann, H., and Schaper, W. (1975). *Tetrahedron Lett.* 3511; also see Potter, S. E., and Sutherland, I. O. *Chem. Commun.* 520 (1973).

Biellmann, J. F., and Ducep, J. B. (1968). *Tetrahedron Lett.* 5629.

Biellmann, J. F., and Ducep, J. B. (1969). *Tetradedron Lett.* 3707.

Biellmann, J. F., and Ducep, J. B. (1971). *Tetrahedron* **27**, 5861.

Biellmann, J. F., and Schmitt, J. L. (1973). *Tetrahedron Lett.* 4615.

Biellmann, J. F., and Vicens, J. J. (1974). *Tetrahedron Lett.* 2915.

Block, E. (1977). *In* "Organic Compounds of Sulphur, Selenium and Tellurium" (Specialist Periodical Reports) (D. R. Hogg, ed.), Vol. IV. Chemical Society of London.

Boekelheide, V., Anderson, P. H., and Hylton, T. A. (1974). *J. Am. Chem. Soc.* **96**, 1558.

Bordwell, F. G. (1968). *In* "Organosulfur Chemistry" (M. J. Janssen, ed.). Wiley, New York.

Bordwell, F. G. (1970). *Accounts Chem. Res.* **3**, 281.

Bordwell, F. G., and Doomes, E. (1974). *J. Org. Chem.* **39**, 2531 and references therein.

Bordwell, F. G., Matthews, W. S., and Vanier, N. R. (1975a). *J. Am. Chem. Soc.* **97**, 442.

Bordwell, F. G., Vanier, N. R., Matthews, W. S., Hendrickson, J. B., and Skipper, P. L. (1975b). *J. Am. Chem. Soc.* **97**, 7160.

Bordwell, F. G., Van Der Puy, M., and Vanier, N. R. (1976). *J. Org. Chem.* **41**, 1883, 1885.

Bordwell, F. G., Bares, J. E., Bartmess, J. E., Drucker, G. E., Gerhold, J., McCollum, G. J., Van Der Puy, M., Vanier, N. R., and Matthews, W. S., (1977). *J. Org. Chem.* **42**, 326.

Bory, S., and Marquet, A. (1973). *Tetrahedron Lett.* 4155.

Bory, S., Luche, M. J., Moreau, B., Lavielle, S., and Marquet, A. (1975). *Tetrahedron Lett.* 827; see also Marquet, A., *Pure Appl. Chem.* **49**, 183 (1977).

Bos, H. J. T., Brandsma, L., and Arens, J. F. (1967). *Monatsh. Chem.* **98**, 1043.

Brown, C. A. (1974). *J. Org. Chem.* **39**, 3913.

Büchi, G., and Freidinger, R. M. (1974), *J. Am. Chem. Soc.* **96**, 3332.

Campbell, R. V. M., Crombie, L., Findley, D. A. R., King, R. W., Pattenden, G., and Whiting, D. A. (1975). *J. Chem. Soc. Perkin Trans* **1**, 897; also see Hendrickson, J. B., Giga, A., and Wareing, J., *J. Am. Chem. Soc.* **96**, 2275 (1974).

Carlson, R. G., and May, K. D. (1975). *Tetrahedron Lett.* 947.

Carpino, L. A., McAdams, L. V., Rynbrandt, R. H., and Spiewak, J. W. (1971). *J. Am. Chem. Soc.* **93**, 476.

Cilento, G. (1960). *Chem. Rev.* **60**, 146.

Coffen, D. L. (1969). *Rec. Chem. Progr.* **30**, 275.

Coffen, D. L., Grant, B. D., and Williams, D. L. (1971). *Int. J. Sulfur Chem. A* **1**, 113.

Corey, E. J. (1967). *Pure Appl. Chem.* **14**, 19.

Corey, E. J., and Block, E. (1969). *J. Org. Chem.* **34**, 1233.

Corey, E. J., and Chaykovsky, M. (1965). *J. Am. Chem. Soc.* **87**, 1345.

Corey, E. J., and Durst, T. (1966). *J. Am. Chem. Soc.* **88**, 5656.

Corey, E. J., and Erickson, B. W. (1971). *J. Org. Chem.* **36**, 3553.

Corey, E. J., and Lowry, T. H. (1965a). *Tetrahedron Lett.* 793.

Corey, E. J., and Lowry, T. H. (1965b). *Tetrahedron Lett.* 803.

Corey, E. J., and Seebach, D. (1965). *Angew. Chem. Int. Ed. English* **4**, 1075, 1077.

Corey, E. J., and Seebach, D. (1966). *J. Org. Chem.* **31**, 4097.

Corey, E. J., König, H., and Lowry, T. H. (1962). *Tetrahedron Lett.* 515.

Corey, E. J., Erickson, B. W., and Noyori, R. (1971). *J. Am. Chem. Soc.* **93**, 1724; also see Corey, E. J., and Noyori, R. *Tetrahedron Lett.* 311 (1970); Erickson, B. W. *Org. Synth.* **54**, 19 (1974).
Corey, E. J., Danheiser, R. L., and Chandrasekaran, S. (1976). *J. Org. Chem.* **41**, 260.
Coulson, C. A. (1969). *Nature (London)* **221**, 1106.
Craig, D. P., and Thirunamachandran, T. (1965). *J. Chem. Phys.* **43**, 4183; also see Craig, D. P., and Magnusson, E. A. *J. Chem. Soc.* 4895 (1956).
Cristol, S. J., Harrington, J. K., and Singer, M. S. (1966). *J. Am. Chem. Soc.* **88**, 1529.
Dagonneay, M. (1973). *C. R. Acad. Sci. Paris C* **276**, 1683.
Dodson, R. M., Hammen, P. D., and Fan, J. Y. (1971). *J. Org. Chem.* **36**, 2703 and previous papers in series.
Doering, W. v. E., and Hoffmann, A. K. (1955). *J. Am. Chem. Soc.* **77**, 521.
Doering, W. v. E., and Levy, L. K. (1955). *J. Am. Chem. Soc.* **77**, 509.
Durst, T. (1971). *Tetrahedron Lett.* p. 4171.
Durst, T., and Molin, M. (1975). *Tetrahedron Lett.* 63.
Durst, T., and Tin, K.-C. (1970). *Can. J. Chem.* **48**, 845.
Durst, T., and Viau, R. (1973). *Intra-Sci. Chem. Rep.* **7**, 63.
Durst, T., Viau, R., Van Den Elzen, R., and Nguyen, C. H. (1971). *Chem. Commun.* 1334.
Eliel, E. L., Hartmann, A. A., and Abatjoglou, A. G. (1974). *J. Am. Chem. Soc.* **96**, 1807.
Epiotis, N. D., Yates, R. L., Bernardi, F., and Wolfe, S. (1975). *J. Am. Chem. Soc.* **98**, 5435.
Evans, D. A., and Andrews, G. C. (1974). *Accounts Chem. Res.* **7**, 147.
Fischli, A., and Mayer, H. (1975). *Helv. Chim. Acta* **58**, 1492, 1586.
Fischli, A., Branca, Q., and Daly, J. (1976). *Helv. Chim. Acta* **59**, 2443.
Florey, J. B., and Cusachs, L. C. (1972). *J. Am. Chem. Soc.* **95**, 3040.
Fraser, R. R., and Ng, L. K. (1976). *J. Am. Chem. Soc.* **98**, 4334.
Fromm, E. (1889). *Justus Liebigs Ann. Chem.* **253**, 135.
Fukuyama, T., and Kishi, T. (1976). *J. Am. Chem. Soc.* **98**, 6723.
Gilman, H., and Webb, F. J. (1940). *J. Am. Chem. Soc.* **62**, 987.
Gresser, M. (1969). *Mech. Reactions Org. Sulfur Compounds* **4**, 29.
Grieco, P. A., and Masaki, Y. (1975). *J. Org. Chem.* **40**, 150.
Grieco, P. A., Masaki, Y., and Boxler, D. (1975). *J. Org. Chem.* **40**, 2261.
Gröbel, B.-T., and Seebach, D. (1977). *Synthesis*, p. 357.
Gronowitz, S. (1975). *In* "Organic Sulphur Chemistry" (C. J. M. Stirling, ed.). Butterworth, London.
Grossert, J. S., Buter, J., Asveld, E. W. H., and Kellogg, R. M. (1974). *Tetrahedron Lett.* 2805.
Hauser, C. R., Kantor, S. W., and Brasen, W. R. (1953). *J. Am. Chem. Soc.* **75**, 2660.
Hayashi, T., and Baba, H. (1975). *J. Am. Chem. Soc.* **97**, 1608.
Henderson, J. W. (1973). *Quart. Rev. Chem. Soc. (London)* **2**, 397.
Herrmann, J. L., Richman, J. E., and Schlessinger, R. H. (1973). *Tetrahedron Lett.* 3275 and earlier papers in series; see also Damon, R. E., Schlessinger, R. H., and Blount, J. F., *J. Org. Chem.* **41**, 3773 (1976) and Paulsen, H., Koebernick, W., and Koebernick, H., *Tetrahedron Lett.* 2297 (1976).
Hine, J., and Dalsin, P. D. (1972). *J. Am. Chem. Soc.* **94**, 6998.
Hine, J., and Porter, J. J. (1960). *J. Am. Chem. Soc.* **82**, 6178.
Hirai, K., Matsuda, H., and Kishida, Y. (1971). *Tetrahedron Lett.* 4359; also see Hirai, K., and Kishida, Y., *Tetrahedron Lett.* 2117 (1972).
House, H. O., and Larson, J. K. (1968). *J. Org. Chem.* **33**, 61.
Johnson, A. W. (1975). *In* "Organic Compounds of Sulphur, Selenium and Tellurium" (D. H. Reid, ed.), Vol. III and earlier volumes. Chemical Society of London.
Johnson, C. R., and Katekar, G. F. (1970). *J. Am. Chem. Soc.* **92**, 5753.

Johnson, C. R., Nakanishki, A., Nakanishi, N., and Tanaka, K. (1975). *Tetrahedron Lett.* 2865.

Jones, M., Temple, P., Thomas, E. J., and Whitham, G. H. (1974). *J. Chem. Soc. Perkin I* 433.

Julia, M., and Arnould, D. (1973). *Bull. Soc. Chim. Fr.* 743, 746, 3065.

Julia, M., and Guy-Roualt, A. (1967). *Bull. Soc. Chim. Fr.* 1411.

Julia, M., and Saussine, L. (1974). *Tetrahedron Lett.* 3443.

Kattenberg, J., de Waard, E. R., and Huisman, H. O. (1975). *Rec. Trav. Chim. Pays-Bas* **94**, 89.

Kienzle, F., and Minder, R. E. (1976). *Helv. Chim. Acta* **59**, 439.

Kimball, G. E. (1940). *J. Chem. Phys.* **8**, 188.

King, J. F., and Beatson, R. P. (1975). *Tetrahedron Lett.* 973.

Kishi, Y., Fukuyama, T., and Nakatsuka, S. (1973). *J. Am. Chem. Soc.* **95**, 6490, 9492, 6493.

Koch, H. P., and Moffitt, W. E. (1951). *Trans. Faraday Soc.* **47**, 7; also see Bernardi (1975, footnote 60).

Kodama, M., Matsuki, Y., and Ito, S. (1976). *Tetrahedron Lett.* 1121; (1975) 3065.

Koenig, K. E., Felix, R. A., and Weber, W. P. (1974). *J. Org. Chem.*, **39**, 1539.

Kondo, K., and Tunemoto, D. (1975). *Tetrahedron Lett.* 1007.

Kondo, K., Negishi, A., Matsui, K., Tunemoto, D., and Masamune, S. (1972). *Chem. Commun.* 1311.

Lehn, J.-M., and Wipff, G. (1976). *J. Am. Chem. Soc.* **98**, 7498.

Lett, R., Chassaing, G., and Marquet, A. (1976). *J. Organomet. Chem.* **111**, C17.

Lever, O. W., Jr. (1976). *Tetrahedron* **32**, 1943.

Manas, A-R. B., and Smith, R. A. J. (1975). *Chem. Commun.* 216.

Manchand, P. S., Rosenberger, M., Saucy, G., Wehrli, P. A., Wong, H., Chambers, L., Ferro, M. P., and Jackson, W. (1976). *Helv. Chim. Acta* **59**, 387; also see Fischli, A., Mayer, H., Simon, W., and Stoller, H-J., *ibid.* **59**, 397 (1976).

Marshall, J. A., and Belletire, J. L. (1971). *Tetrahedron Lett.* 871.

Marshall, J. A., and Seitz, D. E. (1974). *J. Org. Chem.* **39**, 1814.

Martel, J., and Huynh, C. (1967). *Bull. Soc. Chim. Fr.* 982.

Martel, H. J. J-B., and Rasmussen, M. (1971). *Tetrahedron Lett.* 3843.

Maryanoff, C. A., Maryanoff, B. A., Tang, R., and Mislow, K. (1973). *J. Am. Chem. Soc.* **95**, 5839.

Matthews, W. S., Bares, J. E., Bartmess, J. E., Bordwell, F. G., Cornforth, F. J., Drucker, G. E., Margolin, Z., McCallum, R. J., McCollum, G. J., and Vanier, N. R. (1975). *J. Am. Chem. Soc.* **97**, 7006.

McMahon, T. B., and Kebarle, P. (1976). *J. Am. Chem. Soc.* **98**, 3399.

Meyers, A. I., and Ford, M. E. (1976). *J. Org. Chem.* **41**, 1735.

Meyers, C. Y., Malte, A. M., and Matthews, W. S. (1969). *J. Amer. Chem. Soc.* **91**, 7510; also see Meyers, C. Y., Matthews, W. S., and Malte, A. M., U.S. Pat.3,830,862 (May 20, 1974) and 3,949,001 (May 6, 1976).

Mislow, K. (1967). *Rec. Chem. Progr.* **28**, 4.

Mitchell, K. A. R. (1969). *Chem. Rev.* **69**, 157.

Mitchell, F. H., Otsubo, T., and Boekelheide, V. (1975). *Tetrahedron Lett.* 219; Mitchell, R. H., and Carruthers, R. J., *ibid.* 4331 (1975).

Montanari, F. (1975). *In* "Organic Sulphur Chemistry" (C. J. M. Stirling, ed.). Butterworth, London.

Mukaiyama, T., Narasaka, K., Maekawa, K., and Furusato, M. (1971). *Bull. Chem. Soc, Soc. Jpn.* **44**, 2285.

Mukaiyama, T., Narasaka, K., and Furusato, M. (1972). *J. Am. Chem. Soc.* **94**, 8641.
Mura, A. J. Jr., Majetich, G., Grieco, P. A., and Cohen, T. (1975). *Tetrahedron Lett.* 4437.
Musher, J. I. (1972). *J. Am. Chem. Soc.* **94**, 1370; *Adv. Chem. Ser.* **110**, Chapter 3 (1972).
Mutterer, F., and Fluery, J. P. (1974). *J. Org. Chem.* **39**, 640.
Narasaka, K., Hayashi, M., and Mukaiyama, T. (1972). *Chem. Lett.* (*Jpn.*) 259.
Neureiter, N. P. (1965). *J. Org. Chem.* **30**, 1313.
Normant, H., and Castro, B. (1964). *C. R. Acad. Sci. Paris* **269**, 830.
Oae, S., Tagaki, W., and Ohno, A. (1964). *Tetrahedron* **20**, 417, 427.
Oshima, K., Takahaski, H., Yamamoto, H., and Nozaki, H. (1973a). *J. Am. Chem. Soc.* **95**. 2693.
Oshima, K., Shimoji, K., Takahashi, H., Yamamoto, H., and Nozaki, H. (1973b). *J. Am. Chem. Soc.* **95**, 2694; see also Muthkrishnan, R., and Schlosser, M. *Helv. Chim. Acta* **59**, 13 (1976).
Oshima, K., Yamamoto, H., and Nozaki, H. (1973c). *J. Am. Chem. Soc.* **95**, 7926.
Paquette, L. A. (1968). *Accounts Chem. Res.* **1**, 209; Paquette, L. A. (1968). *In* "Mechanisms of Molecular Migrations" (B. S. Thyagarajan, ed.), Vol. I, p. 121. Wiley (Interscience), New York.
Paquette, L. A., and Philips, J. C. (1969). *J. Am. Chem. Soc.* **91**, 3973.
Paquette, L. A., Meisinger, R. H., and Wingard, Jr., R. E. (1973). *J. Am. Chem. Soc.* **95**, 2230.
Paquette, L. A., Oku, M., Heyd, W. E., and Meisinger, R. H. (1974). *J. Am. Chem. Soc.* **96**, 5815.
Parham, W. E., and Motter, R. F. (1959). *J. Am. Chem. Soc.* **81**, 2146; also see Parham, W. E., Kalmins, M. A., and Theissen, D. R., *J. Org. Chem.* **27**, 2698 (1962).
Paulsen, H., Sinnwell, V., and Stadler, P. (1972). *Angew. Chem. Int. Ed. English* **11**, 149.
Photis, J. M., and Paquette, L. A. (1974). *J. Am. Chem. Soc.* **96**, 4715.
Peterson, D. J. (1967). *J. Org. Chem.* **32**, 1717.
Peterson, D. J. (1972). *Organometal. Chem. Rev. Sect. A* **7**, 295.
Price, C. C., and Oae, S. (1962). "Sulfur Bonding." Ronald Press, New York.
Ramberg, L., and Bäcklund, B. (1940). *Arkiv. Kemi. Mineral Geol.* **13A**, No. 27; *Chem. Abstr.* **34**, 4725.
Rauk, A., Buncel, E., Moir, R. Y., and Wolfe, S. (1965). *J. Am. Chem. Soc.* **87**, 5498.
Rauk, A., Wolfe, S., and Csizmadia, I. G. (1969). *Can. J. Chem.* **47**, 113.
Rayner, D. R., Gordon, A. J., and Mislow, K. (1968). *J. Am. Chem. Soc.* **90**, 4854.
Reece, C. A., Rodin, J. O., Brownlee, R. G., Duncan, W. G., and Silverstein, R. M. (1968). *Tetrahedron* **24**, 4249.
Reich, H. J. (1975). *J. Org. Chem.* **40**, 2570.
Roitman, J. N., and Cram, D. J. (1971). *J. Am. Chem. Soc.* **93**, 2225.
Russell, G. A., and Weiner, S. A. (1966). *J. Am. Chem. Soc.* **31**, 248.
Salmond, W. G. (1968). *Quart. Rev.* **22**, 253.
Schank, K., Wilmes, R., and Ferdinand, G. (1973). *Int. J. Sulfur Chem.* **8**, 397.
Schaumann, E., and Walter, W. (1974). *Chem. Ber.* **107**, 3562.
Schöllkopf, V., Woerner, F. P., and Wiskott, E. (1966). *Chem. Ber.* **99**, 806.
Schönberg, A., Černik, D., and Urban, W. (1931). *Chem. Ber.* **64**, 2577.
Seebach, D. (1969). *Angew. Chem., Intl. Ed. English* **8**, 639.
Seebach, D., and Beck, A. K. (1971). *Org. Synth.* **51**, 76.
Seebach, D., and Beck, A. K. (1972). *Chem. Ber.* **105**, 3892.
Seebach, D., and Beck, A. K. (1974). *Angew. Chem. Int. Ed. English* **13**, 806.
Seebach, D., and Burstinghaus, R. (1975). *Angew. Chem. Int. Ed. English* **14**, 57.
Seebach, D., and Corey, E. J. (1975). *J. Org. Chem.* **40**, 231.

Seebach, D., and Kolb, M. (1974). *Chem. Ind.* (*London*) 687.

Seebach, D., and Peleties, N. (1972). *Chem. Ber.* **105**, 511.

Seebach, D., Erickson, B. W., and Singh, G. (1966). *J. Org. Chem.*, **31**, 4303.

Seebach, D., Geiss, K-H., Beck, A. K., Graf, B., and Daum, H. (1972). *Chem. Ber.* **105** 3280.

Seebach, D., Kolb, M., and Gröbel, B-T. (1973). *Angew. Chem. Int. Ed. English* **12**, 69.

Semmelhack, M. F., and Ryono, L. S. (1975). *J. Am. Chem. Soc.* **97**, 3873.

Shatenshtein, A. I., and Gvozdeva, H. A. (1969). *Tetrahedron* **25**, 2749.

Shirley, D. A., and Reeves, B. J. (1969). *J. Organomet. Chem.* **16**, 1; also see Dolak, T. M., and Bryson, T. A., *Tetrahedron Lett.* 1961 (1977).

Sowerby, R. L., and Coates, R. M. (1972). *J. Am. Chem. Soc.* **94**, 4758, also see McMurray, J. E., and von Beroldingen, L. A. (1974). *Tetrahedron* **30**, 2027.

Stotter, P. L., and Hornish, R. E. (1973). *J. Am. Chem. Soc.* **95**, 4444.

Streitwieser, A., Jr., and Ewing, S. P. (1975). *J. Am. Chem. Soc.* **97**, 190.

Streitwieser, A., Jr., and Holtz, D. (1967). *J. Am. Chem. Soc.* **89**, 692.

Streitwieser, A., Jr., and Mares, F. (1968). *J. Am. Chem. Soc.* **90**, 2444.

Streitwieser, A., Jr., and Williams, J. E., Jr. (1975). *J. Am. Chem. Soc.* **97**, 191; also see Fabian, J., Schönfeld, P., and Mayer, R., *Phosphorus and Sulfur* **2**, 151 (1976).

Tang, C. S. F., Morrow, C. J., and Rapoport, H. (1975). *J. Am. Chem. Soc.* **97**, 759.

Trost, B. M., and Miller, C. H. (1975). *J. Am. Chem. Soc.* **97**, 7182.

Trost, B. M., Kelley, D., and Bogdanowicz, M. J. (1973). *J. Am. Chem. Soc.* **95**, 3068; also see Trost, B. M., and Keeley, D. E., *ibid.* **96**, 1252 (1974); **98**, 248 (1976).

Trost, B. M., Preckel, M., and Leichter, L. M. (1975). *J. Am. Chem. Soc.* **97**, 2224.

Trost, B. M., Arndt, H. C., Strege, P. E., and Verhoeven, T. R. (1976). *Tetrahedron Lett.* 3477.

Trost, B. M., Keeley, D. E., Arndt, H. C., Rigby, J. H., and Bogdanowicz, M. J. (1977). *J. Am. Chem. Soc.* **99**, 3080.

Truce, W. E., and Roberts, F. E. (1963). *J. Org. Chem.* **28**, 961.

Tsuchihashi, G., Iriuchijima, S., and Maniwa, K. (1973). *Tetrahedron Lett.* 3389.

Tsuchihashi, G., Mitamura, S., and Ogura, K. (1976). *Tetrahedron Lett.* 855.

van Tamelen, E. E., McCurry, P., and Huber, N. (1971). *Proc. Nat. Acad. Sci. U.S.* **68**, 1294.

van Tamelen, E. E., Holton, R. A., Hopla, R. E., and Konz, W. E. (1972). *J. Am. Chem. Soc.* **94**, 8228.

Uneyama, K., and Torii, S. (1976). *Tetrahedron Lett.* 443.

Vedejs, E., and Fuchs, P. L. (1971). *J. Org. Chem.* **36**, 366.

Wallace, T. J., Pobiner, H., Hofmann, J. E., and Schriesheim, A. (1965). *J. Chem. Soc.* 1271.

Weinges, K., and Klessing, K. (1976). *Chem. Ber.* **109**, 793.

Wolfe, S. (1972). *Accounts Chem. Res.* **5**, 102.

Wolfe, S. (1976). Personal communication.

Wolfe, S., Rauk, A., and Csizmadia, I. G. (1967). *J. Am. Chem. Soc.* **89**, 5710.

Wolfe, S., Rauk, A., and Csizmadia, I. G. (1969). *J. Am. Chem. Soc.* **91**, 1567.

Wolfe, S., Rauk, A., Tel, L. M., and Csizmadia, I. G. (1970). *Chem. Commun.* 96.

Wolinsky, J., Marhenke, R. L., and Lau, R. (1972). *Synth. Commun.* **2**, 165.

Wright, A., and West, R. (1974). *J. Am. Chem. Soc.* **96**, 3222.

Yamamoto, S., Shiono, M., and Mukaiyama, T. (1973). *Chem. Lett.* (*Jpn.*) 961; see also Negishi, E., Yoshida, T., Siveira, A., Jr., and Chiour, B. L., *J. Org. Chem.* **40**, 814 (1975); Negishi, E., Chiu, K-W., and Yoshida, T., *J. Org. Chem.* **40**, 1677 (1975); Hughes, R. J., Pelter, A., Smith, K., Negishi, E., and Yoshida, T., *Tetrahedron Lett.* **87** (1976).

Chapter 3 | Sulfur Ylides

Sulfur ylides are zwitterions in which a carbanion is attached directly to a positively charged sulfur (Trost and Melvin, 1975; A. W. Johnson, 1966).† Although stabilized sulfur ylides were prepared as early as 1930 (Ingold and Jessop, 1930), it was only with the demonstration by Corey and Chaykovsky in 1962 of the facile preparation and highly selective methylene transfer capabilities of dimethylsulfonium and dimethylsulfoxonium methylides (**3-1** and **3-2**, respectively) that the chemistry of sulfur ylides came of age (Corey,

$$(CH_3)_2\underset{+}{S}{-}\underset{-}{C}H_2 \qquad\qquad (CH_3)_2\underset{+}{S}(O){-}\underset{-}{C}H_2$$

3-1 **3-2**

1962a, b; Corey and Chaykovsky, 1965). The rich chemistry characterizing sulfur ylides will be outlined below.

3.1 Preparation

General

Methods for the preparation of sulfur ylides are summarized in Scheme 3-1. These methods include deprotonation of sulfonium salts, reaction of sulfur compounds with carbenes, benzynes, and various active methylene compounds and electrophilic substances, and electrochemical reduction of sulfonium salts.

Deprotonation of Sulfonium Salts

Deprotonation of sulfonium salts, the most general approach to sulfur ylides, merits further discussion because in addition to the desired reaction there can be significant interference from competing processes such as β-elimination (or α,β'-elimination) and reaction at sulfur (Eq. 3-1). For example, treatment of 1,2,2,4-tetramethylthietanonium fluoroborate with

† Wittig explains that the name "ylide" was chosen because the "onium" atom is bound to the neighboring carbon in a homopolar fashion (yl) yet the linkage is simultaneously classified as ionic (ide) (Wittig, 1974).

Scheme 3-1. Preparation of Sulfur Ylides

Proton abstraction

$(CH_3)_3S^+ + n\text{-}C_4H_9Li \xrightarrow{THF} (CH_3)_2\overset{+}{S}\text{—}\overset{-}{C}H_2$	Corey and Chaykovsky (1965
$(CH_3)_3SO^+ + NaH \xrightarrow{THF} (CH_3)_2\overset{+}{S}(O)\overset{-}{C}H_2$	Corey and Chaykovsky (1965
$Ph_2\overset{+}{S}CH(CH_3)_2 \xrightarrow[THF]{Cl_2CHLi \text{ or } t\text{-}C_4H_9Li} Ph_2\overset{+}{S}\overset{-}{C}(CH_3)_2$	Corey *et al.* (1967)
(cyclopropyl, H)$\text{—}\overset{+}{S}Ph_2 + KOH \xrightarrow{DMSO}$ (cyclopropyl anion)$\text{—}\overset{+}{S}Ph_2$	Trost and Bogdanowicz (1973)
$(CH_3)_2\overset{+}{S}(O)N(CH_3)_2 + CH_3S(O)CH_2Na \longrightarrow CH_3\overset{+}{S}(N(CH_3)_2)\overset{-}{C}H_2$ (with $S{=}O$)	Johnson *et al.* (1970)
$(C_2H_5)_2\overset{+}{S}CH_2\overset{+}{S}(C_2H_5)_2 \xrightarrow[CH_3OH]{KOH} (C_2H_5)_2\overset{+}{S}\overset{-}{C}H\overset{+}{S}(C_2H_5)_2$	Lillya (1966)

Reaction of sulfur compounds with carbenes

$R_2S + (CH_3O_2C)_2CN_2 \xrightarrow[\Delta/Cu]{h\nu \text{ or}} R_2\overset{+}{S}\text{—}\overset{-}{C}(CO_2CH_3)_2$	Ando *et al.* (1972); Ando (19
$(CH_3)_2SO + (CH_3O_2C)_2CN_2 \xrightarrow{h\nu} (CH_3)_2\overset{+}{S}(O)\overset{-}{C}(CO_2CH_3)_2$	Ando *et al.* (1972)
$t\text{-}C_4H_9SS\text{-}t\text{-}C_4H_9 + Cl_2C{:} \longrightarrow t\text{-}C_4H_9\overset{+}{S}(\overset{-}{C}Cl_2)S\text{-}t\text{-}C_4H_9$	Searles and Wann (1965)
$PhCH_2SPh + PhSCH{:} \longrightarrow PhCH_2\overset{+}{S}(Ph)\overset{-}{C}HSPh$	Julia *et al.* (1972)
$PhSC(CH_3)_2CH_2\ddot{C}Ph \longrightarrow$ Ph—$\overset{+}{S}$ (four-membered ring with $C(CH_3)_2$ and $\overset{-}{C}Ph$)	Kondo and Ojima (1975)

Reaction of sulfides with benzyne

$(CH_2{=}CHCH_2)_2S$ + (benzyne) $\longrightarrow CH_2{=}CHCH_2\overset{+}{S}(Ph)\overset{-}{C}HCH{=}CH_2$	Hellmann and Eberle (1963)

Reaction of sulfur ylides with electrophilic compounds

$2\,(CH_3)_2\overset{+}{S}(O)\overset{-}{C}H_2 + C_2H_5O_2CCl \longrightarrow (CH_3)_2\overset{+}{S}(O)\overset{-}{C}HCO_2C_2H_5$	Nozaki *et al.* (1967)

(continued)

Scheme 3-1 (*continued*)

$(CH_3)_2\overset{+}{S}(O)\overset{-}{C}H_2 + PhC{\equiv}CC(O)Ph \longrightarrow$ [cyclic product: 3,5-diphenyl ring with $\overset{-}{C}$, $\overset{+}{S}$(=O)CH_3]	Hortmann (1965)
Michael addition to a vinyl sulfonium salt	
$CH_2{=}CH\overset{+}{S}(OCH_3)Ph + NaOCH_3 \longrightarrow CH_3OCH_2\overset{-}{C}H\overset{+}{S}(OCH_3)Ph$	Johnson and Phillips (1967)
$CH_2{=}CHCH{=}CH\overset{+}{S}(CH_3)_2 + NaOCH_3 \longrightarrow CH_3OCH_2CH{=}CH\overset{-}{C}H\overset{+}{S}(CH_3)_2$	Braun *et al.* (1973)
Reaction of sulfoxides with active methylene compounds	
$(CH_3)_2SO + CH_2(CN)_2 \longrightarrow (CH_3)_2\overset{+}{S}{-}\overset{-}{C}(CN)_2$	Middleton *et al.* (1965)
Electroreduction of sulfonium salts	
$(CH_3)_3S^+ \xrightarrow[DMSO]{e^-} (CH_3)_3S\cdot \xrightarrow{-H\cdot} (CH_3)_2\overset{+}{S}{-}\overset{-}{C}H_2$	Shono *et al.* (1973)

n-butyllithium at $-78°$ affords *n*-butylmethyl sulfide and 1,1,2-trimethylcyclopropane by way of a "σ-sulfurane" or tetracoordinate sulfur intermediate (Eq. 3-2; Trost *et al.*, 1971b).

$$-\underset{H}{C}-\underset{H}{C}-\overset{+}{\ddot{S}}\Big< \xrightarrow[B^-]{base} \begin{cases} -\underset{H}{C}-\overset{-}{C}-\overset{+}{\ddot{S}}\Big< & \text{Ylide} \\ >C{=}C\Big<_H + :\ddot{S}\Big< & \beta\text{-elimination} \\ >CH-CH-\overset{B}{\ddot{S}}\Big< & \text{"}\sigma\text{-sulfurane"} \end{cases} \qquad (3\text{-}1)$$

$$\text{2,2-(CH}_3)_2\text{-4-CH}_3\text{-1-CH}_3\text{-thietanium } S^+ \xrightarrow[-78°]{n\text{-}C_4H_9Li} [\text{σ-sulfurane: } S(C_4H_9)(CH_3)\text{ ring with } CH_3, CH_3, CH_3] \longrightarrow \text{1-CH}_3\text{-2,2-(CH}_3)_2\text{-cyclopropane} + n\text{-}C_4H_9SCH_3 \qquad (3\text{-}2)$$

The difficulties associated with sulfonium salt deprotonation reactions are well illustrated by Corey's preparation of diphenylsulfonium isopropylide (Corey *et al.*, 1967; Corey and Jautelat, 1967). It should first be noted that the ease of deprotonation of a sulfonium ion R_2S^+CHXY (and the stability of the resulting ylide $R_2S^+C^-XY$) vary markedly with the nature of groups X and Y. If X and/or Y are electron acceptors, such as cyano, carbonyl, or sulfonio (R_2S^+—), deprotonation requires only a very mild base. If X and Y are hydrogen, a strong base, such as *t*-butoxide or methylsulfinyl carbanion, is required and the ylide becomes thermally unstable, necessitating working at temperatures below 0°C. Sulfonium *n*-alkylides (X = H, Y = *n*-alkyl) are even less stable than sulfonium methylides (X = Y = H) but can be satisfactorily prepared at −78° in tetrahydrofuran solution using the poorly nucleophilic strong bases *t*-butyllithium or lithium diisopropylamide. More nucleophilic alkyllithiums such as methyl, *n*-butyl- or phenyllithium are unsatisfactory, leading to the formation of substantial quantities of by-products through β-elimination or σ-sulfurane routes. When both of the ylide substituents X and Y are alkyl groups as in the case of diphenylsulfonium isopropylide, deprotonation of the sulfonium salt precursor becomes particularly difficult. Corey succeeded in generating this latter ylide by methylation of a dimethoxyethane solution of diphenylsulfonium ethylide followed by treatment at −70° with dichloromethyllithium (generated *in situ* with methylene chloride and lithium diisopropylamide) (Eq. 3-3; Corey *et al.*, 1967; Corey and Jautelat, 1967).

$$Ph_2\overset{+}{S}—\overset{-}{C}HCH_3 \xrightarrow[\text{DME}]{CH_3I} Ph_2\overset{+}{S}—CH(CH_3)_2I^- \xrightarrow[\text{DME, } -70^\circ]{LiCHCl_2} Ph_2\overset{+}{S}—\overset{-}{C}(CH_3)_2 \quad (3\text{-}3)$$

In Section 2.6 we have considered the differential reactivity of diastereotopic methylene protons adjacent to the pyramidal sulfoxide group. Sulfonium sulfur is also pyramidal and configurationally stable under ordinary conditions. Thus, the report by Fava and co-workers (Barabarella *et al.*, 1975) that the methylene protons in conformationally rigid sulfonium salt **3-3** show differential kinetic acidity is not too surprising. However, the degree of stereospecificity is surprising: the relative exchange rates in D_2O are

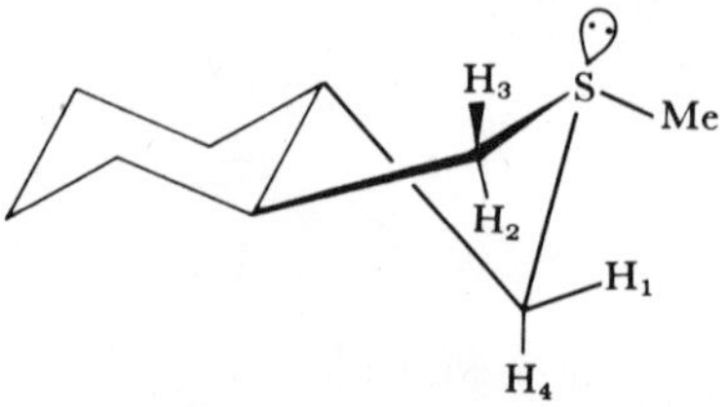

H3 S Me H2 H1 H4

3-3

$H_1 < 1$, $H_2 = 200$, $H_3 = 3$, $H_4 = 3$, with k_{H_2}/k_{H_1} estimated to be as large as 1300. Fava suggests that this is one of the most stereospecific nonenzymatic reactions yet reported but also concludes that the exchange data are inconsistent with any theory (e.g., the gauche effect theory) based on the idea that the all-important factor in determining ylide stability is the C_α—S dihedral angle.

3.2 σ- and π-Sulfuranes

A compound containing a sulfur atom singly bonded to four substituents has been termed a "σ-sulfurane" (Trost, 1973) in analogy to the hypothetical sulfurane molecule, H_4S. While σ-sulfuranes in which sulfur is bonded to one or more fluorine atoms or oxygen groups in addition to carbon are reasonably stable, as illustrated by the isolation of **3-4** (Martin and Perozzi, 1974), **3-5** (Kapovits and Kalman, 1971), and **3-6** (Denney *et al.*, 1973), σ-sulfuranes containing four carbon–sulfur bonds (e.g., the product of nucleophilic attack of an alkyllithium on sulfonium sulfur) readily revert to dicoordinate sulfides (Eq. 3-4; Sheppard, 1971). Using the sulfurane nomenclature, a compound containing a sulfur atom singly bonded to two ligands and doubly bonded to

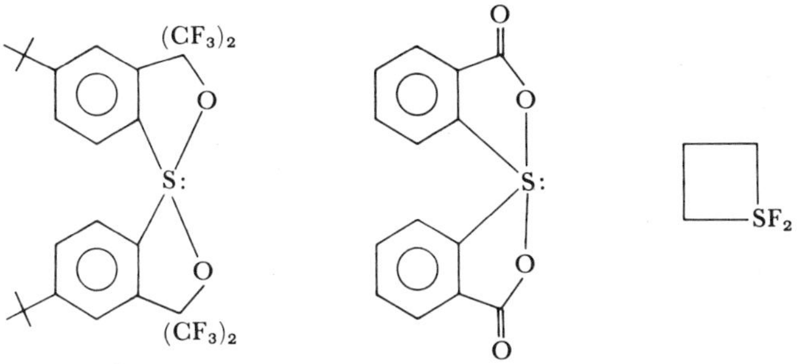

(CF₃)₂ O S: O (CF₃)₂ O O S: O O SF₂

3-4 **3-5** **3-6**

$$(C_6F_5)_4S: \xrightarrow{>0°C} C_6F_5—C_6F_5 \ (70\%) + (C_6F_5)_2S \ (71\%) \qquad (3\text{-}4)$$

one ligand, e.g., a sulfonium ylide in its fully covalent resonance form, is termed a π-sulfurane (Trost, 1973). Alternatively, resonance forms **3-7** and **3-8** are described as ylide and ylene representations, respectively. A recent

$$>\overset{+}{S}—\overset{-}{C}< \longleftrightarrow >S=C<$$

3-7 **3-8**

ab initio study of phosphonium ylides, thiocarbanions, and related species (Whangbo *et al.*, 1975) indicates that the *C*-heteroatom bond length does not correlate with the overlap population of the bond but correlates very well with the ionic bond orders, defined by the coulombic term $-q_A q_B/r_{AB}$ for a general bond A–B. The implication of this result is that the observed C-heteroatom bond shortening seen on deprotonation of onium salts to give ylides need not be ascribed to an ylene formulation such as **3-8** involving p–d π conjugation, but rather can be attributed to coulombic effects in **3-7**. In view of the fact that the accumulating body of experimental and theoretical studies on sulfur ylides indicates the absence of appreciable double-bond character as implied by **3-8** (Trost and Melvin, 1975), the use of the "π-sulfurane" nomenclature seems to this author to be misleading and undesirable.

3.3 Thiabenzenes

The generation and characterization of thiabenzenes and other thiaarenes represent a fascinating chapter in organosulfur chemistry which, as we shall see, is appropriately included in this survey of sulfur ylides. Structures **3-9**–**3-13** ($R_1 = R_2$ = Ph in **3-13**) were originally described by Price and coworkers as stable planar molecules possessing cyclic aromatic ring currents (Price *et al.*, 1963, 1971; Polk *et al.*, 1969; Suld and Price, 1961). Very recently Mislow established that **3-9**–**3-11** as prepared by Price are actually oligomeric materials of presently undetermined structure (Maryanoff *et al.*, 1975). Both Hortmann and Mislow have concluded that substituted thiaarenes such as **3-12**–**3-14**, which are monomeric in solution, are best depicted as nonaromatic, unstable *cyclic sulfonium ylides* (Hortmann and Harris, 1970; Hortmann *et al.*, 1974; Maryanoff *et al.*, 1975) which are pyramidal at sulfur

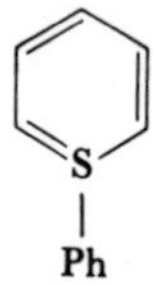

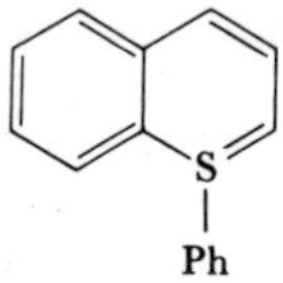

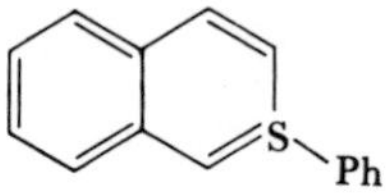

S Ph **3-9** S Ph **3-10** S Ph **3-11**

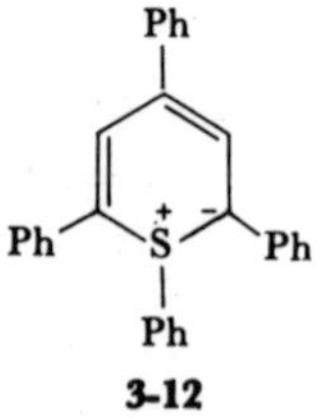

Ph Ph S⁺ Ph Ph **3-12** S⁺ R₂ R₁ **3-13** Ph Ph S⁺ CH₃ **3-14**

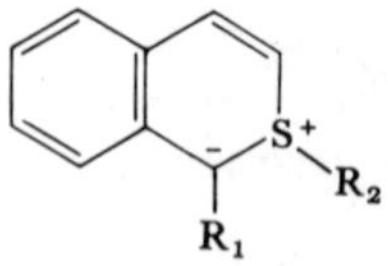

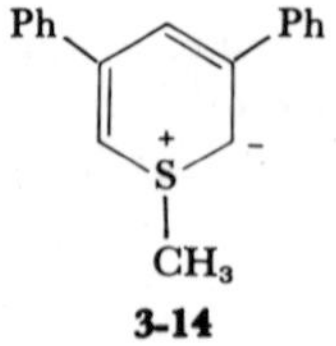

(Maryanoff *et al.*, 1975). Evidence supporting the characterization of **3-12**–**3-14** as ylides includes

(1) NMR data indicating diastereotopic isopropyl methyl groups in **3-13** $R_1 = i\text{-}C_3H_7$, R_2 = Ph (requires that sulfur be pyramidal with a substantial

barrier to inversion (Maryanoff *et al.*, 1975)) and indicating the substantial shielding of the 2 and 6 protons in **3-14** (δ4.03;† Hortmann and Harris, 1970; Hortmann *et al.*, 1974) and the 3 proton in **3-13**, R_1 = Ph, R_2 = CH_3 (δ4.47;† Hortmann *et al.*, 1974; Maryanoff, 1975);

(2) the facile Stevens rearrangement of **3-13**, R_1 = Ph, R_2 = CH_3, to **3-15** (Hortmann *et al.*, 1974; Maryanoff *et al.*, 1975); and

(3) solvent and substituent effects characteristic of charged (or zwitterionic) species (Hortmann *et al.*, 1974; Maryanoff *et al.*, 1975).

Hortmann has reported a novel synthesis of a thiabenzene 1-oxide which,

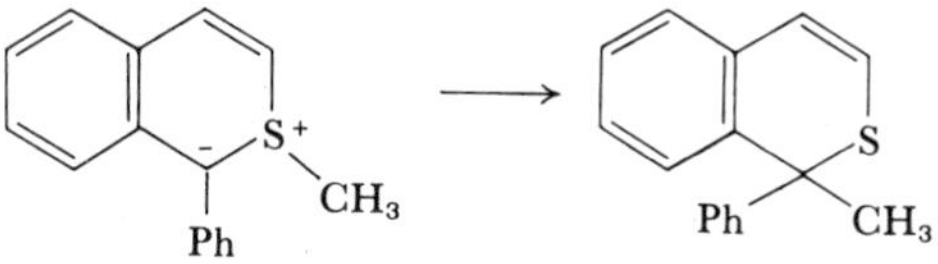

Ph S^+ CH_3 → Ph S CH_3

3-13 **3-15**

in accord with its description as a *cyclic sulfoxonium ylide*, is substantially more stable than the related thiabenzene **3-14** (Eq. 3-5; Hortmann, 1965; Hortmann and Harris, 1971):

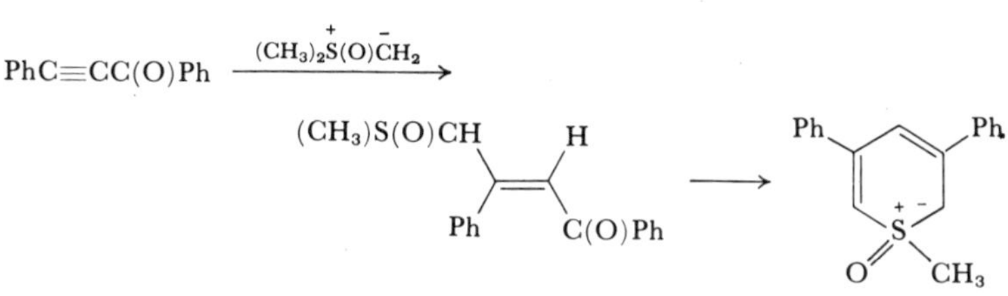

$$PhC{\equiv}CC(O)Ph \xrightarrow{(CH_3)_2\overset{+}{S}(O)\overset{-}{C}H_2} (CH_3)S(O)CH(Ph)C{=}CH\,C(O)Ph \longrightarrow \text{Ph, Ph, }\overset{+}{S}\overset{-}{}(O)CH_3 \quad (3\text{-}5)$$

3.4 Thiocarbonyl Ylides (Thione Ylides)

Thiocarbonyl ylides, species containing two trivalent carbon atoms bonded to sulfur, may be viewed as 1,3-dipoles, although contributions from other resonance structures are possible (see **3-16**) (Kellogg, 1976). Approaches

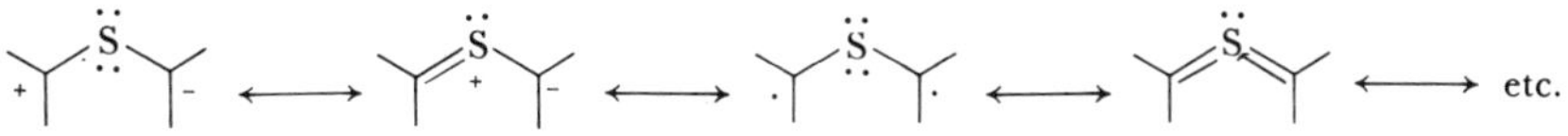

⟷ ⟷ ⟷ ⟷ etc.

3-16

to these interesting, reactive species include carbene addition to thiocarbonyl compounds and other processes in which a disubstituted carbon becomes attached to a thiocarbonyl sulfur (Eqs. 3-6 and 3-7), fragmentation of Δ^3-1,3,4-thiadiazolines (Eq. 3-8), deprotonation of sulfonium compounds

† For comparison, the 2,5 protons of thiophene appear at δ7.30.

(Eq. 3-9), photocyclization of divinyl sulfides (Eq. 3-10), and ring-opening of episulfides (Eq. 3-11).

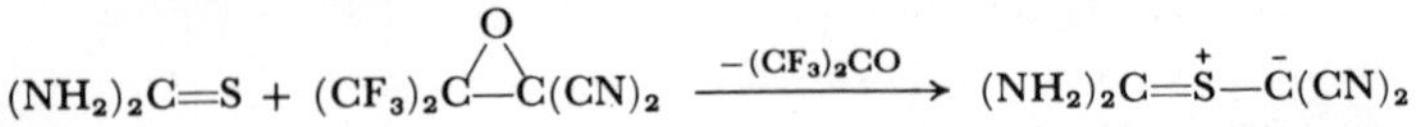

$$(NH_2)_2C{=}S + (CF_3)_2C\overset{O}{-}C(CN)_2 \xrightarrow{-(CF_3)_2CO} (NH_2)_2C{=}\overset{+}{S}{-}\overset{-}{C}(CN)_2 \quad (3\text{-}6)$$

(Middleton, 1966)

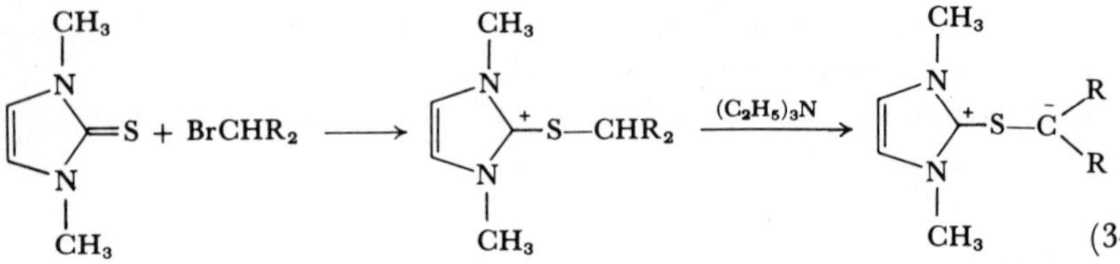

(3-7)

(Arduengo and Burgess, 1976)

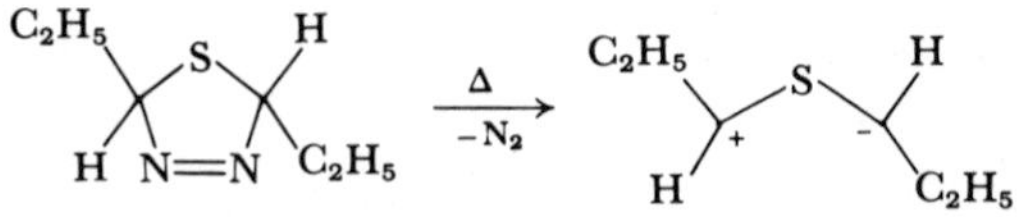

(3-8)

(Buter *et al.*, 1972)

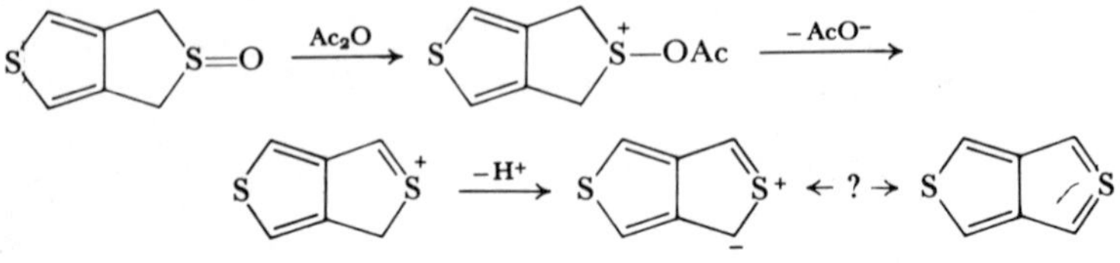

(3-9)

thieno[3,4-C]thiophene

(Cava and Lakshmikantham, 1975)

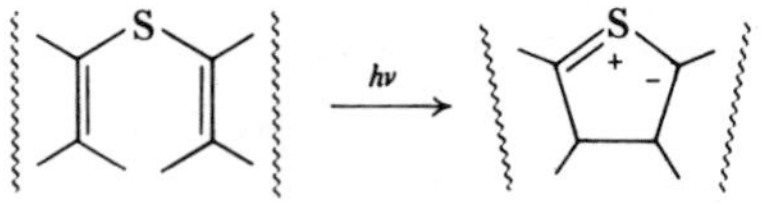

(3-10)

(Block, 1969; Schultz and DeTar, 1976)

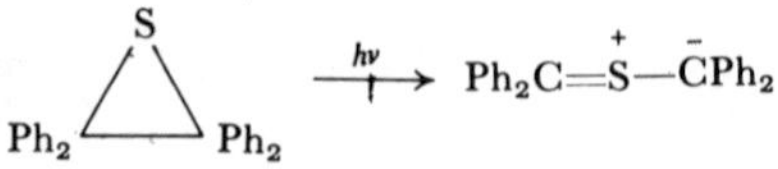

$$\xrightarrow{h\nu} Ph_2C{=}\overset{+}{S}{-}\overset{-}{C}Ph_2 \quad (3\text{-}11)$$

(Becker *et al.*, 1970)

One of the characteristic reactions of thiocarbonyl ylides is closure to the valence tautomeric episulfides. In agreement with the qualitative analogy between the molecular orbital descriptions of thiocarbonyl ylides and the allyl anion, *conrotatory* closure is observed with substituted thiocarbonyl ylides (Eq. 3-12; Buter *et al.*, 1972). Thiocarbonyl ylides also undergo stereospecific cycloaddition with 1,3-dipolarophiles (Buter *et al.*, 1972; Kellogg, 1973). Burgess has recently shown by an x-ray study that the "allyl anion" bonding model of a thiocarbonyl ylide, which requires a planar ylide geometry, is not followed by heavily substituted ylides such as that in Eq. 3-7 (Arduengo and Burgess, 1976).

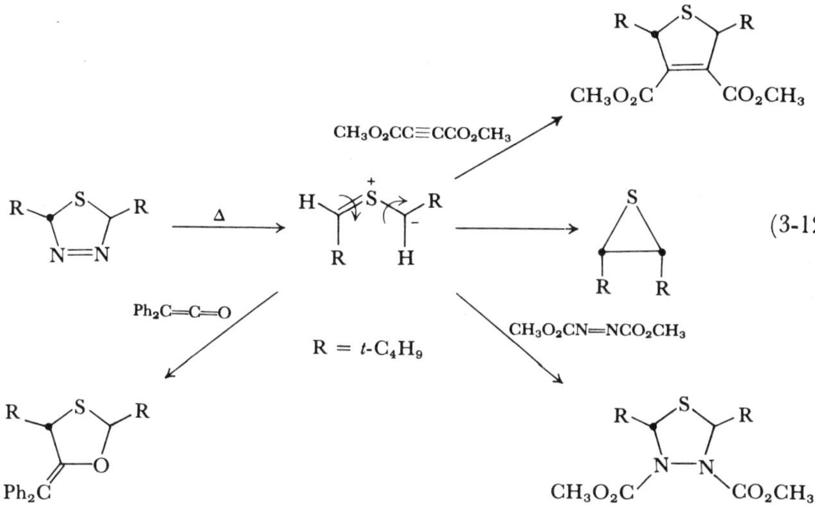

$R = t\text{-}C_4H_9$ (3-12)

Closure to episulfides is occasionally followed by spontaneous elimination of sulfur (Eq. 3-13). Barton has devised a clever "twofold extrusion" approach to olefin synthesis shown in Eq. 3-14 which involves pyrolysis of Δ^3-1,3,4-thiadiazolines (available in good yields from ketones or thioketones

(3-13)

(Middleton, 1966)

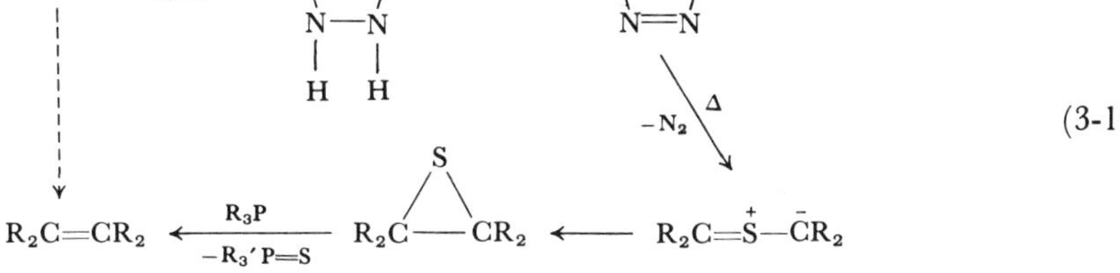

(3-14)

Table 3-1. Olefin Synthesis via Twofold Extrusion

Ketone (or thioketone)	Olefin product	Overall yield (%)	Reference
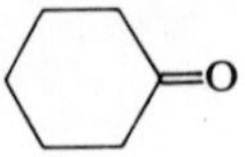	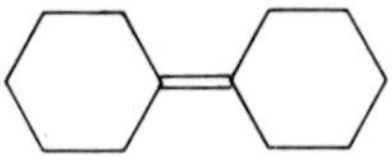	42	Barton and Willis (1972)
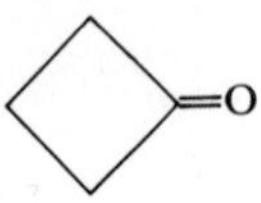	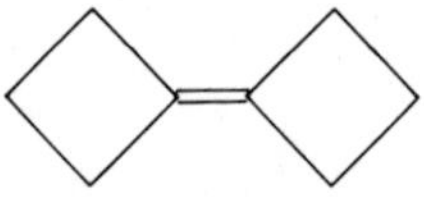	69	Everett and Garratt (1972)
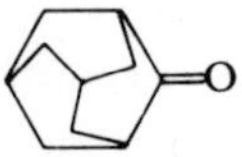	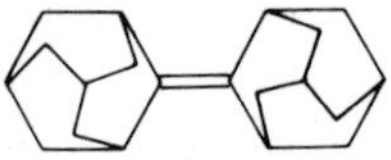	65	Schaap and Faler (1973)
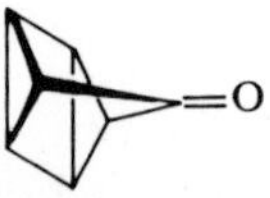	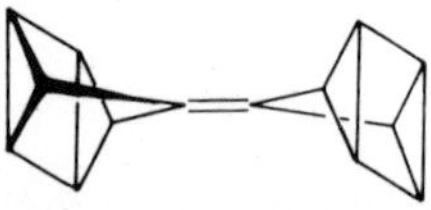	20	Sauter *et al.* (1973)
$(t\text{-}C_4H_9)_2C{=}S$[a]	$(t\text{-}C_4H_9)_2C{=}CPh_2$	68[b]	Barton *et al.* (1974)
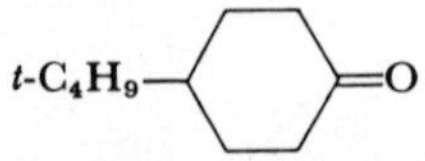 $t\text{-}C_4H_9$–(cyclohexanone)=O	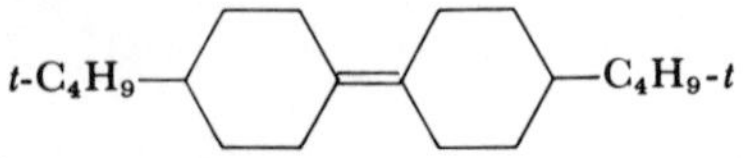 $t\text{-}C_4H_9$–(cyclohexylidene)=(cyclohexylidene)–$C_4H_9\text{-}t$	90[b]	Kellogg *et al.* (1975)

[a] Coreactant is diphenyldiazomethane.
[b] From *cis*-Δ^3-1,3,4-thiadiazoline.

as shown) in the presence of phosphines, reagents known to extrude sulfur from episulfides (Barton and Willis, 1972; Barton *et al.*, 1974). Applications of this reaction appear in Table 3-1.

Two other reactions seen with thiocarbonyl ylides are *protonation* followed by capture of the intermediate alkylidene sulfonium ion by a nucleophile (Eq. 3-15) and, in the case of cyclic thiocarbonyl ylides, [1,4]-suprafacial hydrogen shifts (Eq. 3-16).

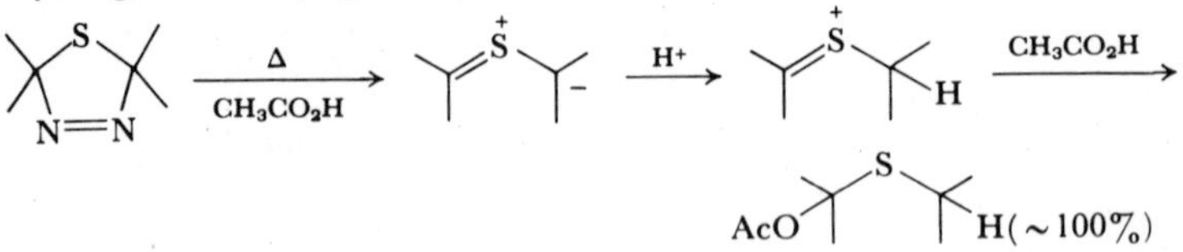

S, N=N; $\xrightarrow[CH_3CO_2H]{\Delta}$ $\overset{+}{S}$ $\xrightarrow{H^+}$ $\overset{+}{S}$ H $\xrightarrow{CH_3CO_2H}$ AcO S H(~100%) (3-15)

(Buter *et al.*, 1974)

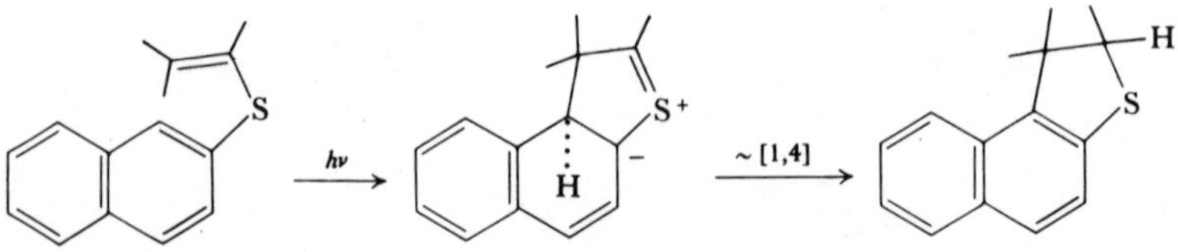

S $\xrightarrow{h\nu}$ S+, H $\xrightarrow{\sim[1,4]}$ H, S (87%) (3-16)

(Schultz and DeTar, 1976)

3.5 Reactions of Sulfur Ylides

The reactions of sulfur ylides include carbonyl addition (and addition to other carbon–heteroatom multiple bonds), Michael addition (cyclopropanation), alkylation (and other displacements), elimination (α and α,β'), fragmentation, and rearrangement ([1,2]-, [1,4]-, and [2,3]sigmatropic). Representative examples of each of these classes of sulfur ylide reactions are presented below. More detailed discussions and many additional examples of these several reactions may be found in the recent monograph by Trost and Melvin (1975) and chapters by Johnson (1970, 1973, 1975) and Block (1977).

Carbonyl Addition

While phosphorus ylides react with carbonyl compounds giving olefins via the well-known Wittig reaction, sulfur ylides under analogous conditions afford *epoxides* (Eq. 3-17).

The relative weakness of the sulfur–oxygen double bond compared to the phosphorus–oxygen double bond† along with the superior leaving group ability of sulfides (or sulfoxides and sulfonamides, considering the ylides in Table 3-2) compared to phosphines can account for the contrasting results following carbonyl addition. In addition, if phosphorus–oxygen and sulfur–oxygen single bond strengths parallel the respective double bond strengths, oxaphosphetanes such as Eq. 3-17 might be expected to be more stable than

$Ph_3\overset{+}{P}\overset{-}{C}H_2$; $\overset{+}{P}Ph_3$; O^-; PPh_3; **3-17**; $=CH_2 + Ph_3PO$; $=O$; $(CH_3)_2\overset{+}{S}\overset{-}{C}H_2$; O^-; $+S(CH_3)_2$; CH_3; CH_3; **3-18**; $-CH_3\overset{O}{\overset{\|}{S}}CH_3$; $=CH_2$; CH_2; $+ CH_3SCH_3$ (3-17)

† The sulfur–oxygen bond strength in $Ph_2S{=}O$ is 89 kcal/mol while the phosphorus-oxygen bond strength in $Ph_3P{=}O$ is 128.4 kcal/mol (Astrologes and Martin, 1976).

Table 3-2. Epoxide Formation with 4-*tert*-Butylcyclohexanone

Entry	Ylide	Product(s)	Ratio	Reference
1.	$(CH_3)_2\overset{+}{S}—\overset{-}{C}H_2$		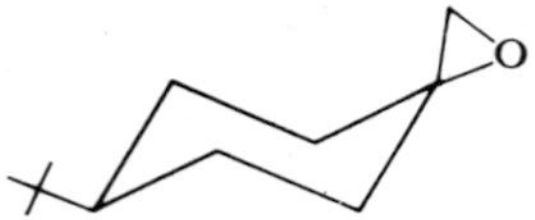  5:1	Corey and Chaykovsky (1965)
2.	$Ph_2\overset{+}{S}—\overset{-}{C}HCH{=}CH_2$		 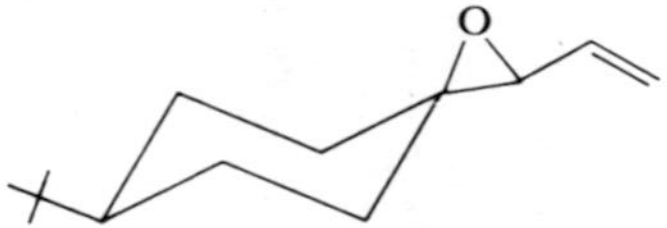 1:4	LaRochelle *et al.* (1971)
3.	$(CH_3)_2\overset{+}{S}—\overset{-}{C}HCO_2^-$		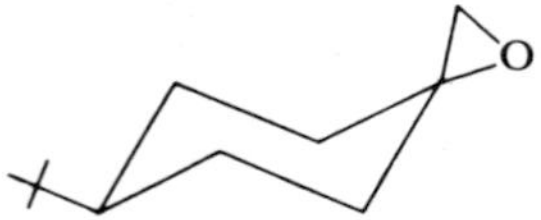  3:97	Adams *et al.* (1970)
4.	$(CH_3)_2\overset{+}{S}(O)CH_2^-$			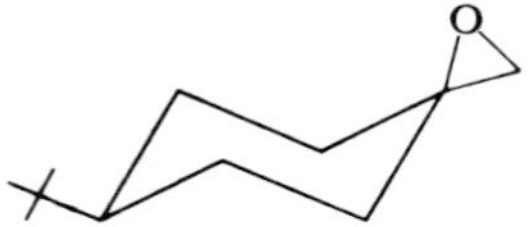 Corey and Chaykovsky (1965)
5.	$Ph\overset{+}{S}(O)CH_2^-$ with $N(CH_3)_2$ on S			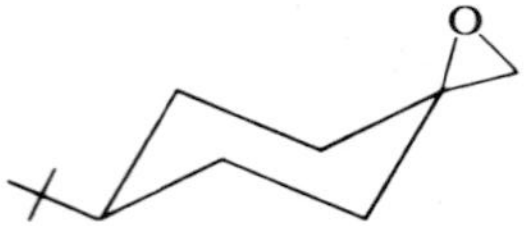 Johnson and Schroeck (1973a)
6.	$Ph_2\overset{+}{S}$—cyclopropyl anion			 Trost and Bogdanowicz (1973)

their sulfur counterpart (A. W. Johnson, 1966; certain oxaphosphetanes have actually been isolated at low temperatures (Vedejs, 1973)).

In view of the importance of epoxides as synthetic intermediates and the ready availability of carbonyl compounds, epoxide formation using sulfur ylides is a very useful procedure that has been studied in great detail both from a mechanistic and preparative standpoint. The generality of the reaction is demonstrated by the lack of interference from enol ethers, acetals, amides, nitriles, divalent sulfur, and in some cases, esters, hydroxyl, and amino groups (Trost and Melvin, 1975).

The stereochemical course of epoxide formation with a variety of sulfur ylides is best illustrated with studies on the conformationally immobile 4-*tert*-butylcyclohexanone as summarized in Table 3-2.

The observed stereochemical preferences can be rationalized by arguing that dimethylsulfonium methylide (and to a lesser extent the other sulfonium ylides) give products dictated by kinetic control whereas sulfoxonium ylides give products dictated by thermodynamic considerations (Johnson (1973); Trost and Melvin, 1975). The reversibility of sulfoxonium ylide addition and the irreversibility of sulfonium ylide addition are nicely demonstrated by the following results (Eqs. 3-18 and 3-19; Johnson, 1973).

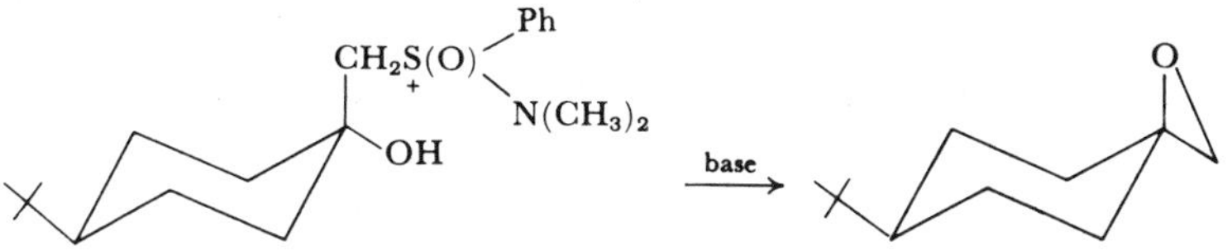

$CH_2\overset{+}{S}(O)(Ph)N(CH_3)_2$ / OH $\xrightarrow{\text{base}}$ (3-18)

$CH_2\overset{+}{S}(CH_3)Ph$ / OH $\xrightarrow{\text{base}}$ (3-19)

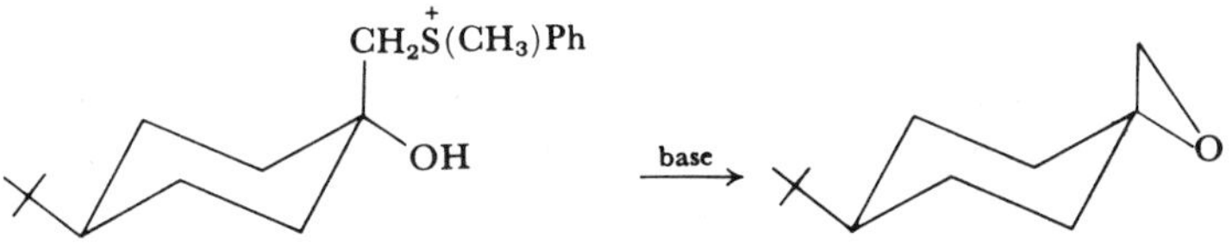

The sulfoxonium ylides, being more stable than sulfonium ylides, would be expected to be better leaving groups allowing reversible betaine formation. If a sulfur ylide is made so stable that its nucleophilicity is substantially diminished, as in the case of $(R_2S^+)_2CH^-$ and $R_2S^+CH^-C(O)R$, then

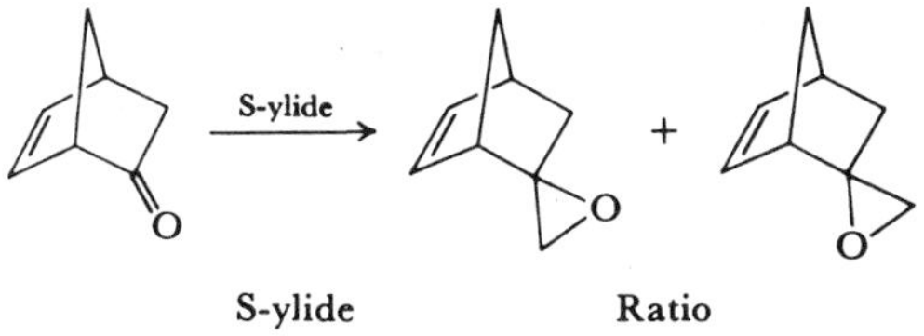

$\xrightarrow{\text{S-ylide}}$ + (3-20)

S-ylide	Ratio
$(CH_3)_2\overset{+}{S}(O)\overset{-}{C}H_2$	2.3:1
$(CH_3)_2\overset{+}{S}\overset{-}{C}H_2$	1:15.7

no carbonyl addition at all is observed. The differences in epoxide stereochemistry obtained with sulfonium and sulfoxonium ylides is of obvious synthetic utility as is further illustrated in Eq. 3-20 (Bly *et al.*, 1968). Asymmetric syntheses of epoxides may be realized using optically active sulfur ylides (Eq. 3-21; Johnson and Schroeck, 1973). Sulfur ylides are useful for

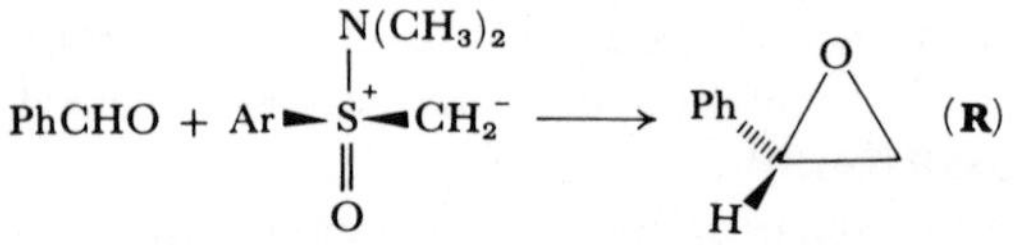

$$\text{PhCHO} + \text{Ar}-\overset{\overset{\text{N(CH}_3)_2}{|}}{\underset{\underset{\text{O}}{\|}}{\text{S}^+}}-\text{CH}_2^- \longrightarrow \text{Ph(H)-epoxide} \quad (\mathbf{R}) \tag{3-21}$$

(**R**) (60% yield; 20% optical purity)

preparing labeled epoxides for biosynthetic studies, as shown by the example in Eq. 3-22 (Corey *et al.*, 1968). The use of cyclopropyl sulfur ylides is the

$$\text{Ph}_2\overset{+}{\text{S}}-\overset{-}{\text{C}}\text{HCH}_3 \xrightarrow{^{14}\text{CH}_3\text{I}} \text{Ph}_2\overset{+}{\text{S}}-\overset{-}{\text{C}}(\text{CH}_3)_2^*$$

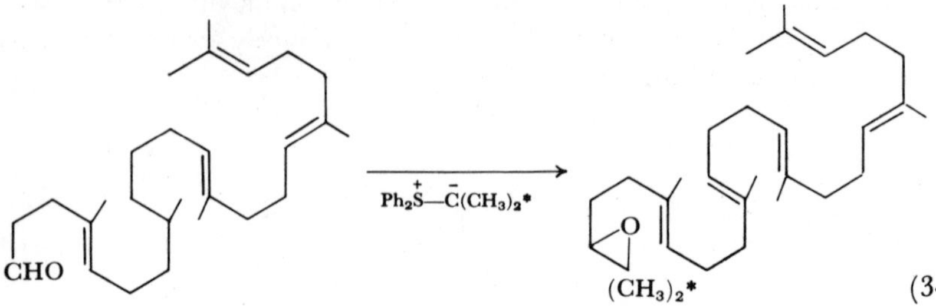

CHO $\xrightarrow{\text{Ph}_2\overset{+}{\text{S}}-\overset{-}{\text{C}}(\text{CH}_3)_2^*}$ O (CH$_3$)$_2$* (3-22)

basis for the novel process of "spiroannelation" (Eq. 3-23; Trost and Bogdanowicz; 1973; Trost, 1974; Johnson, 1973). When nucleophilic

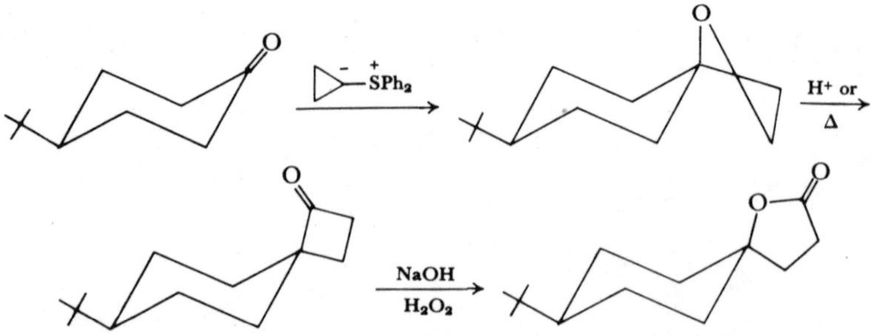

$\xrightarrow{\triangleright^- \overset{+}{\text{S}}\text{Ph}_2}$ O $\xrightarrow[\Delta]{\text{H}^+ \text{ or}}$

O $\xrightarrow[\text{H}_2\text{O}_2]{\text{NaOH}}$ O O (3-23)

centers are adjacent to carbonyl functions, ylide addition may be diverted by processes that may or may not involve oxiranes, e.g., Eq. 3-24 (Bravo *et al.*, 1969). Finally, methylene transfer to other carbon–heteroatom double

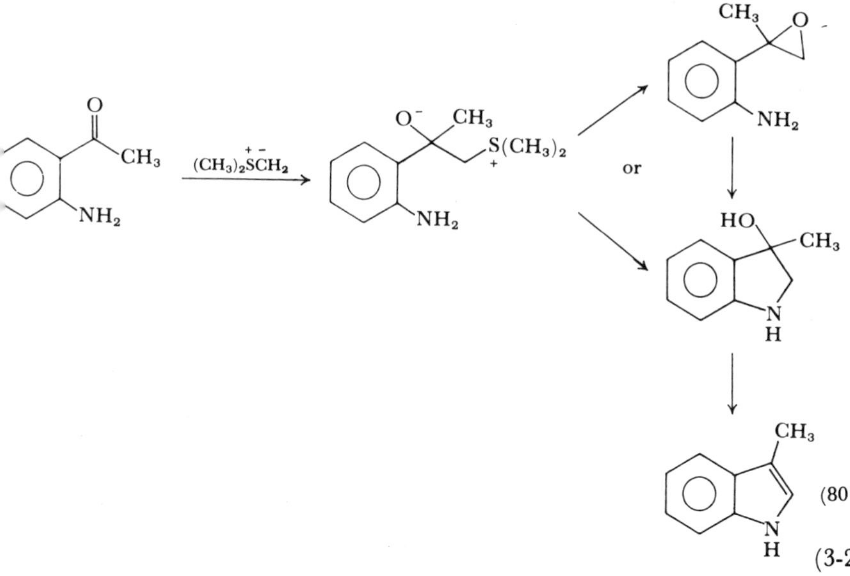

$$\text{o-aminoacetophenone} \xrightarrow{(CH_3)_2\overset{+}{S}\overset{-}{C}H_2} \text{alkoxide intermediate with } CH_2\overset{+}{S}(CH_3)_2 \;\rightarrow\; \text{epoxide (or) 3-hydroxy-3-methylindoline} \rightarrow \text{3-methylindole } (80\%) \qquad (3\text{-}24)$$

bonds has been used effectively in organic synthesis as in Eqs. 3-25 and 3-26.

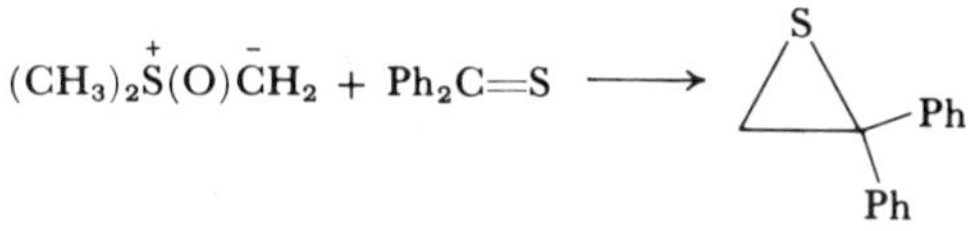

$$(CH_3)_2\overset{+}{S}(O)\overset{-}{C}H_2 + Ph_2C{=}S \longrightarrow \text{2,2-diphenylthiirane} \qquad (3\text{-}25)$$

(Corey, 1965)

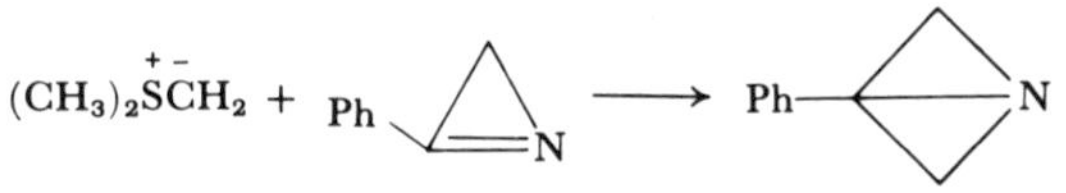

$$(CH_3)_2\overset{+}{S}\overset{-}{C}H_2 + \text{3-phenyl-2H-azirine} \longrightarrow Ph\text{-1-azabicyclo[1.1.0]butane} \qquad (3\text{-}26)$$

(Hortmann and Robertson, 1967)

Michael Addition (Cyclopropanation)

Sulfonium and sulfoxonium ylides readily add to such Michael acceptors as α,β-unsaturated ketones, esters, nitriles, isonitriles, sulfones, sulfoxides, sulfonamides, sulfonates, and nitro compounds to afford cyclopropanes (Trost and Melvin, 1975), e.g., Eqs. 3-27 to 3-29. These reactions are all nucleophilic cyclopropanations which proceed best with electron deficient olefins in contrast to addition of α-thiocarbenes/-carbenoids (cf. Chapter 6) and other carbenes/carbenoids that require *electron-rich* olefins.

It is known that biosynthesis of the cyclopropane ring in a number of natural products occurs by transfer of a methylene group from the methyl group of *S*-adenosylmethionine (**3-19**), presumably by way of a sulfonium ylide **3-20**, to an unactivated olefin such as an oleic ester (Law, 1971). While

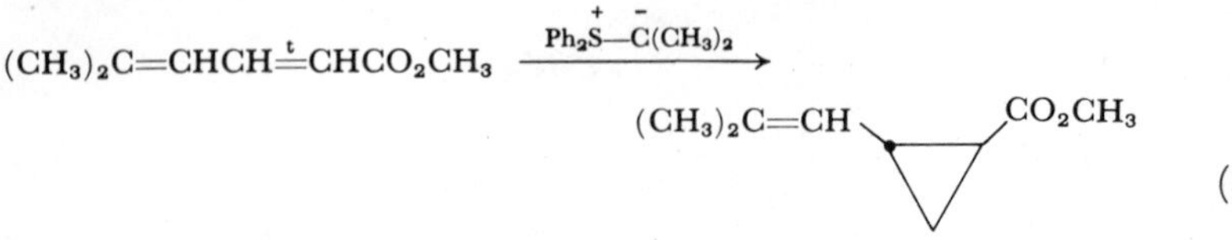

(3-27)

methyl chrysanthemate

(Corey and Jautelat, 1967b)

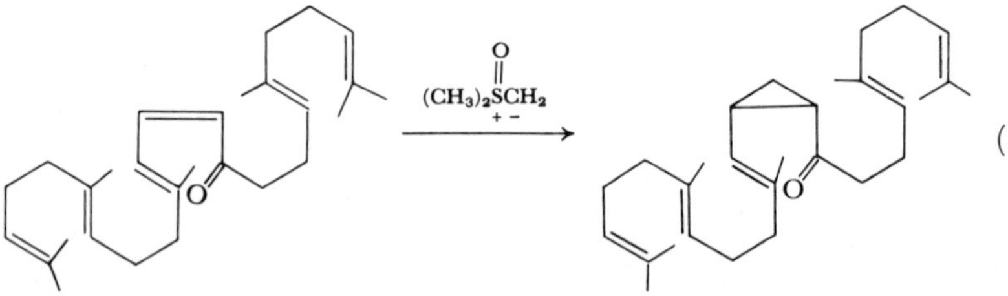

(3-28)

(Corey and Ortiz de Montellano, 1968a)

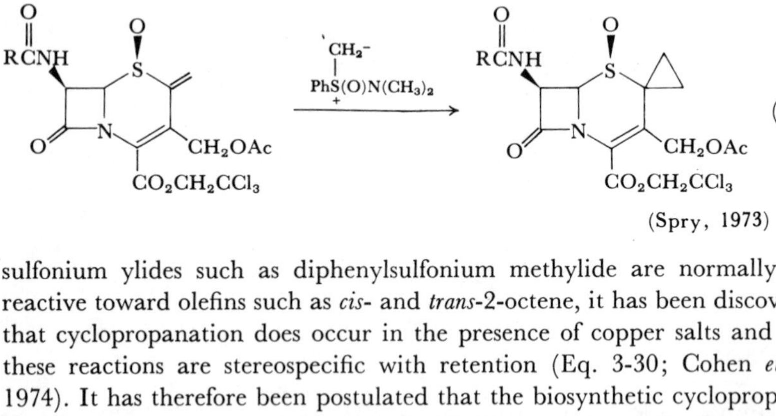

(3-29)

(Spry, 1973)

sulfonium ylides such as diphenylsulfonium methylide are normally unreactive toward olefins such as *cis*- and *trans*-2-octene, it has been discovered that cyclopropanation does occur in the presence of copper salts and that these reactions are stereospecific with retention (Eq. 3-30; Cohen *et al.*, 1974). It has therefore been postulated that the biosynthetic cyclopropanations involve activation of the ylide derived from *S*-adenosylmethionine via a transition metal chelate such as that shown for copper(I) in **3-21** (Cohen

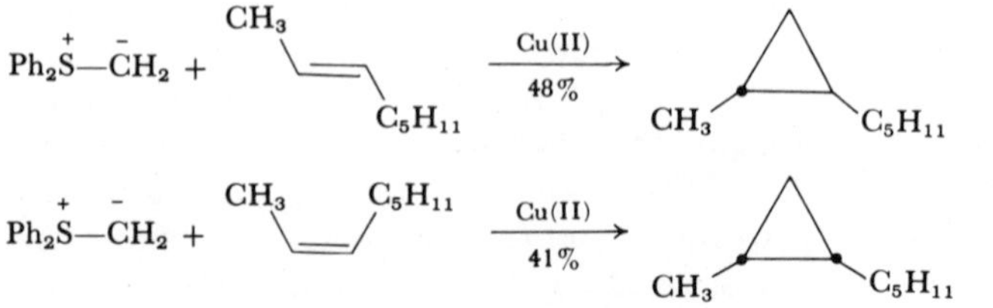

(3-30)

Table 3-3. Addition of Sulfur Ylides to Various Cyclohexenones

Entry	Cyclohexenone	Ylide	Product	Reference
1	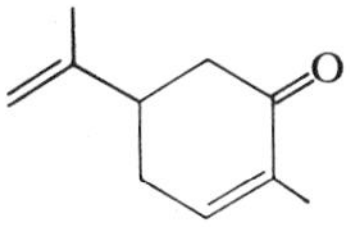	$(CH_3)_2\overset{+}{S}\overset{-}{C}H_2$	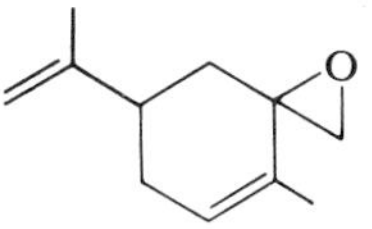	Corey and Chaykovsky (1965)
2	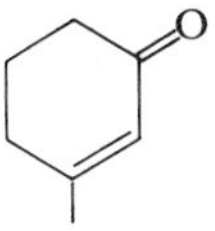	$Ph_2\overset{+}{S}\overset{-}{C}(CH_3)_2$	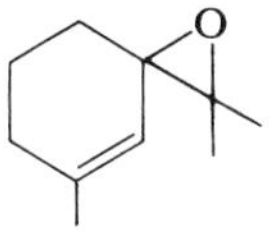	Corey and Jautelat (1967)
3	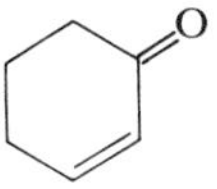	$Ph_2\overset{+}{S}\overset{-}{C}(CH_3)_2$	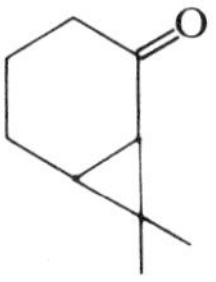	Corey and Jautelat (1967)
4	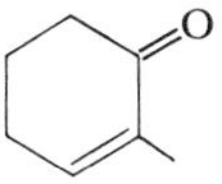	$Ph_2\overset{+}{S}\overset{-}{C}(CH_3)_2$	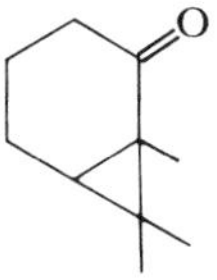	Corey and Jautelat (1967)
5	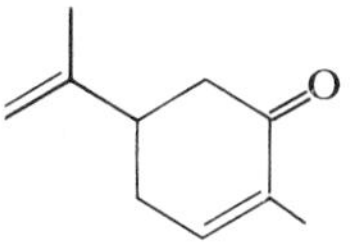	$(CH_3)_2\overset{+}{S}(O)CH_2^-$	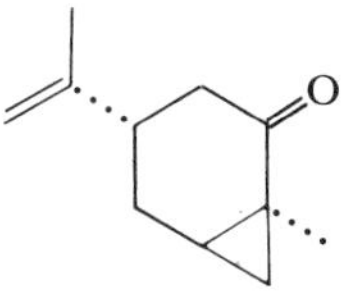	Corey and Chaykovsky (1965)
6	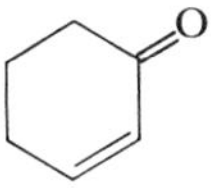	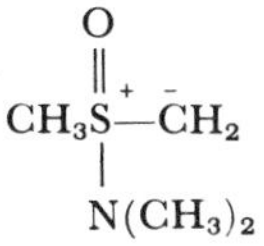 $CH_3\overset{+}{S}(=O)(N(CH_3)_2)-\overset{-}{C}H_2$	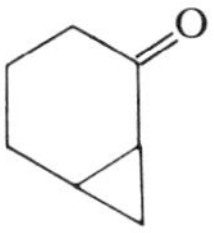	Johnson (1973)
7	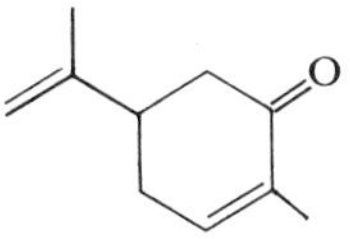	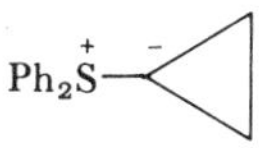 $Ph_2\overset{+}{S}-\overset{-}{C}$ (cyclopropylide)	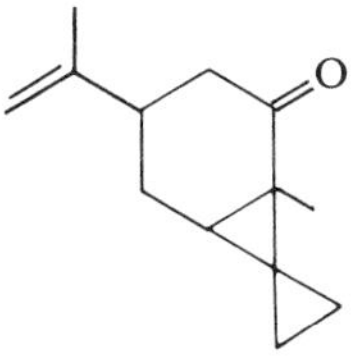	Trost and Bogdanowicz (1973)

NH_3^+ CH_3 $^-O_2CCH(CH_2)_2\overset{+}{S}CH_2$ adenosyl (**3-19**) $\xrightarrow[-H^+]{\text{base}}$ $^-O_2CCH(CH_2)_2\overset{+}{S}CH_2$ adenosyl, NH_3^+ CH_2^- (**3-20**) $\xrightarrow{Cu(I)}$ **3-21**

3-19

3-20

Cu(I)

3-21

n-C_8H_{17} $(CH_2)_7CO_2R$ H H

$[\,\geq Cu{=}CH_2\,] \leftarrow ? —$

n-C_8H_{17} $(CH_2)_7CO_2R$

enzyme—S—R′

R OPP → R $\overset{+}{S}$ Enz R′

base

(3-31)

CO_2R''

3-22

Chrysanthemic acid

$=CHCO_2R''$

(R=H)

R $\overset{-}{}$ $\overset{+}{S}$ Enz R′

Cu^+

R CH_2OPP

(R = geranyl)

H CH_2OPP R RCH_2

3-23

Presqualene pyrophosphate

et al., 1974). Methylene transfer may then be preceded by formation of a metal–carbene complex (examples of stable transition metal–carbene complexes have been described by Fischer and are discussed further in Chapter 6).

Other sulfur ylides related to **3-20** may be involved in the biosynthesis of presqualene pyrophosphate **3-23** (Cohen *et al.*, 1974) and chrysanthemic acid **3-22** (Pattenden and Storer, 1973) (Eq. 3-31).

The addition of sulfur ylides to eneones presents an interesting situation since addition to either the carbonyl group (1,2-addition) or to the olefin (1,4- or Michael addition) may occur depending on the nature of the ylide and enone. Table 3-3 illustrates this point.

In rationalizing the varying results obtained, for example, with carvone (Table 3-3, entries 1, 5, 7) and several different ylides or with diphenylsulfonium isopropylide and various cyclohexenones (Table 3-3, entries 2–4), it can be argued that the reversibility of betaine formation plays a major role in product determination. Betaine formation is thought to be essentially irreversible (i.e., $k_2 > k_{-1}$) with dimethylsulfonium methylide so that kinetic control prevails here, while in the case of the more stable ylides (or with hindered enones) reversible betaine formation (i.e., $k_{-1} > k_2$) allows thermodynamic control to operate (Eq. 3-32; Trost and Melvin, 1975). As

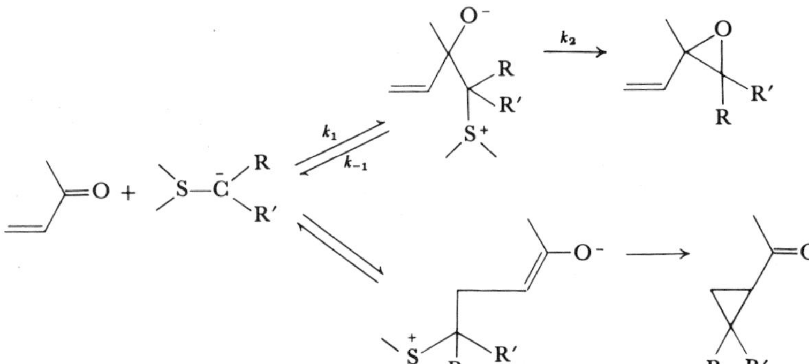

(3-32)

in the case of oxirane formation, use of optically active ylides permits the synthesis of optically active cyclopropanes (Eq. 3-33; Johnson and Schroeck, 1973b). With Michael acceptors containing nucleophilic groups, diversion

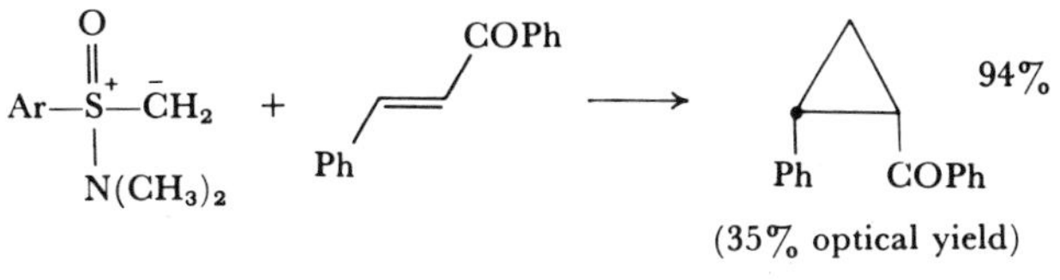

94% (3-33)

(35% optical yield)

$$(CH_3)_2\overset{+}{S}(O)\overset{-}{C}H_2 + CH_2{=}CHCONHR \longrightarrow (CH_3)_2\overset{+}{S}(O)\text{—}CH_2CH_2CH_2C(=NR)O^- \longrightarrow \text{N-R pyrrolidin-2-one} \qquad (3\text{-}34)$$

of the initial Michael adduct is sometimes observed (Eq. 3-34; König *et al.*, 1966). While highly stabilized ylides of low nucleophilicity such as $(CH_3)_2S^+C^-(CN)_2$ and $((CH_3)_2S^+)_2CH^-$ fail to undergo Michael addition, cyclopropane formation can be realized with some stabilized ylides such as $(CH_3)_2S^+CH^-CN$, $(CH_3)_2S^+CH^-CO_2C_2H_5$, $(CH_3)_2S^+CH^-SPh$ and the novel lactone ylide shown in Eq. 3-35 (Trost and Melvin, 1975).

$$(CH_3)_2\overset{+}{S}\text{—(}\alpha\text{-carbanion of }\gamma\text{-butyrolactone)} \xrightarrow{CH_2{=}CHCN} \text{spiro cyclopropane lactone (CN on cyclopropane)} \qquad (3\text{-}35)$$

Alkylation (and other Displacements)

Sulfur ylides participate in a variety of displacement reactions, in keeping with their carbanionic character (Eq. 3-36).

$$Ph_2\overset{+}{S}\text{—}\overset{-}{C}HCH_3 \xrightarrow{CH_3I} Ph_2\overset{+}{S}\text{—}CH(CH_3)_2$$ (Corey *et al.*, 1967)

$$Ph_2\overset{+}{S}CH_2CH_2CH_2Cl \xrightarrow[THF]{NaH} Ph_2\overset{+}{S}\overset{-}{C}HCH_2CH_2Cl \longrightarrow Ph_2\overset{+}{S}\text{—}C_3H_5\ (\text{cyclopropyl})$$ (Trost and Bogdanowicz, 1973)

(3-36)

$$(CH_3)_2\overset{+}{S}(O)\overset{-}{C}H_2 \xrightarrow{(CH_3)_3SiCl} (CH_3)_2\overset{+}{S}(O)CH_2Si(CH_3)_3$$ (Schmidbaur and Kapp, 1972)

$$(CH_3)_2\overset{+}{S}(O)\overset{-}{C}H_2 \xrightarrow{CH_3SO_2Cl} (CH_3)_2\overset{+}{S}(O)CH_2SO_2CH_3$$ (Truce and Madding, 1966)

α-Elimination

A limited number of examples of α-elimination reactions of sulfur ylides have been described. These reactions produce carbenes which undergo such characteristic reactions as Wolff rearrangement, insertion into C—H bonds, addition to C=C and formation of "dimers" as indicated by the following representative examples (Eqs. 3-37 through 3-41). It has been demonstrated that decomposition of stable sulfur ylides to give "dimer" is significantly catalyzed by "electron deficient" disulfides such as dimethoxy disulfide

Wolff Rearrangement:

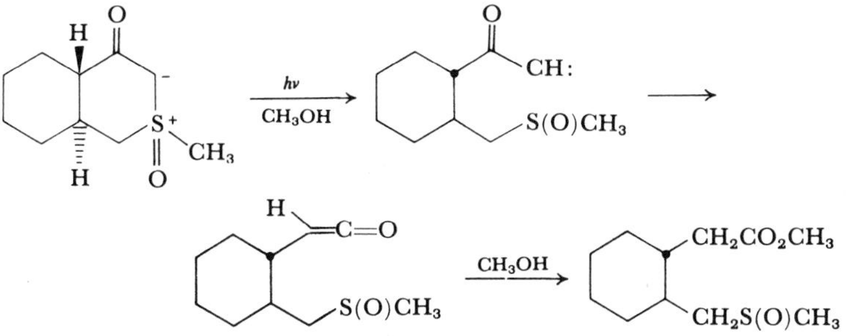

$\xrightarrow[CH_3OH]{h\nu}$ CH: S(O)CH$_3$ $\longrightarrow$

C=O S(O)CH$_3$ $\xrightarrow{CH_3OH}$ $CH_2CO_2CH_3$ $CH_2S(O)CH_3$ (3-37)

(Corey and Chaykovsky, 1964)

Insertion/Addition

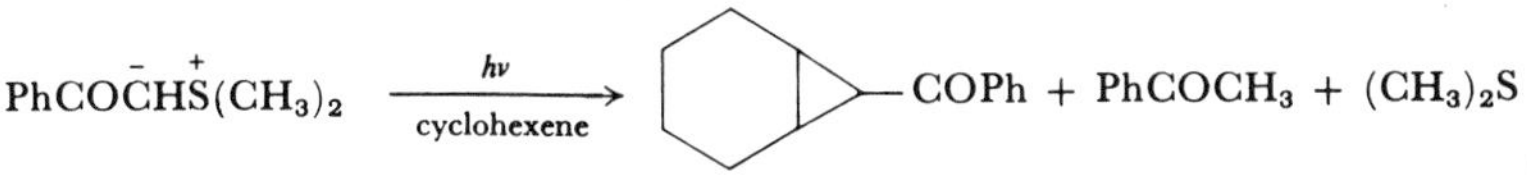

$PhCO\overset{-}{C}H\overset{+}{S}(CH_3)_2 \xrightarrow[\text{cyclohexene}]{h\nu}$ —COPh + $PhCOCH_3$ + $(CH_3)_2S$ (3-38)

(Trost, 1967)

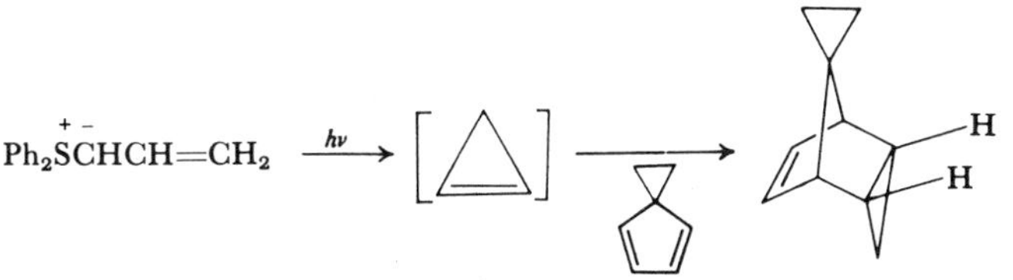

$Ph_2\overset{+}{S}\overset{-}{C}HCH{=}CH_2 \xrightarrow{h\nu}$ [ ] $\longrightarrow$ (3-39)

(Trost, 1967)

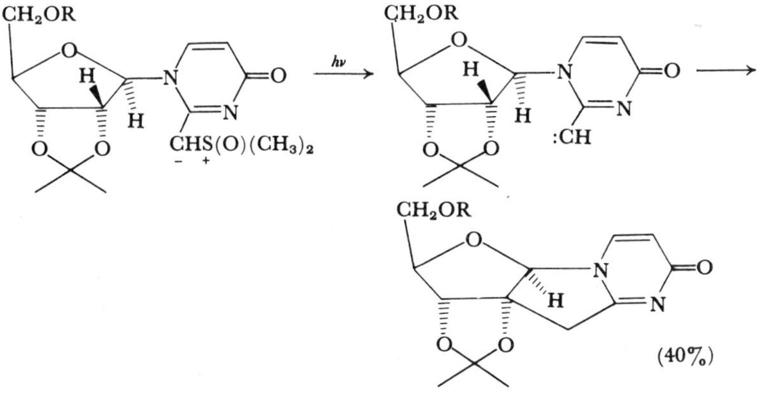

CH$_2$OR ... CHS(O)(CH$_3$)$_2$ $\xrightarrow{h\nu}$ CH$_2$OR ... :CH $\longrightarrow$

CH$_2$OR ... (40%) (3-40)

(Kunieda and Witkop, 1971)

"Dimerization"

$(CH_3)_2\overset{+}{S}{-}\overset{-}{C}HSPh \xrightarrow{\Delta} PhSCH: \xrightarrow{Me_2\overset{+}{S}\overset{-}{C}HSPh} PhSCH{=}CHSPh$ (70%) (3-41)

(Hayasi and Nozaki, 1971)

($CH_3OSSOCH_3$), bis (trifluoromethyl) disulfide, or thiocyanogen (Matsuyama *et al.*, 1973).

α,β′-Elimination

In addition to undergoing decomposition by α-elimination, ylides bearing β-hydrogens can undergo α,β′-elimination, as shown for example by the early work of Franzen and Mertz (1960) (Eq. 3-42)† and Searles and Wann (1965) (Eq. 3-43):

$$(CH_3CD_2)_3S^+ \xrightarrow{Ph_3C^-} CH_3CD_2\overset{+}{S}(CD_2)(\overset{-}{C}DCH_3)\ [\text{H transfer from } CH_2] + Ph_3C{-}D\ (75\%\ d_1) \longrightarrow CH_3CD_2SCDHCH_3 + CH_2{=}CD_2 \tag{3-42}$$

$$(CH_3)_3CSSC(CH_3)_3 + Cl_2C: \longrightarrow (CH_3)_3CS\overset{+}{S}(\overset{-}{C}Cl_2){-}C(CH_3)_2CH_2{-}H \longrightarrow (CH_3)_3CSSCHCl_2 + (CH_3)_2C{=}CH_2 \tag{3-43}$$

The α,β′-elimination reaction of oxysulfonium ylides **3-25** (formed by deprotonation of oxysulfonium salts **3-24**) is a key step in a variety of syn-

$$(CH_3)_2SO + X{-}CH\!< \xrightarrow{a} \; >\!CH{-}O\overset{+}{S}(CH_3)_2\ (\textbf{3-24}) \; \xleftarrow{b} >\!CH{-}OH + (CH_3)_2\overset{+}{S}{-}Y$$

$$\textbf{3-24} \xrightarrow{-H^+} >\!C(H){-}O{-}\overset{+}{S}CH_3(\overset{-}{C}H_2)\ (\textbf{3-25}) \longrightarrow\ >\!C{=}O + (CH_3)_2S \tag{3-44}$$

† Note that triethylsulfonium iodide has been converted to the corresponding ylide (with lithium diethylamide) which is reported to be "fairly stable" at −76° and which can be trapped with cyclohexanone (Corey and Oppolzer, 1964).

thetically important selective oxidations of alcohols or alcohol derivatives (Eq. 3-44). The oxysulfonium salt intermediate **3-24** may be prepared through nucleophilic displacement by sulfoxide oxygen as in the Kornblum oxidation of alkyl halides and tosylates (Eq. 3-44(a); Kornblum *et al.*, 1959). The high temperature needed with DMSO alone can be avoided through activation of the displacement process with $AgBF_4$ (Eqs. 3-45(b) and 3-46;

$$CH_3(CH_2)_7X + (CH_3)_2S{=}O \xrightarrow[\text{or b}]{\text{a}} CH_3(CH_2)_7OS(CH_3)_2 \longrightarrow CH_3(CH_2)_6CHO \qquad (3\text{-}45)$$

conditions a: X=OTS, $NaHCO_3$ 150°; yield 78%
conditions b: X=Br, $AgBF_4$ then $(C_2H_5)_3N$, 25°; yield 83%

$$BrCH_2C(CH_3){=}C(CH_3)CH_2Br\ (\textit{trans}) \xrightarrow[\text{(2) }(C_2H_5)_3N]{\text{(1) DMSO, }AgBF_4\text{, 25°}} \underset{60\%}{OHCC(CH_3){=}C(CH_3)CHO(\textit{trans})} \qquad (3\text{-}46)$$

Ganem and Boeckman, 1974). An alternative means of forming the oxysulfonium salt **3-24** involves nucleophilic displacement at sulfur by an alcohol (Eq. 3-44b). In the Pfitzner–Moffatt oxidation, DMSO is activated through condensation with dicyclohexylcarbodiimide (DCC) in the presence of a

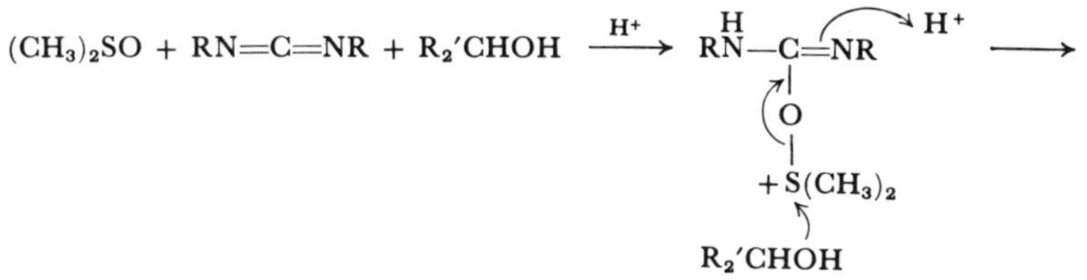

$$(CH_3)_2SO + RN{=}C{=}NR + R_2'CHOH \xrightarrow{H^+} RNH{-}C({=}NR)({-}O{-}\overset{+}{S}(CH_3)_2) \ [+\ R_2'CHOH,\ H^+] \longrightarrow RNHC(O)NHR + R_2'CHO\overset{+}{S}(CH_3)_2 \qquad (3\text{-}47)$$

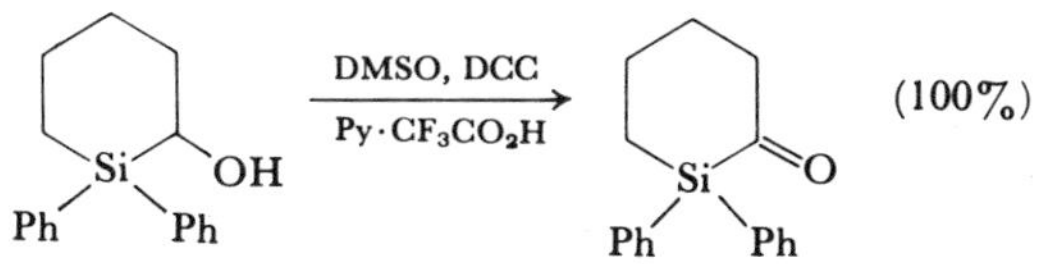

2,2-Diphenyl-2-silacyclohexan-1-ol (Si, Ph, Ph, OH) $\xrightarrow[\text{Py}\cdot CF_3CO_2H]{\text{DMSO, DCC}}$ 2,2-diphenyl-2-silacyclohexanone (Si, Ph, Ph, =O) (100%) (3-48)

(Brook and Pierce, 1965)

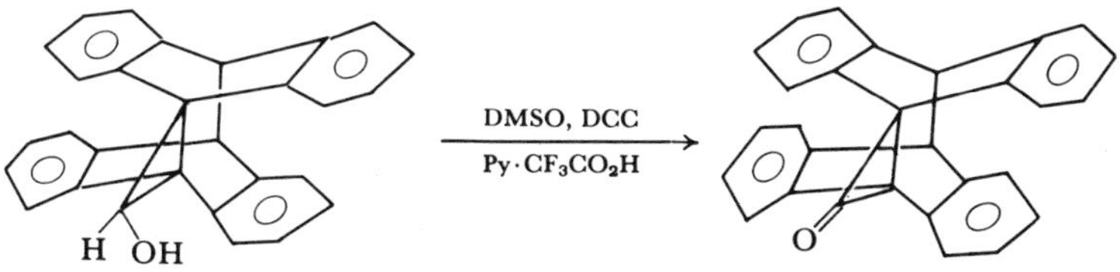

H OH $\xrightarrow[\text{Py}\cdot CF_3CO_2H]{\text{DMSO, DCC}}$ O (3-49)

(Weinshenker and Greene, 1968)

proton donor (pyridinium trifluoroacetate or phosphoric acid) (Eq. 3-47; Moffatt, 1971; Butterworth and Hanessian, 1971). Typical applications are indicated in Eqs. 3-48 to 3-50. A variety of other DMSO activating agents

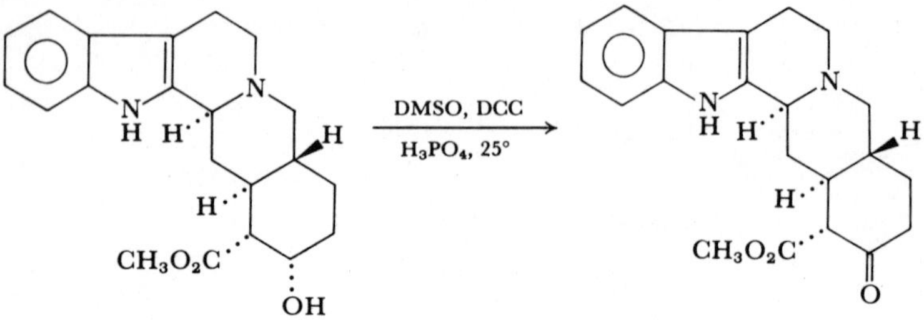
DMSO, DCC
H_3PO_4, 25°
N
H H
N
H
H
CH_3O_2C
OH
N
H H
N
H
H
CH_3O_2C
O

(3-50)

(80%)

(Albright and Goldman, 1965)

favoring displacement at sulfur have been examined as substitutes for DCC, such as acid anhydrides (Eqs. 3-51 and 3-52) and pyridine-SO_3 (Eq. 3-53), among others.

$$\text{ArCOCH(OH)Ar} \xrightarrow[\text{(a) Ac}_2\text{O, 25°; (b) Ms}_2\text{O, }-20°]{\text{DMSO}} \text{ArCOCOAr} \qquad (3\text{-}51)$$

(a) Ar = $pCH_3OC_6H_4$, 89%
(b) Ar = Ph, 99%

((a) Van Dyke and Pritchard, 1967)
((b) Albright, 1974)

$$\text{PhCH(OH)C(CH}_3)_3 \xrightarrow[\text{Ac}_2\text{O}]{\text{DMSO}} \text{PhCOC(CH}_3)_3 \qquad (3\text{-}52)$$

(98%)

(Clement *et al.*, 1969)

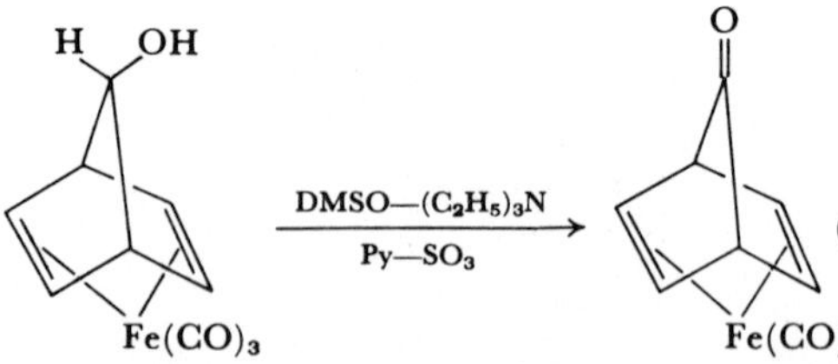
H OH
DMSO—$(C_2H_5)_3N$
Py—SO_3
$Fe(CO)_3$
O
$Fe(CO)_3$

(70%) (3-53)

(Landesberg and Sieczkowski, 1971)

A number of novel sulfonium and sulfoxonium reagents of the type R_2S^+X and $R_2S^+(O)X$ that have also proven useful in alcohol oxidations have been recently reported by Corey and Kim (Eqs. 3-54 to 3-56). It is significant that vicinal diols can be oxidized without complications from C—C bond cleavage (the case with most other oxidants!). Oxidations of the type shown in Eq. 3-56 can also be achieved in excellent yield using polystyrene containing chlorosulfonium groups (e.g., polymer-C_6H_4—$S^+(Cl)CH_3$; Crosby *et al.*, 1975).

This procedure using an insoluble polymer support may be useful in industrial processes.

Deuterium labeling experiments (Eq. 3-57 to 3-59) have provided support for the base-initiated collapse of oxysulfonium salts to carbonyl compounds according to the mechanism of Eq. 3-44. In these reactions the evidence

$$\text{C}_6\text{H}_{11}\text{—CH}_2\text{OH} \xrightarrow[\text{(2) } (C_2H_5)_3N]{\text{(1) } (CH_3)_2\overset{+}{S}(O)Cl,\ -45^\circ} \text{C}_6\text{H}_{11}\text{—CHO } (>94\%) \qquad (3\text{-}54)$$

(Corey and Kim, 1973a)

$$\xrightarrow{Ph\overset{+}{S}(Cl)CH_3} \quad (80\%) \qquad (3\text{-}55)$$

(Corey and Kim, 1974)

$$\text{—CH}_2\text{OH} \xrightarrow[\text{then } (C_2H_5)_3N]{Ph\overset{+}{S}(Cl)CH_3,\ -25^\circ} \text{—CHO} \quad (93\%) \qquad (3\text{-}56)$$

(Corey and Kim, 1973b)

$$Ph\overset{+}{S}(OCD_3)CH_3 \xrightarrow{\text{base}} Ph\overset{+}{S}(OCD_3)\overset{-}{C}H_2 \longrightarrow PhSCH_2D + CD_2{=}O$$

$$Ph\overset{+}{S}(OCH_3)CD_3 \xrightarrow{\text{base}} Ph\overset{+}{S}(OCH_3)\overset{-}{C}D_2 \longrightarrow PhSCHD_2 + CH_2{=}O \qquad (3\text{-}57)$$

(Johnson and Phillips, 1967)

$$(CH_2)_{11}\text{C(D)OH} \xrightarrow[\text{NCS—}(C_2H_5)_3N]{\text{DMSO}} (CH_2)_{11}\text{C=O} + CH_3SCH_2D \qquad (3\text{-}58)$$

(McCormick, 1974)

indicates that the ylide is very efficient in acting as an internal base and is not reprotonated to form the starting oxysulfonium salt (Johnson and Phillips, 1967). Occasional deviation from the mechanism of Eq. 3-44 can

$$CH_3CH_2CH_2CD_2OH \xrightarrow[DCC]{DMSO} CH_3SCH_2D + C_3H_7CDO \qquad (3\text{-}59)$$

(Fenselau and Moffatt, 1966)

occur when the protons α to the oxygen (in oxysulfonium salt **3-24**) are particularly acidic. Thus in Eq. 3-60 the new C═O bond is formed by a β-rather than α,β'-elimination (Torssell, 1966). A second deviation from the

$$ArCOCH_2Br \xrightarrow{DMSO} ArCOCH_2O\overset{+}{S}(CH_3)_2 \xrightarrow{(C_2H_5)_3N}$$

$$\underset{\text{base} \curvearrowright H}{ArCOCH}\text{—}O\text{—}\overset{+}{S}(CH_3)_2 \longrightarrow ArCOCHO \qquad (3\text{-}60)$$

mechanism of Eq. 3-44 occurs in more polar media, viz., the conversion of an alcohol into its methylthiomethyl ether as illustrated in Eq. 3-61 (Corey and Kim, 1972). Thus, the ratio of **3-26** to **3-27** varies from >99:1 in toluene

$$C_6H_{11}CH_2OH + \text{(succinimido)}N\text{—}\overset{+}{S}(CH_3)_2\overset{-}{Cl} \longrightarrow C_6H_{11}CH_2O\overset{+}{S}(CH_3)_2$$

$$C_6H_{11}CH_2\text{—}O\text{—}\overset{+}{S}(CH_3)\text{—}\overset{-}{C}H_2 \dashrightarrow \left[C_6H_{11}CH_2\overset{-}{O} \quad CH_2{=}\overset{+}{S}CH_3\right]$$

$$\underset{\textbf{3-26}}{C_6H_{11}CHO + (CH_3)_2S} \qquad \underset{\textbf{3-27}}{C_6H_{11}CH_2OCH_2SCH_3} \qquad (3\text{-}61)$$

to 4.6:1 in methylene chloride to ~1:1 in 1:1 methylene chloride–dimethyl sulfoxide (Corey and Kim, 1972), presumably reflecting the increasing stabilization of the developing charge separation in the transition state leading to formation of the ion $CH_3SCH_2^+$ which leads to **3-27** by subsequent reaction with the nucleophilic alcohol. The reaction leading to ether **3-27** resembles the Pummerer rearrangement, which will be considered in Chapter 4.

In all of the above discussion of the reactions of oxysulfonium salts, the role of the counter anion has been ignored. In the absence of added base, a number of oxysulfonium halides can be converted into halides under very mild conditions (Eq. 3-62; Corey *et al.*, 1972). This reaction is of synthetic utility for the selective conversion of allylic and benzylic alcohols to halides under neutral conditions.

$$(CH_3)(HOCH_2CH_2)C{=}CH(CH_2OH) + (CH_3)_2\overset{+}{S}{-}N(\text{succinimide})\ Cl^- \xrightarrow[-20^\circ]{CH_2Cl_2}$$

$$(CH_3)(HOCH_2CH_2)C{=}CH(CH_2O\overset{+}{S}(CH_3)_2Cl^-) \xrightarrow{0^\circ} (CH_3)(HOCH_2CH_2)C{=}CH(CH_2Cl) + (CH_3)_2SO$$

(87%) (3-62)

Fragmentation

A limited number of reactions which fall into the category of "fragmentation reactions" have been reported for sulfonium ylides (Eq. 3-63 to 3-65).

$$\text{(tetrahydrothiophenium ylide)}\ \overset{+}{S}{-}R \longrightarrow C_2H_4 + RSCH{=}CH_2 \qquad (3\text{-}63)$$

$R = CH_3, C_6F_4H$

(Weygand and Daniel, 1961; Brewer *et al.*, 1968)

$$\text{(benzo-fused cyclic sulfonium)}\ \overset{+}{S}{-}CH_3 \xrightarrow[(C_2H_5)_2O]{PhLi} \text{(ylide)}\ \overset{+}{S}{-}CH_3 \longrightarrow$$

$$[\text{o-xylylene with } SCH_3] \longrightarrow \text{(dibenzocyclooctadiene bearing } CH_3S,\ SCH_3) + \text{(spiro dimer bearing two } SCH_3) \qquad (3\text{-}64)$$

(Bornstein *et al.*, 1967)

$$\text{(2-}CH_3\text{-3-}(CH_3)_2CH\text{-thiirane)} + C_2H_5O_2CCHN_2 \xrightarrow[(C_2H_5)_2O]{Cu(II)}$$

$$\text{(thiiranium ylide)}\ \overset{+}{S}{-}\overset{-}{C}HCO_2C_2H_5 \longrightarrow$$

$$\text{H(CH}_3\text{)C{=}C(H)CH(CH}_3\text{)}_2 + S{=}CHCO_2C_2H_5 \qquad (3\text{-}65)$$

(72%) (> 99.5% *cis*) (Hata *et al.*, 1975)

Reaction 3-65 represents a novel means of stereospecifically desulfurizing thiiranes.

Sigmatropic Rearrangements: The Stevens and Sommelet–Hauser Reactions

Sulfur ylides are known to undergo [1,2] sigmatropic (Stevens), [2,3] sigmatropic (Sommelet–Hauser), and [1,4] sigmatropic rearrangements.† The first two "name reactions" have been studied extensively with respect to mechanism (Lepley and Giumanini, 1971; Stevens and Watts, 1973) and are of demonstrated synthetic utility. In a number of cases, sulfur ylides undergo two of the three types of rearrangement at the same time, as illustrated in Eqs. 3-66 to 3-68.

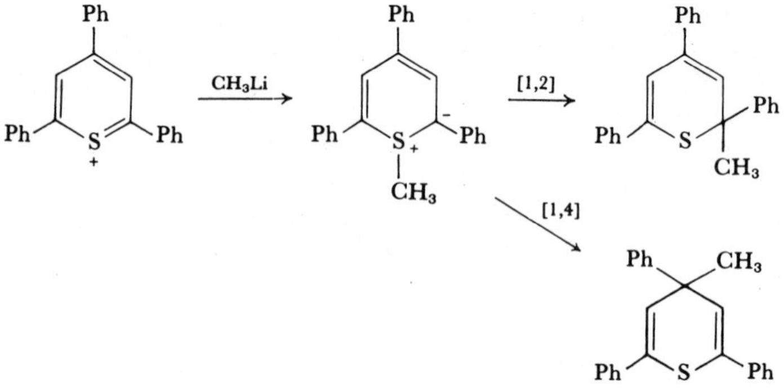

(3-66)

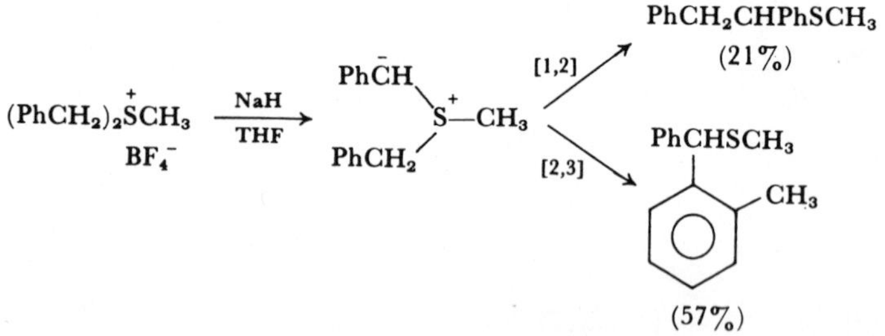

(3-67)

$(PhCH_2)_2\overset{+}{S}CH_3\ BF_4^-$ $\xrightarrow[THF]{NaH}$ $Ph\overset{-}{C}H(PhCH_2)\overset{+}{S}{-}CH_3$ → [1,2] $PhCH_2CHPhSCH_3$ (21%); [2,3] $PhCHSCH_3$ with ortho CH_3 on benzene ring (57%)

(Mitchell and Boekelheide, 1974)

The Stevens ([1,2] sigmatropic) rearrangement of sulfonium ylides has been observed only with stabilized sulfonium ylides (cf. the examples in Eq. 3-66 to 3-68). In competition with [2,3] sigmatropic rearrangement, the [1,2] sigmatropic rearrangement seems to be of higher activation energy

† Sigmatropic rearrangements and other pericyclic processes are discussed more fully in Chapter 7.

(Baldwin and Hackler, 1969). While the accumulated evidence demonstrates that there is a free radical component to the Stevens rearrangement, the possibility of a concerted component should not be discounted (Dewar and

KO-*t*-Bu [2,3] SPh Ph SPh (3-68)

SPh (11%) SPh SPh SPh SPh PhS SPh SPh SPh

(Baldwin and Hackler, 1969)

Ramsden, 1974; Ollis *et al.*, 1975; Dolling *et al.*, 1975; see Chapter 5 for more details on free radical intermediates in the Stevens rearrangement). Some notable examples of the application of the Stevens rearrangement in the synthesis of cyclophanes have been reported by Boekelheide and others (Eqs. 3-69 and 3-70; see Chapter 5 for related synthetic approaches to cyclophanes).

Although the first example of a Sommelet–Hauser ([2,3] sigmatropic) rearrangement of a sulfur ylide was reported as early as 1938 (Eq. 3-71), only in recent years has the reaction proved to be of synthetic importance, as illustrated, for example, by Eqs. 3-72 to 3-77. Examples of [1,2] and [2,3] rearrangement reactions in cyclic sulfur ylides formed by intramolecular carbene capture will be described in Chapter 6. Repetitive use of [2,3]

methylation $\longrightarrow$ CH_3—S^+ ^+S—CH_3 $\xrightarrow{\text{base}}$

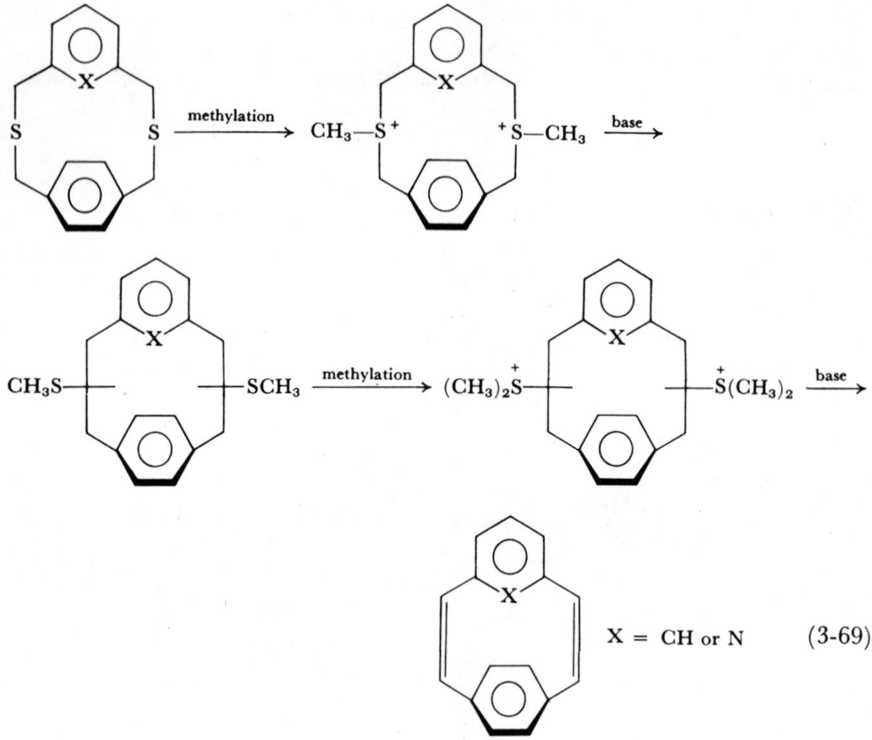

CH_3S SCH_3 $\xrightarrow{\text{methylation}}$ $(CH_3)_2\overset{+}{S}$ $\overset{+}{S}(CH_3)_2$ $\xrightarrow{\text{base}}$

X = CH or N (3-69)

(Boekelheide *et al.*, 1974a,b)

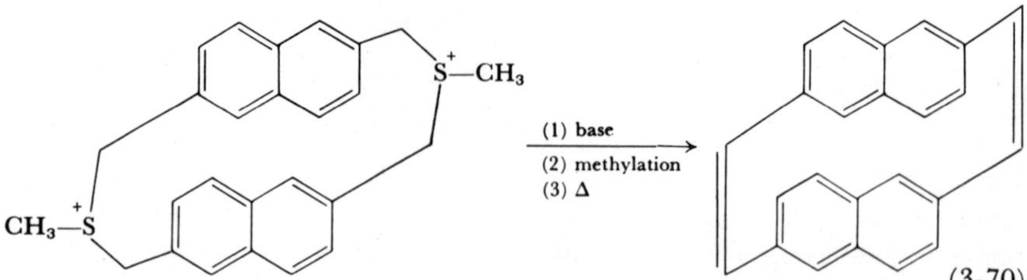

$\overset{+}{S}$—CH_3

CH_3—$\overset{+}{S}$

(1) base
(2) methylation
(3) Δ

(3-70)

(Haenel and Staab, 1973)

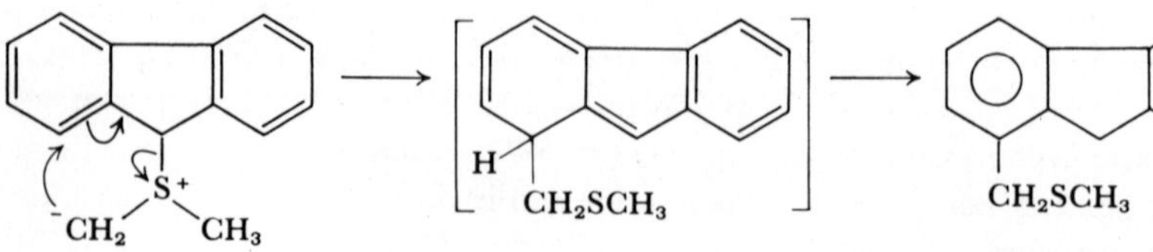

S^+ $^-CH_2$ CH_3 H CH_2SCH_3 CH_2SCH_3 (3-71)

(Pinck and Hilbert, 1938)

"PhS̈CH"

$\equiv$ (3-72)

(Julia *et al.*, 1972)

CHCl$_3$ / NaOH; SiO$_2$ (3-73)

(53%)

(Andrews and Evans, 1972)

n-C$_4$H$_9$Li, −70°

H$_2$C=C=C(Ph)CHC≡CPh (SC$_2$H$_5$) (3-74)

(Baldwin *et al.*, 1968)

BF$_4^-$; n-C$_4$H$_9$Li (3-75)

(Corey and Walinsky, 1972)

$LiC{\equiv}CC(CH_3)_2Cl$ ($\equiv$:C=C=C(CH$_3$)$_2$)

SCH$_3$ Hg(II)

O

(3-76)

Artemisia ketone

(Michelot *et al.*, 1974)

base SR (1) methylation (2) H_2O

R

HO

(3-77)

Yomogi alcohol

(Trost *et al.*, 1971b)

sigmatropic shifts forms the basis of a novel "ring-growing reaction" (Eq. 3-77) described by Vedejs and Hagen (1975). Clever utilization of the Sommelet–Hauser rearrangement of ylides derived from *azasulfonium salts* by Gassman and co-workers (cf. Eq. 3-79) has formed the basis for a useful

$PhCOCHN_2$ S$^+$···$\bar{C}$HCOPh

COPh + *trans* Isomer $Ph_3P{=}CH_2$ Ph $(CH_3O_2C)_2CN_2$

Ph (3-78)

$(CH_3O_2C)_2C^-$ $(CO_2CH_3)_2$

general synthetic approach to a variety of key organonitrogen compounds (Gassman *et al.*, 1974a, b; Gassman and Drewes, 1974; Gassman and Gruetzmacher, 1974; Gassman and van Bergen, 1974). The requisite azasulfonium salts are prepared either through interaction of *N*-chloroanilines with

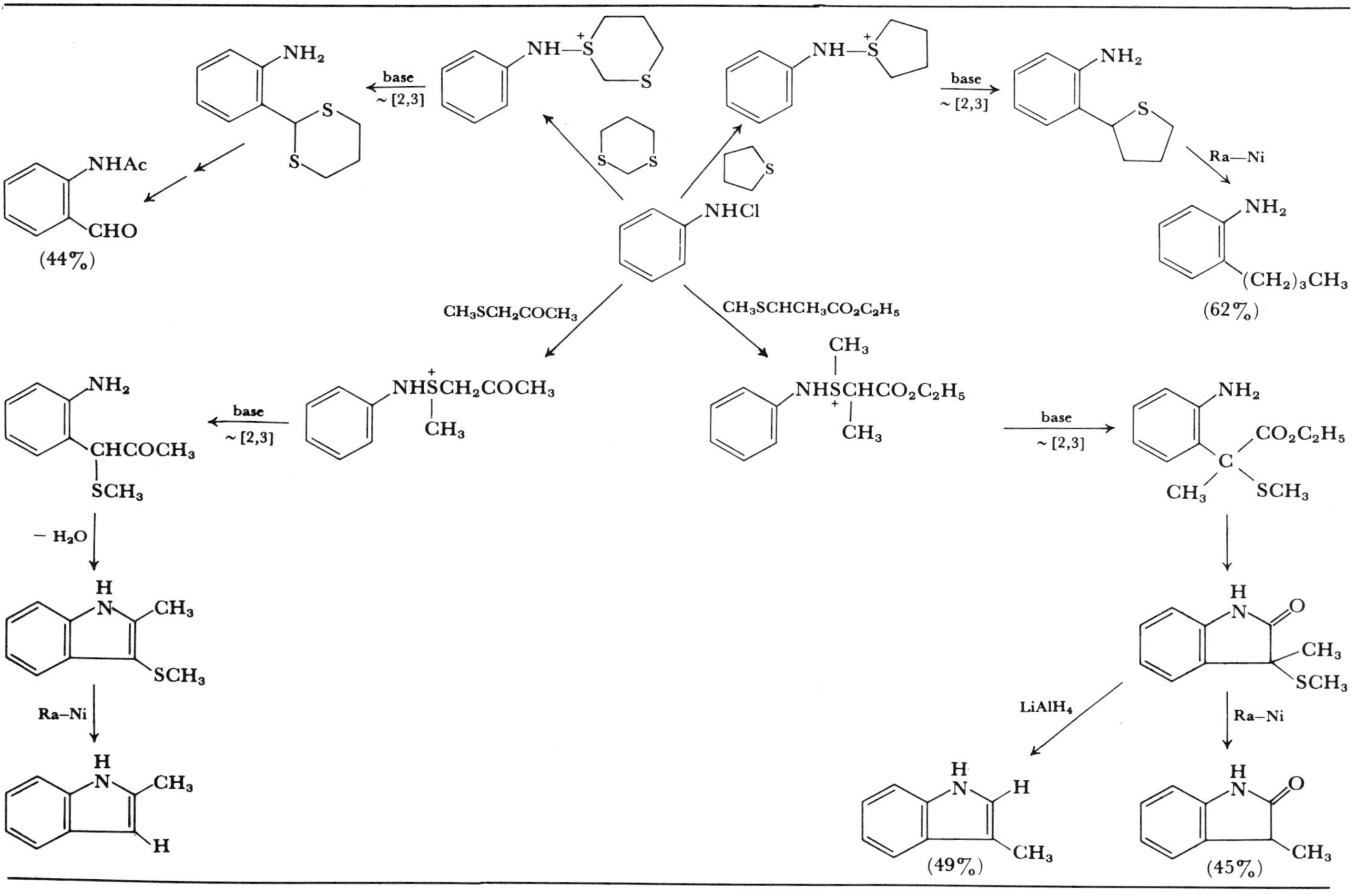
NH–S⁺
base
~[2,3]
NH2
NHAc
CHO
(44%)
NHCl
Ra–Ni
(CH2)3CH3
(62%)
CH3SCH2COCH3
CH3SCHCH3CO2C2H5
NHSCH2COCH3
CH3
NHSCHCO2C2H5
CHCOCH3
SCH3
CO2C2H5
– H2O
H
N
LiAlH4
(49%)
(45%)

sulfides or reaction of chlorosulfonium salts with anilines. As illustrated in Scheme 3-2, use of sulfides, 1,3-dithiane, β-ketosulfides or β-carboalkoxy sulfides followed by base-induced Sommelet–Hauser rearrangement and desulfurization permits the synthesis of ortho-alkylated and ortho-formylated anilines, 2- and 3-substituted indoles and oxindoles. While the reactions are illustrated using *N*-chloroaniline, a variety of other *N*- or ring-substituted anilines may be used as well. Significantly many of the transformations illustrated are "one-pot" reactions.

(3-79)

General References

Block, E. (1977). *In* "Organic Compounds of Sulphur, Selenium and Tellurium" (Specialist Periodical Reports) (D. R. Hogg, ed.), Vol. IV. Chem. Soc. of London.

Johnson, A. W. (1966). "Ylid Chemistry." Academic Press, New York.

Johnson, A. W. (1970, 1973, 1975). *In* "Organic Compounds of Sulphur, Selenium and Tellurium" (D. H. Reid, ed.), Vols. I–III. Chemical Society of London.

Trost, B. M., and Melvin, L. S. Jr. (1975). "Sulfur Ylides." Academic Press, New York.

References

Adams, J., Hoffman, L., and Trost, B. M. (1970). *J. Org. Chem.* **35**, 1600.

Albright, J. D. (1974). *J. Org. Chem.* **39**, 1977.

Albright, J. D., and Goldman, L. (1965). *J. Org. Chem.* **30**, 1107.

Ando, W. (1977). *Accounts Chem. Res.* **10**, 179.

Ando, W., Yagihara, T., Tozune, S., Imai, I., Suzuki, J., Toyama, T., Nakaido, S., and Migita, T. (1972). *J. Org. Chem.* **37**, 1721.

Andrews, G., and Evans, D. A. (1972). *Tetrahedron Lett.* 5121.

Arduengo, A. J., and Burgess, E. M. (1976). *J. Am. Chem. Soc.* **98**, 5020, 5021.

Astrologes, G. W., and Martin J. C. (1976). *J. Am. Chem. Soc.* **98**, 2895.

Baldwin, J. E., and Hackler, R. E. (1969). *J. Am. Chem. Soc.* **91**, 3646.

Baldwin, J. E., Hackler, R. E., and Kelly, D. P. (1968b). *Chem. Commun.* 1083.

Barabarella, G., Garbesi, A., and Fava, A. (1975). *J. Am. Chem. Soc.* **97**, 5883.

Barton, D. H. R., and Willis, B. J. (1972). *J. Chem. Soc. Perkin Trans.* **1**, 305.

Barton, D. H. R., Guziec, F. S., Jr., and Shahak, I. (1974). *J. Chem. Soc. Perkin Trans.* **1**, 1974.

Becker, R. S., Bost, R. O., Kole, J., Bertoniere, N. R., Smith, R. L., and Griffin, G. W. (1970), *J. Am. Chem. Soc.* **92**, 1302.

Block, E. (1969). *Quart. Rev. Sulfur Chem.* **4**, 264.

Bly, R. S., DuBose, C. M., Jr., and Konizer, G. B. (1968). *J. Org. Chem.* **33**, 2188.

Boekelheide, V., Anderson, P. H., and Hylton, T. A. (1974a). *J. Am. Chem. Soc.* **96**, 1558.

Boekelheide, V., Galuszko, K., and Szeto, K. S. (1974b). *J. Am. Chem. Soc.* **96**, 1578.

Bornstein, J., Shields, J. E., and Supple, J. H. (1967). *J. Org. Chem.* **32**, 1499.

Braun, H., Mayer, N., Strobl, G., and Kresze, G. (1973). *Justus Liebigs Ann. Chem.* 1317; Braun, H., Huber, G., and Kresze, G. (1973). *Tetrahedron Lett.* 4033.

Bravo, P., Gaudiano, G., and Umari-Ronchi, A. (1969). *Tetrahedron Lett.* 679.

Brewer, J. P. N., Heaney, H., and Jablonski, J. M. (1968). *Tetrahedron Lett.* 4455.

Brook, A. G., and Pierce, J. B. (1965). *J. Org. Chem.* **30**, 2566.

Buter, J., Wassenaar, S., and Kellogg, R. M. (1972). *J. Org. Chem.* **37**, 4045 and references therein.

Buter, J., Raynolds, P. W., and Kellogg, R. M. (1974). *Tetrahedron Lett.* 2901.

Butterworth, R. F., and Hanessian, S. (1971). *Synthesis* **2**, 70.

Cava, M. P., and Lakshmikantham, M. V. (1975). *Accounts Chem. Res.* **8**, 139 and references therein.

Clement, W. H., Dangieri, T. J., and Tuman, R. W. (1969). *Chem. Ind.* (*London*) 755.

Cohen, T., Herman, G., Chapman, T. M., and Kuhn, D. (1974). *J. Am. Chem. Soc.* **96**, 5627.

Corey, E. J., and Chaykovsky, M. (1962a). *J. Am. Chem. Soc.* **84**, 867.

Corey, E. J., and Chaykovsky, M. (1962b). *J. Am. Chem. Soc.* **84**, 3782.

Corey, E. J., and Chaykovsky, M. (1964). *J. Am. Chem. Soc.* **86**, 1640.

Corey, E. J., and Chaykovsky, M. (1965). *J. Am. Chem. Soc.* **87**, 1353.

Corey, E. J., and Jautelat, M. (1967). *J. Am. Chem. Soc.* **89**, 3912.

Corey, E. J., and Kim, C. U. (1972). *J. Am. Chem. Soc.* **94**, 7586.

Corey, E. J., and Kim, C. U. (1973a). *Tetrahedron Lett.* 919.

Corey, E. J., and Kim, C. U. (1973b). *J. Org. Chem.* **38**, 1233.

Corey, E. J., and Kim, C. U. (1974). *Tetrahedron Lett.* 287.

Corey, E. J., and Oppolzer, W. (1964). *J. Am. Chem. Soc.* **86**, 1899.

Corey, E. J., and Walinsky, S. W. (1972). *J. Am. Chem. Soc.* **94**, 8932.

Corey, E. J., and Ortiz de Montellano, P. R. (1968). *Tetrahedron Lett.* 5113.

Corey, E. J., Jautelat, M., and Oppolzer, W. (1967). *Tetrahedron Lett.* 2325.

Corey, E. J., Ortiz de Montellano, P. R., and Yamamoto, H. (1968). *J. Am. Chem. Soc.* **90**, 6254.

Corey, E. J., Kim, C. V., and Takeda, M. (1972). *Tetrahedron Lett.* 4339.

Crosby, G. A., Weinshenker, N. M., and Uh, H.-S. (1975). *J. Am. Chem. Soc.* **97**, 2232.

Denney, D. B., Denney, D. Z., and Hsu, Y. F. (1973). *J. Am. Chem. Soc.* **95**, 4064.

Dewar, M. J. S., and Ramsden, C. A. (1974). *J. Chem. Soc. Perkin Trans.* **1**, 1839.

Dolling, U. H., Closs, G. L., Cohen, A. H., and Ollis, W. D. (1975). *Chem. Commun.* 545.

Everett, J. W., and Garratt, P. J. (1972). *Chem. Commun.* 642.

Fenselau, A. N., and Moffatt, J. G. (1966). *J. Am. Chem. Soc.* **88**, 1762.

Franzen, V., and Mertz, Ch. (1960). *Chem. Ber.* **93**, 2819.

Ganem, B., and Boeckman, R. K., Jr. (1974). *Tetrahedron Lett.* 917.

Gassman, P. G., and Drewes, H. R. (1974). *J. Am. Chem. Soc.* **96**, 3002.

Gassman, P. G., and Gruetzmacher, G. D. (1974). *J. Am. Chem. Soc.* **96**, 5487.

Gassman, P. G., and van Bergen, T. J. (1974). *J. Am. Chem. Soc.* **96**, 5508.

Gassman, P. G., van Bergen, T. J., Gilbert, D. P., and Cue, B. W. Jr. (1974a) *J. Am. Chem. Soc.* **96**, 5495.

Gassman, P. G., Gruetzmacher, G., and van Bergen, T. J. (1974b). *J. Am. Chem. Soc.* **96**, 5512; also see Gassman, P. G., and Parton, R. L. (1977). *Tetrahedron Lett.* 2055.

Haenel, M., and Staab, H. A. (1973). *Chem. Ber.* **106**, 2203.
Hata, Y., Watanabe, M., Inoue, S., and Oae, S. (1975). *J. Am. Chem. Soc.* **97**, 2553.
Hayasi, Y., and Nozaki, H. (1971). *Bull. Chem. Soc. Jpn.* **45**, 198; see also Wilson, S. R., and Myers, R. S. (1976). *Tetrahedron Lett.* 3413.
Hellmann, H., and Eberle, D. (1963). *Justus Liebigs Ann. Chem.* **662**, 188; also see Blackburn, G. M., Ollis, W. D., Plackett, J. D., Smith, C., and Sutherland, I. O. (1968). *Chem. Commun.* 186; Otsubo, T., and Boekelheide, V. (1975). *Tetrahedron Lett.* 3881.
Hortmann, A. G. (1965). *J. Am. Chem. Soc.* **87**, 4972.
Hortmann, A. G., and Harris, R. L. (1970). *J. Am. Chem. Soc.* **92**, 1803.
Hortmann, A. G., and Harris, R. L. (1971). *J. Am. Chem. Soc.* **93**, 2471.
Hortmann, A. G., and Robertson, D. A. (1967). *J. Am. Chem. Soc.* **89**, 5974; **94**, 2758 (1972).
Hortmann, A. G., Harris, R. L., and Miles, J. A. (1974). *J. Am. Chem. Soc.* **96**, 6119.
Ingold, C. K., and Jessop, J. A. (1930). *J. Chem. Soc.* (*London*) 713.
Johnson, C. R. (1973). *Accounts Chem. Res.* **6**, 341.
Johnson, C. R., and Phillips, W. G. (1967). *J. Org. Chem.* **32**, 1926.
Johnson, C. R., and Schroeck, C. W. (1973a). *J. Am. Chem. Soc.* **95**, 7418.
Johnson, C. R., and Schroeck, C. W. (1973b). *J. Am. Chem. Soc.* **95**, 7424.
Johnson, C. R., Haake, M., and Schroeck, C. W. (1970). *J. Am. Chem. Soc.* **92**, 6594. Also see Johnson, C. R., and Rogers, P. E. (1973). *J. Org. Chem.* **38**, 1793; and Schmidbauer, H., and Kammel, G. (1971). *Chem. Ber.* **104**, 3241, 3252.
Julia, S., Huynh, C., and Michelot, D. (1972). *Tetrahedron Lett.* 3587.
Kapovits, I., and Kálmán, A. (1971). *Chem. Commun.* 649.
Kellogg, R. M. (1973). *J. Org. Chem.* **38**, 844.
Kellogg, R. M. (1976). *Tetrahedron* **32**, 2165.
Kellogg, R. M., Noteboom, M., and Kaiser, J. M. (1975). *J. Org. Chem.* **39**, 2573.
Kondo, K., and Ojima, I. (1975). *Bull. Chem. Soc. Jpn.* **48**, 1490.
König, H., Metzger, H., and Seelert, K. (1966). *Chem. Ber.* **98**, 3712, 3724.
Kornblum, N., Jones, W. J., and Anderson, G. J. (1959). *J. Am. Chem. Soc.* **81**, 4113.
Kunieda, T., and Witkop, B. (1971). *J. Am. Chem. Soc.* **93**, 3487.
Landesberg, J. M., and Sieczkowski, J. (1971). *J. Am. Chem. Soc.* **93**, 972.
LaRochelle, R. W., Trost, B. M., and Krepski, L. (1971). *J. Org. Chem.* **36**, 1126.
Law, J. H. (1971). *Accounts Chem. Res.* **4**, 199.
Lepley, A. R., and Giumanini, A. G. (1971). *In* "Mechanisms of Molecular Migration" (B. S. Thyagarajan, ed.), Vol. 3, p. 297. Wiley (Interscience), New York.
Lillya, C. P. (1966). *J. Am. Chem. Soc.* **88**, 1559.
Martin, J. C., and Perozzi, E. F. (1974). *J. Am. Chem. Soc.* **96**, 3155.
Maryanoff, B. E., Stackhouse, J., Senkler, G. H., Jr., and Misllw, K. (1975). *J. Am. Chem. Soc.* **97**, 2718, and previous papers in series.
Matsuyama, H., Minato, H., and Kobayashi, M. (1973). *Bull. Chem. Soc. Jpn.* **46**, 3158.
McCormick, J. P. (1974). *Tetrahedron Lett.* 1701.
Michelot, D., Linstrumelle, G., and Julia, S. (1974). *Chem. Commun.* 10.
Middleton, W. J. (1966). *J. Org. Chem.* **31**, 3731.
Middleton, W. J., Buhle, E. L., McNally, J. G., Jr., and Zanger, M. (1965). *J. Org. Chem.* **30**, 2384.
Mitchell, R. H., and Boekelheide, V. (1974). *J. Am. Chem. Soc.* **96**, 1547.
Moffatt, J. G. (1971). *In* "Techniques and Applications in Organic Synthesis: Oxidation" (R. Augustine, and D. J. Trecker, ed.), Vol. 2, p. 1. Dekker, New York.
Nozaki, H., Tunemoto, D., Matubara, S., and Kondo, K. (1967). *Tetrahedron* **23**, 545.

Ollis, W. D., Rey, M., Sutherland, I. O., and Closs, G. L. (1975). *Chem. Commun.* 543.
Pattenden, G., and Storer, R. (1973). *Tetrahedron Lett.* 3473.
Pinck, L. A., and Hilbert, G. E. (1938). *J. Am. Chem. Soc.* **60**, 494; *ibid.* **68**, 751 (1946).
Polk, M., Siskin, M., and Price, C. C. (1969). *J. Am. Chem. Soc.* **91**, 1206.
Price, C. C., Hori, M., Parasaran, T., and Polk, M. (1963). *J. Am. Chem. Soc.* **85**, 2278.
Price, C. C., Follweiler, J., Pirelahi, H., and Siskin, M. (1971). *J. Org. Chem.* **36**, 791.
Sauter, H., Horster, H. G., and Prinzbach, H. (1973). *Angew. Chem. Int. Ed. English* **12**, 991.
Schaap, A. P., and Faler, G. R. (1973). *J. Org. Chem.* **38**, 3061.
Schmidbaur, H., and Kapp, W. (1972). *Chem. Ber.* **105**, 1203.
Schultz, A. G., and DeTar, M. B. (1976). *J. Am. Chem. Soc.* **98**, 3564; Schultz, A. G. (1974). *J. Org. Chem.* **39**, 3185.
Searles, S., Jr., and Wann, R. E. (1965). *Tetrahedron Lett.* 2899; Field, L., and Banks, C. H. (1975). *J. Org. Chem.* **40**, 2774.
Sheppard, W. A. (1971). *J. Am. Chem. Soc.* **93**, 5597.
Shono, T., Akazawa, T., and Mitani, M. (1973). *Tetrahedron*, **29**, 817.
Spry, D. O. (1973). *Tetrahedron Lett.* 2413.
Stevens, T. S., and Watts, W. E. (1973). "Selected Molecular Rearrangements." Van Nostrand-Reinhold, Princeton, New Jersey.
Suld, G., and Price, C. C. (1961). *J. Am. Chem. Soc.* **83**, 1770.
Torssell, K. (1966). *Tetrahedron Lett.* 4445.
Trost, B. M. (1967). *J. Am. Chem. Soc.* **89**, 138.
Trost, B. M. (1973). *Fortschr. Chem. Forsch.* **41**, 1.
Trost, B. M. (1974). *Accounts Chem. Res.* **7**, 85.
Trost, B. M., and Bogdanowicz, M. J. (1973). *J. Am. Chem. Soc.* **95**, 5298, 5307, 5311, and 5321.
Trost, B. M., Schinski, W. L., Chen, F., and Mantz, I. B. (1971a). *J. Am. Chem. Soc.* **93**, 676.
Trost, B. M., Conway, P., and Biddlecom, W. G. (1971b). *Chem. Commun.* 1639.
Truce, W. E., and Madding, G. D. (1966). *Tetrahedron Lett.* 3681.
Van Dyke, M., and Pritchard, N. D. (1967). *J. Org. Chem.* **32**, 3204.
Vedejs, E. (1973). *J. Am. Chem. Soc.* **95**, 5778.
Vedejs, E., and Hagen, J. P. (1975). *J. Am. Chem. Soc.* **97**, 6878.
Weinshenker, N. M., and Greene, F. D. (1968). *J. Am. Chem. Soc.* **90** 506.
Weygand, F., and Daniel, H. (1961). *Chem. Ber.* **94**, 3145.
Whangbo, M.-H., Wolfe, S., and Bernardi, F. (1975). *Can. J. Chem.* **53**, 3040.
Whangbo, M.-H., Wolfe, S., and Bernardi, F. (1975). *Can. J. Chem.* **53**, 3040; also see Bernardi, F., Schlegel, H. B., Whangbo, M.-H., and Wolfe, S. (1977). *J. Am. Chem. Soc.* **99**, 5633.
Wittig, G. (1974). *Accounts Chem. Res.* **7**, 6.

Chapter 4 | Sulfur-Containing Carbocations

4.1 Introduction

Carbocations (carbonium ions†) are carbon cations containing a positively charged carbon surrounded by a sextet of bonding electrons. Carbocations, such as the methyl cation **4-1**, are known to be planar at the charged carbon, thereby allowing the concentration of bonding electrons into the more stable sp^2 orbitals and leaving vacant the less stable p orbital. Stabilization of these carbon cations is realized by charge delocalization:

(a) by charge dispersal onto hydrogen (hyperconjugation) as in the *tert*-butyl cation (**4-2**);

(b) by the presence of adjacent carbon–carbon double bonds as in the benzyl and propenyl cations (**4-3** and **4-4**, respectively); and

(c) by the presence of adjacent atoms possessing lone pairs as in the methylthiomethyl, methoxymethyl, and dimethylaminomethyl cations (**4-5**, **4-6**, and **4-7**).

The type of stabilization illustrated by **4-2**, by **4-3** and **4-4**, and by **4-5**–**4-7** involves, respectively, charge delocalization by σ-, π-, and n-type electrons on atoms immediately adjacent to the carbon cation (α-substituents). It is

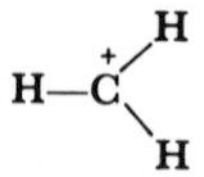

$$\mathrm{H{-}\overset{+}{C}H_2}$$

4-1

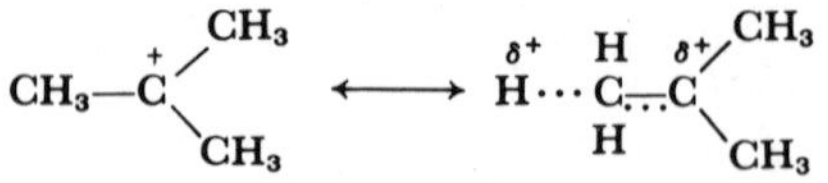

$$\mathrm{CH_3{-}\overset{+}{C}(CH_3)_2} \longleftrightarrow \mathrm{H^{\delta+}\cdots CH_2\cdots\overset{\delta+}{C}(CH_3)_2}$$

4-2

$$\mathrm{C_6H_5{-}CH_2^+} \longleftrightarrow \text{(cyclohexadienyl cation)}{=}\mathrm{CH_2}$$

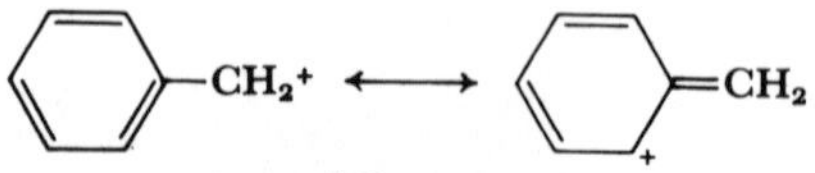

4-3

$$\mathrm{CH_2{=}CH{-}CH_2^+} \longleftrightarrow \mathrm{{}^+CH_2{-}CH{=}CH_2}$$

4-4

† For arguments in favor of using the *carbocation* nomenclature in place of the older *carbonium ion* nomenclature, see Fletcher *et al.* (1974).

well known that charge delocalization involving more distant sources of electron density can also occur, e.g., π-participation in **4-8**, n-participation in **4-9**, and possible σ-participation via the controversial nonclassical carbocations in **4-10** (Capon and McManus, 1976).

$$CH_3—\ddot{\underset{..}{S}}—\overset{+}{C}H_2 \longleftrightarrow CH_3\overset{+}{\underset{..}{S}}{=}CH_2 \qquad CH_3—\ddot{\underset{..}{O}}—\overset{+}{C}H_2 \longleftrightarrow CH_3\overset{+}{\underset{..}{O}}{=}CH_2$$

4-5a **4-5b** **4-6a** **4-6b**

$$(CH_3)_2\ddot{N}—\overset{+}{C}H_2 \longleftrightarrow (CH_3)_2\overset{+}{N}{=}CH_2$$

4-7a **4-7b**

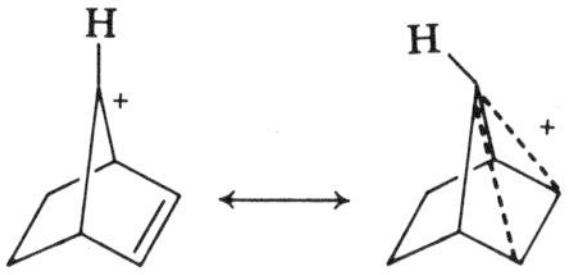

4-8

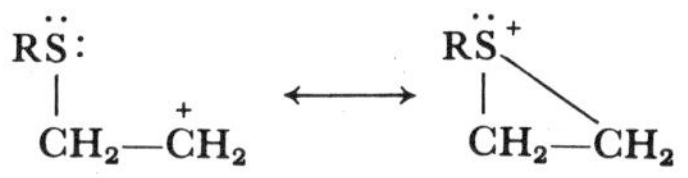

RS:
|
$CH_2—\overset{+}{C}H_2$ ⟷ $R\overset{+}{S}$ (bridged to both carbons) $CH_2—CH_2$

4-9a **4-9b**

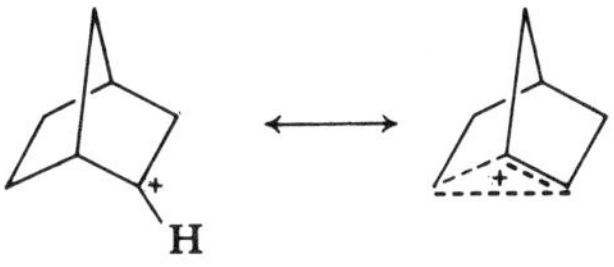

4-10

The stabilization of carbon cations by α- or β-thio groups involves contributions by resonance structures of types **4-5b** and **4-9b** (a methylene sulfonium ion and an episulfonium ion, respectively). Ions **4-5** and **4-9** can be isolated in nonnucleophilic solvents with inert counter anions such as BF_4^- or $SbCl_6^-$. Episulfonium ions also result from the interaction of sulfenium ions (alkylthiyl cations), $R—S^+$, with olefins (Eq. 4-1) (Krimer *et al.*, 1973):

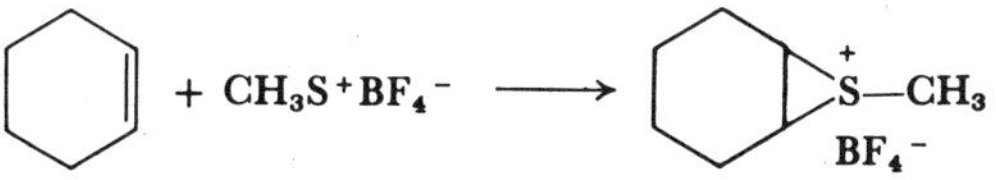

$$\text{cyclohexene} + CH_3S^+BF_4^- \longrightarrow \text{(cyclohexene episulfonium)}\overset{+}{S}—CH_3\ BF_4^- \qquad (4\text{-}1)$$

4.2 Generation of Sulfur-Containing Carbocations

Methods of generating sulfur-containing carbocations are given in Scheme 4-1.

Scheme 4-1. Generation of Sulfur-Containing Carbocations

Heterolytic bond cleavage

$CH_3SCH_2Cl + SbCl_5 \longrightarrow CH_3\overset{+}{S}{=}CH_2 \ \ SbCl_6^-$ Meerwein *et al.* (1965)

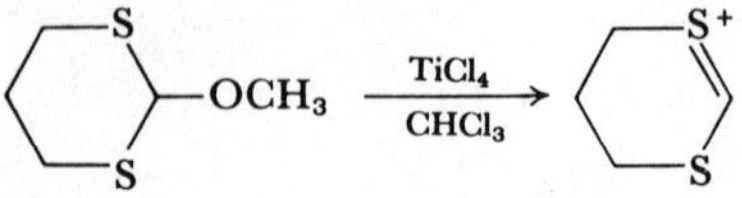

$\xrightarrow[CHCl_3]{TiCl_4}$ Stütz and Stadler (1972)

$CH_3SSCH_2Cl \xrightarrow{dioxane\text{–}H_2O} CH_3SSCH_2^+ \ \ Cl^-$ Block and O'Connor (1974)

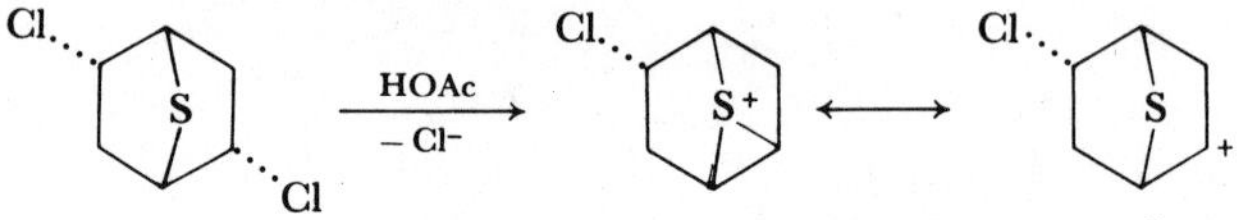

$\xrightarrow[-\,Cl^-]{HOAc}$ $\longleftrightarrow$

Corey and Block (1966)

Hydride transfer

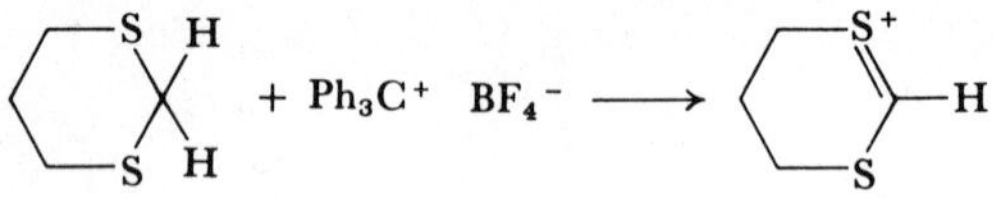

$+ \ Ph_3C^+ \ BF_4^- \longrightarrow$ Corey and Kim (1972)

$+ \ 2Ph_3C^+ \ BF_4^- \longrightarrow$

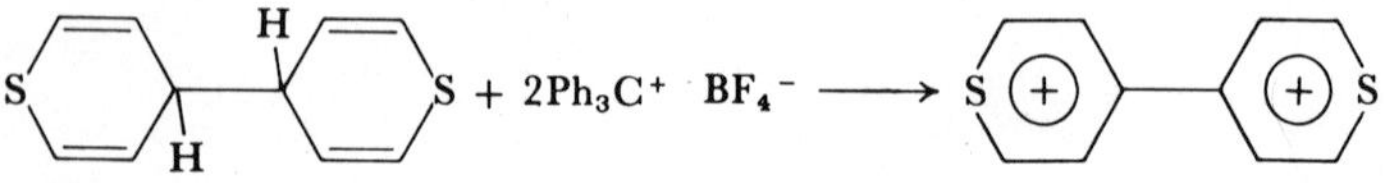

Yoshida (1971)

Electrophilic addition

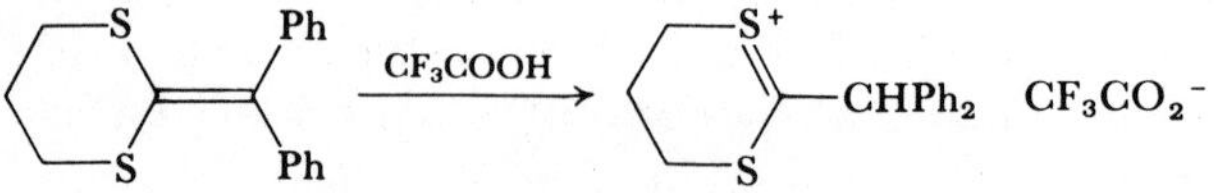

$\xrightarrow{CF_3COOH}$ $-CHPh_2 \ \ CF_3CO_2^-$

Carey and Court (1972)

$\longrightarrow$ $\longleftrightarrow$ Cl^-

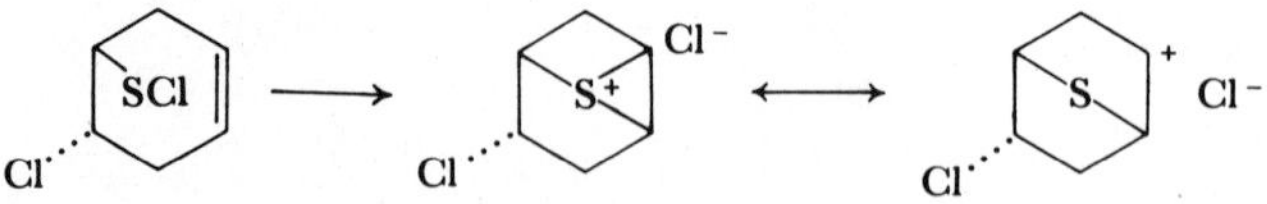

Corey and Block (1966)

$(CH_3S)_2C{=}S + (CH_3)_3O^+ \ BF_4^- \longrightarrow (CH_3S)_3C^+ \ \ BF_4^-$

Tucker and Roof (1967)

Loss of neutral molecule from sulfonium or oxosulfonium ion

$CH_3\overset{+}{S}(Cl)CH_3 \xrightarrow{-\,HCl} CH_3\overset{+}{S}{=}CH_2$ Tuleen and Stephens (1969)

$RS^{+}(=O)(Cl)-CH_2R' \xrightarrow{-HCl} RS^{+}(=O)=CHR'$ Durst *et al.* (1973)

$RS(=O)-S^{+}(R')CH_2H \xrightarrow{-RSOH} R'\overset{+}{S}=CH_2$ Kice and Morkved (1963)

Fragmentation

$\xrightarrow{SOCl_2}$ $\xrightarrow{-SO_2}$ $\overset{+}{S}-C_2H_5$

Ohno *et al.* (1966)

Protonation of a carbene

$\xrightarrow{CF_3COOH}$ Hartzler (1973)

Oxidation of radical

$PhSCH_2CO_2Na \xrightarrow{electrol} PhSCH_2CO_2\cdot \xrightarrow{-CO_2} PhSCH_2\cdot$

$PhSCH_2CO_2\cdot \xrightarrow[-e]{electrol} PhSCH_2CO_2^{+}$; $PhSCH_2\cdot \xrightarrow[-e]{electrol} PhSCH_2^{+}$

$PhSCH_2CO_2^{+} \xrightarrow{-CO_2} PhSCH_2^{+} \xrightarrow{CH_3OH} PhSCH_2OCH_3$ (34%)

Uneyama *et al.* (1971)

Rearrangement of sulfonium ion

$\xrightarrow{h\nu}$ Hogeveen *et al.* (1973)

Electron impact processes

$CH_3SCH_3 + e^- \longrightarrow CH_3SCH_2^{+} + H\cdot + 2e^-$ Taft *et al.* (1965)

4.3 Assessment of Stability of Sulfur-Containing Carbocations

The kinetic expression for carbocation formation according to Eq. 4-2 is

$$\mathrm{RX} \underset{k_{-1}}{\overset{k_1}{\rightleftharpoons}} \mathrm{R}^+ + \mathrm{X}^- \xrightarrow[k_2]{\mathrm{Y}^-} \mathrm{RY} \tag{4-2}$$

given by Eq. 4-3 (McManus and Pittman, 1973). To the extent that the

$$\text{Rate} = \frac{k_1 k_2[\mathrm{RX}][\mathrm{Y}^-]}{k_{-1}[\mathrm{X}^-] + k_2[\mathrm{Y}^-]} \tag{4-3}$$

term $k_{-1}[X^-]$ is small compared to $k_2[Y^-]$ (for example in the determination of the initial rate of solvolysis of a dilute solution of RX in solvent Y (or Y^-)), the rate expression reduces to k_1[RX]. The rate of reaction reflects the energy difference between the starting material RX and the transition state leading to R^+. It has been common practice to use the relative rate constants for solvolysis under identical conditions of a series of structurally related compounds RX as a measure of the relative stabilities of carbocations R^+. In using this approach it is necessary to assume that

1. for all of these reactions, the transition state is reached at the same point along the reaction coordinate and is similarly close in energy to the carbocation;
2. the reactants all have closely similar energies; and
3. differential solvation effects for RX and R^+ are similar in all cases,

assumptions which cannot be easily accepted.

For example, solvation effects by common polar solvents would be expected to be more substantial with "harder" carbocations, e.g., oxonium ions derived from α-haloethers, than with "softer" carbocations, e.g., sulfonium ions derived from α-halosulfides. Despite the limitations of the solvolysis rate data in providing a realistic comparison of the cation-stabilizing abilities of heteroatoms, it is still of some predictive value for cation-generating reactions conducted in polar protic solvents. Table 4-1 summarizes solvolysis rate data for several α-chlorosulfides, α-chloroethers, and related compounds. While S_N1 mechanisms are assumed to be involved in all cases, it may be that the less reactive chlorides are being consumed by S_N2 routes which under the reaction conditions are seen as pseudo-S_N1 processes.

In solution the order of cation-stabilizing ability indicated by solvolysis data is α-amino ≫ α-oxy > α-thio > β-thio ~ δ-thio > γ-thio. The effect

$$\mathrm{ArXCH_2Cl} + (\mathrm{C_2H_5})_4\mathrm{N}^{+\,36}\mathrm{Cl}^- \underset{}{\overset{\mathrm{CH_3CN}}{\rightleftharpoons}} \left[{}^{36}\overset{\delta-}{\mathrm{Cl}}\text{-----}\overset{\overset{\mathrm{XAr}}{|\,\delta+}}{\mathrm{C}}\text{-----}\overset{\delta-}{\mathrm{Cl}} \right] \rightleftharpoons$$

X = O or S

$$\mathrm{ArXCH_2{}^{36}Cl} + (\mathrm{C_2H_5})_4\mathrm{N}^+\,\mathrm{Cl}^- \tag{4-4}$$

Table 4-1. Solvolysis Rate Data for α-Chlorosulfides, α-Chloroethers, and Related Compounds

Entry	Compound	k_{rel}	Conditions	Reference
1	n-$C_6H_{13}Cl$	~1.0	CH_3OH	Bordwell and Brannen (1964)
2	$PhSCH_2Cl$	3.3×10^4		Bordwell and Brannen (1964)
3	$PhS(CH_2)_2Cl$	150		Bordwell and Brannen (1964)
4	$PhS(CH_2)_3Cl$	1.0		Bordwell and Brannen (1964)
5	$PhS(CH_2)_4Cl$	130		Bordwell and Brannen (1964)
6	$PhS(CH_2)_5Cl$	4.3		Bordwell and Brannen (1964)
7	p-NO_2-$C_6H_4SCH_2Cl$	333		Bordwell and Brannen (1964)
8	$PhSCH_2Cl$	1.00	90% aq. dioxane, 25°	Böhme *et al.* (1949)
9	$PhCH_2SCH_2Cl$	17		Böhme *et al.* (1949)
10	CH_3SCH_2Cl	220		Böhme *et al.* (1949)
11	CH_3OCH_2Cl	3.6×10^{5} [a]		Böhme (1941)
12	Cl, CH_2	1.0	96% C_2H_5OH, 29°	Krabbenhoft *et al.* (1974)
13	Cl, S	0.11		Krabbenhoft *et al.* (1974)
14	Cl, O	0.32		Krabbenhoft *et al.* (1974)
15	Cl, CH_3, N	8.1×10^6		Krabbenhoft *et al.* (1974)

[a] Extrapolated value.

of sulfur-containing groups other than the sulfide group will be discussed elsewhere in this chapter. The greater reactivity of α-chloroethers compared to α-chlorosulfides is also seen in S_N2 reactions, such as the reaction shown in Eq. 4-4 (Hayami *et al.*, 1971). In this reaction, where there is the development of partial positive charge on the α-heterocarbon, oxygen is found to accelerate the reaction by a factor of 10^5–10^6 and sulfur by 10^2–10^3 as compared to 2-arylethyl systems. Since the reaction is retarded by electron-withdrawing substituents and accelerated by electron donors on the aromatic ring, the possibility can be eliminated of simple enhanced electron deficiency at the reaction center due to attachment of electronegative heteroatoms.

Comparative kinetic data on the cation stabilizing abilities of sulfur and oxygen have also been obtained from solvolysis studies of substituted cumyl chlorides (Eq. 4-5; Brown *et al.*, 1958) and from studies of the reactivity of sulfur or oxygen substituted benzenes in electrophilic aromatic substitution (protodesilylation and protodetritiation, Eqs. 4-6 and 4-7, respectively, Bailey and Taylor, 1971). Data from Eqs. 4-5 and 4-7 are collected in Table 4-2.

$$R{-}C_6H_4{-}C(CH_3)_2Cl \xrightarrow[10\%\ water]{90\%\ acetone} \left[R{-}C_6H_4{-}\overset{+}{C}(CH_3)_2 \longleftrightarrow \overset{+}{R}{=}C_6H_4{=}C(CH_3)_2\right] \longrightarrow R{-}C_6H_4{-}C(CH_3)_2OH \qquad (4\text{-}5)$$

$$R{-}C_6H_4{-}Si(CH_3)_3 \xrightarrow[H_2O,\ 50^\circ]{HClO_4{-}CH_3OH} \left[R{-}\overset{+}{C_6H_5}{<}^{H}_{Si(CH_3)_3}\right] \searrow R{-}C_6H_4{-}H \qquad (4\text{-}6)$$

$$R{-}C_6H_4{-}T \xrightarrow[70^\circ]{CF_3COOH} \left[R{-}\overset{+}{C_6H_5}{<}^{H}_{T}\right] \xrightarrow{-T^+} R{-}C_6H_4{-}H \qquad (4\text{-}7)$$

The data in Table 4-2 are consistent with the ability of sulfur to provide substantial stabilization for an adjacent positive charge involving 2p–3p π overlap. Arguments for the superior cation stabilizing ability of nitrogen compared to oxygen and sulfur refer to the ionic character of α-haloamines (they are salts) compared to the covalent character of α-haloethers and α-halosulfides (Grob and Ide, 1974), the smaller electron-withdrawing effect of nitrogen compared to sulfur and oxygen, (Grob and Ide, 1974; Taft *et al.*, 1965) and the larger value of Brown's substituent constants $\sigma_p{}^+$ (−1.7 for

Table 4-2. Relative Rate Constants for Some Cation-Generating Reactions Involving Heterosubstituted Benzene Derivatives[a]

Reaction	Rate constant, R = H	CH_3	PhS	PhO	CH_3S	CH_3O	$\frac{kCH_3O}{kCH_3S}$	Reference
4–5	1.0	420	9830	31,000	81,900	190,000	2.3	Brown *et al.* (1958)
4–6	1.0	21.1	10.7	88.5	65.2	1270	19.5	Bailey and Taylor (1971)
4–7	1.0	—	—	—	553.0	3360	6.1	Bailey and Taylor (1971)

[a] See Eqs. 4–5 to 4–7 for conditions.

$(CH_3)_2N$, −0.778 for CH_3O, −0.604 for CH_3S) for nitrogen (Brown *et al.*, 1958).

The greater effectiveness of oxygen compared to sulfur in α-cation stabilization is attributed to the superiority of 2p–2p π bonding compared to 2p–3p π bonding which more than compensates for the greater electron-withdrawing effect of oxygen compared to sulfur. The reduced reactivity of arylthio derivatives can be attributed to the reduction of electron density on sulfur due to delocalization into the aryl ring, e.g., Eq. 4-8. The curious

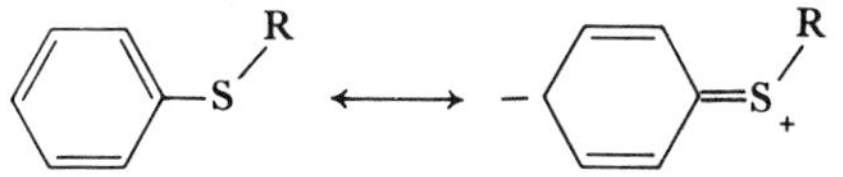

$$C_6H_5\text{–}S\text{–}R \longleftrightarrow {}^{-}C_6H_5{=}S^{+}\text{–}R \qquad (4\text{-}8)$$

results in entries 12–15 in Table 4-1 (Krabbenhoft *et al.*, 1974) have been rationalized by postulating that conjugative electron release by π bond formation is sterically inhibited so that inductive effects come to the fore. The data for the nitrogen-bridged system suggest that conjugative electron release has not been entirely suppressed. It seems that the order of the solvolytic α-heteroatom effect N > 0 > S has been maintained but displaced with respect to the carbon standard because of the increased importance given to inductive effects.

Another possible example of the displacement of electron-donating properties of sulfur and oxygen by steric effects is noted in F NMR studies by Goodman and Taft (1965). The decreased shielding in going from **4-11** to **4-12** paralleling a decrease in electron density at the fluorine is consistent

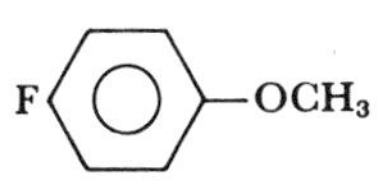

F–C₆H₄–OCH_3 **4-11**   F–C₆H₄–SCH_3 **4-12**   F–C₆H₂(CH_3)₂–OCH_3 **4-13**

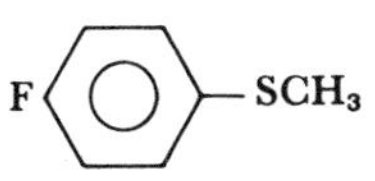

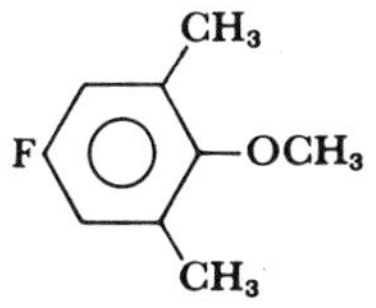

4-14 **4-15**

with the superior conjugative electron-releasing ability for oxygen compared to sulfur. The 3,5-methyl groups, by a steric twisting effect, substantially reduce the electron-donating properties of the bracketed heteroatom. In comparing **4-13** and **4-14** with **4-11** and **4-12**, respectively, it is found that the shielding of fluorine by the *p*-heteroatom has been reduced by about the same extent in each case ($\sim$6 ppm) and in comparing **4-14** with **4-15** it is seen that the electron-accepting properties of sulfur now dominate.

One noteworthy feature of the data in Tables 4-1 and 4-2 is the variation in rate ratios for processes assumed to involve α-oxy and α-thio cations. While the rate ratio for solvolysis of $MeOCH_2Cl$ and $MeSCH_2Cl$ is 1600 (Boehme, 1941; Boehme *et al.*, 1949), substantially smaller ratios are indicated for reactions 1–3 in Table 4-2 suggesting that the relative cation-stabilizing abilities of oxygen and sulfur may vary significantly with the substrate and reaction conditions.

A particularly dramatic illustration of this point is seen in the stabilization energies of substituted methyl cations generated by electron impact in the gas phase as summarized in Table 4-3 (Taft and Rakshys, 1965). In support of Taft's work, more recent studies employing chemical ionization mass spectrometry conclude that "the methylthiomethyl cation is formed in chemical ionization much more readily than is the methoxymethyl cation" and that "the relative stabilizing effect of CH_3S and CH_3O on a charge center are different in the gas phase and in solution" (Field and Weeks, 1970).

On the basis of the data in Table 4-2, Taylor concludes that sulfur has a high capability to respond to demands for transition state resonance stabilization. It is argued that "the weaker electronegativity of the second-row elements is much more important than has been realized . . . under conditions

Table 4-3. Relative Stabilization Energies of Monosubstituted Methyl Cations

Ion	S. E. (kcal $\pm$ 3)	Ion	S. E. (kcal $\pm$ 3)
CH_3^+	(0)	$HSCH_2^+$	64
$CH_3CH_2^+$	35	$CH_3OCH_2^+$	69
$PhCH_2^+$	55	$CH_3SCH_2^+$	74
$HOCH_2^+$	60	$(CH_3)_2NCH_2^+$	106

Table 4-4. Anchimeric Assistance by Sulfur

Entry	Class	Compound	Relative rate of solvoylsis		Reference
			H_2O, 50°	aq. dioxane, 100°	
1	—	$CH_3CH_2CH_2Cl$	1.00	1.00[a]	Blandamer *et al.* (1969)
2	RS-3	$CH_3SCH_2CH_2Cl$	2.93×10^5	2.7×10^3[b]	Böhme and Sell (1948)
3	RS-3	$PhSCH_2CH_2Cl$	—	89	Böhme and Sell (1948)
4	RO-3	$C_2H_5OCH_2CH_2Cl$	—	0.18	Böhme and Sell (1948)
5	RO-5	$CH_3O(CH_2)_4Cl$	40.4	—	Böhme and Sell (1948)
			80% C_2H_5OH, 118.2°		
6	—	cyclohexyl chloride	1.00		Goering and Howe (1957)
7	RS-3	SPh, Cl	7.2×10^4		Goering and Howe (1957)
8	RS-3	SPh, Cl	0.16		Goering and Howe (1957)
			80% aq-$(CH_3)_2CO$, 50°		
9	—	OPNB	1.00		Ikegami *et al.* (1974)
10	RS-4	S, OPNB	0.10		Ikegami *et al.* (1974)

[a] *n*-Butyl chloride.
[b] 2-Chloroethylethyl sulfide.

(continued)

Table 4-4. *(Continued)*

Entry	Class	Compound	Relative rate of solvolysis	Reference
11	RS-3	S; OPNB	5.8×10^6	Ikegami *et al.* (1974)
			AcOH, 25°	
12	RS-3	S; Cl	1.0	Tabushi *et al.* (1975)
13	RS-3	S; Cl	$\geqslant 4.7 \times 10^9$	Tabushi *et al.* (1975)
14	—	CH_2; Cl	2×10^3	Tabushi *et al.* (1975)
15	RO-3	O; Cl	0.14^c	Tabushi *et al.* (1975)
16	RO-3	O; Cl	4.4×10^{-4c}	Tabushi *et al.* (1975)
			80% aq. C_2H_5OH, 77.0°	
17	—	t-C_4H_9—; —OTs	1.00	Ireland and Smith (1959)

18	RS-4	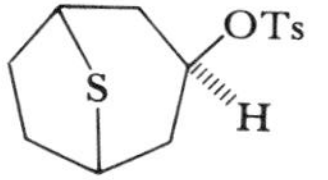 S, OTs, H	0.46	Ireland and Smith (1959)
19	RS-4	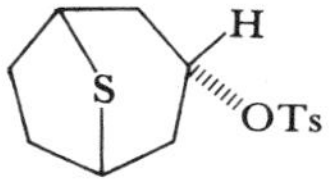 S, H, OTs	0.21	Ireland and Smith (1959)
		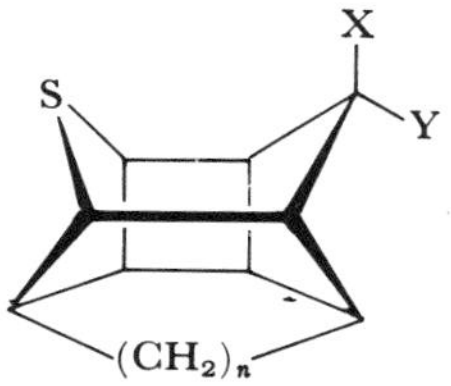	AcOH, 25°	
		S, X, Y, $(CH_2)_n$		
20	RS-4	$n = 1$, X = OTs, Y = H	8.4	Paquette *et al.* (1969)
21	RS-4	$n = 1$, X = H, Y = OTs	32.2	Paquette *et al.* (1969)
22	RS-4	$n = 2$, X = OTs, Y = H	1.8	Paquette *et al.* (1969)
23	RS-4	$n = 2$, X = H, Y = OTs	5564	Paquette *et al.* (1969)
		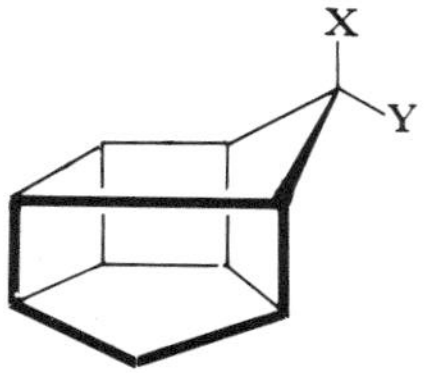 X, Y		
24	—	X = OTs, Y = H	1.0	Paquette *et al.* (1969)
25	—	X = H, Y = OTs	4.9	Paquette *et al.* (1969)

(*continued*)

[c] Solvolyzed in 50% aqueous dioxane with rate comparison to solvolysis of *exo*-2-chloro-7-thiabicyclo[2.2.1]heptane under these same conditions.

Table 4-4. *(Continued)*

Entry	Class	Compound	Relative rate of solvolysis	Reference
			75% aq. dioxane, 148.8°	
26	—	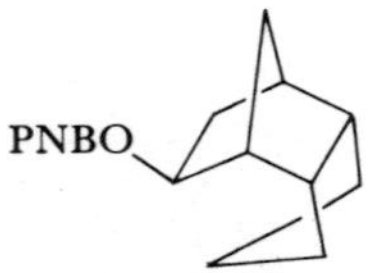 PNBO	1.0	Gratz and Wilder (1970)
27	RS-5	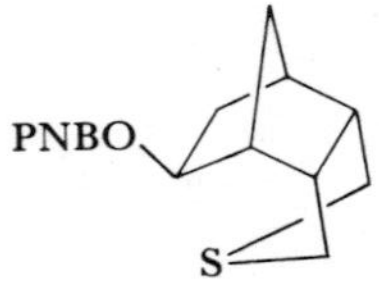 PNBO, S	752	Gratz and Wilder (1970)
28	RS-5	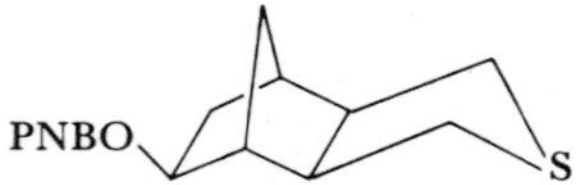 PNBO, S	0.8	Gratz and Wilder (1970)

of very high electron demand, the second-row elements can show large polarizability effects" (Bailey and Taylor, 1971).

In an important theoretical study, Bernardi, Csizmadia, Schlegel, and Wolfe have performed *ab initio* calculations on CH_2O, CH_2S, CH_2OH^+, and CH_2SH^+ (Bernardi *et al.*, 1975). While, as expected, the π-overlap population in formaldehyde is found to be greater than the π-overlap population in thioformaldehyde, the situation is reversed with the cations (the π-overlap population in CH_2SH^+ is however less than that in CH_2S). Furthermore, there is substantially more positive charge on sulfur in CH_2SH^+ than on oxygen in CH_2OH^+. Consistent with the stronger π bond between carbon and sulfur in CH_2SH^+ than between carbon and oxygen in CH_2OH^+, the C—S rotational barrier in CH_2SH^+ is greater than the C—O rotational barrier in CH_2OH^+. In explaining these results, it is noted that charge transfer between orbitals (e.g., $2p_C$ and $2p_O$ or $2p_C$ and $3p_S$) is inversely proportional to the energy difference between the interacting orbitals and carbon 2p orbitals are closer in energy to sulfur 3p than oxygen 2p orbitals in the cation (Bernardi *et al.*, 1975). In view of these theoretical calculations and the gas phase studies of Taft and Field, it must be concluded that solvation effects dominate the condensed phase processes leading to heterostabilized cations.

4.4 Neighboring Group Participation by Divalent Sulfur

The data in Table 4-1 also indicate the effectiveness of β- and δ-thio groups in stabilizing carbocations. While β-oxy groups retard solvolysis due to the strong inductive effect of oxygen, δ-oxy groups are known to provide modest anchimeric assistance in solvolysis, though to a lesser degree than the δ-thio group (see Table 4-4). The greater effectiveness of sulfur compared to oxygen as a neighboring group may be attributed to

1. the greater polarizability of the sulfur 3p electrons than the oxygen 2p electrons which would tend to make the sulfur electrons somewhat more available (solvation effects may also be important here: see footnote † on page 22)
2. the longer bond lengths associated with sulfur, which should favor intramolecular interactions, and
3. the lesser destabilization of nearby developing positive charges by the more weakly electron-withdrawing sulfur atom compared to the oxygen atom.

The data in Table 4-4 are classified according to the method suggested by Winstein whereby the size of the thiacyclic (or oxacyclic) ring whose formation would result from S (or O) participation is indicated by RS-*n* (or RO-*n*): thus β-participation is designated by RS-3, γ-participation by RS-4, etc. (Winstein *et al.*, 1958). The data in Table 4-4 clearly indicate that the effectiveness of sulfur in providing anchimeric assistance is dependent not only on solvent and substituent (e.g., alkyl versus aryl) effects but to a major extent on the molecular geometry, particularly the alignment of the sulfur 3p orbital with the developing electron-deficient center. Thus a steady increase in the effectiveness of sulfur as a neighboring RS-3 group (compared to an appropriate reference compound) is seen in going from the acyclic compound in entry 3 to the more rigid exocyclic structure in entry 7 to the thiacycle in entry 11 to the thiabicycle in entry 13. The last case represents the largest value ever reported for the effect of sulfur participation!

In the case of RS-4 participation, the preferred conformations of thiacyclohexane (entry 10), 8-thiabicyclo[3.2.1]octane (entry 19), and the polycycle in entry 21 do not place the sulfur in a favorable enough position to allow anchimeric assistance. Only with the compound in entry 23 does RS-4 assistance seem to be significant. RS-5 participation has been realized for both acyclic (cf. Table 4-1, entry 5) and polycyclic substrates (Table 4-4, entry 27) while RS-6 participation is seen with acyclic systems (Table 4-1, entry 6).

In the examples in Table 4-4, and in related cases, sulfur participation controls product formation. Thus solvolysis of 3-thiacyclohexyl *p*-nitrobenzoate affords the same mixture of products as obtained from 2-carbinylthiacyclopentane *p*-nitrobenzoate indicating a common intermediate (Eq. 4-9; Ikegami *et al.*, 1974). On the other hand acetolysis of *endo*-2-

OPNB S → [S+] ← S OPNB

OH S 82% OH S 18% (4-9)

chloro-7-thiabicyclo[2.2.1]heptane (entry 13) afforded the corresponding *endo*-2-acetoxy derivative while *exo*-2-chloro-7-thiabicyclo[2.2.1]heptane (entry 12) gave only the rearranged product *exo*-3-acetoxy-2-thiabicyclo-

[2.2.1]heptane indicating a different course for the two reactions (Eq. 4-10) (Tabushi *et al.*, 1975). Solvolysis of the 3-*endo*-tosylate of 8-thiabicyclo[3.2.1]-

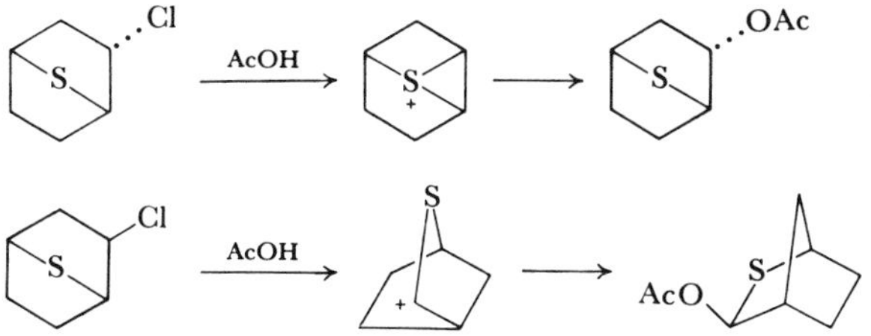

(4-10)

octane gives exclusively *endo* alcohol in 80% yield, suggesting the intervention of a sulfonium ion after the initial ionization step (since anchimeric assistance is not observed) (Eq. 4-11; Ireland and Smith, 1959). An unusual form of

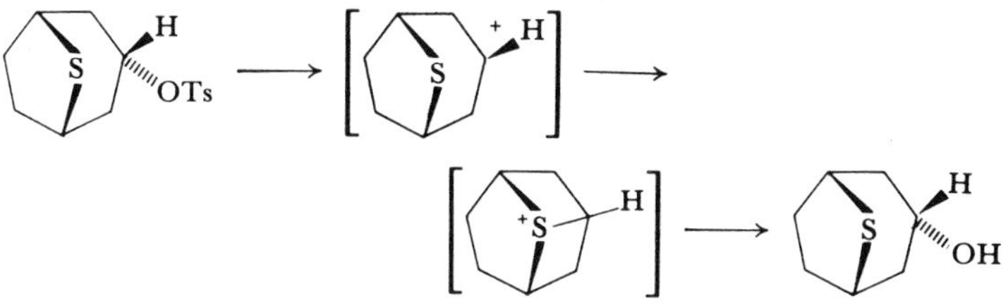

(4-11)

sulfur participation occurs with the polycyclic structure in Table 4-4, entry 21, leading to an olefin via fragmentation (Eq. 4-12; Paquette *et al.*, 1969).

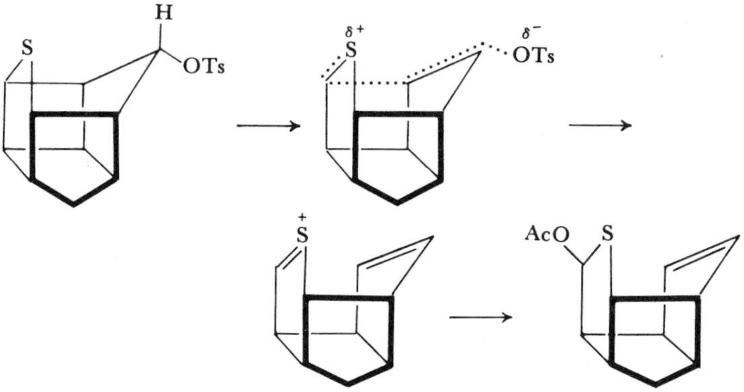

(4-12)

Additional examples of fragmentation reactions involving sulfur as the "electrofugal" (electron-releasing) group will be considered elsewhere in this chapter.

Anchimeric assistance of type RS-3 is also seen in the solvolysis of β-alkythiovinyl and β-arylthiovinyl sulfonates as indicated by the rate data in Table 4-5 (Burighel *et al.*, 1972). The probable intermediate in these

Table 4-5. Solvolysis Rate Data for Some β-Thiovinyl Sulfonates

Compound	k_{rel} (25°; nitromethane–methanol)	E_a (kcal)	$\Delta S^{\ddagger}$ (e.u.), 25°
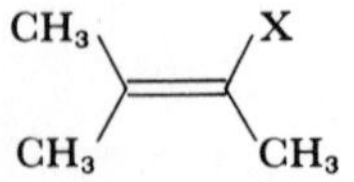 CH_3, X CH_3, CH_3	1.00	25.3	−12
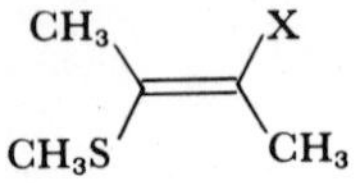 CH_3, X CH_3S, CH_3	3.8×10^4	22.3	−1
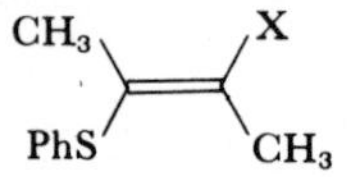 CH_3, X PhS, CH_3	3.5×10^3	—	—

solvolyses where assistance by sulfur is indicated is a thiirenium ion **4-16**. Molecular orbital calculations indicate that this bridged structure is somewhat more stable (1–14 kcal/mol) than the open structure, $RS(R)C{=}CR^+$

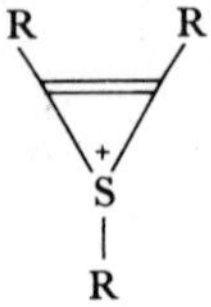

R R
S^+
R

4-16

(Csizmadia *et al.*, 1974), while solvolysis studies on unsymmetrically substituted systems require a bridged ion at least as a transition state in order to explain the formation of products in which a group RS has migrated (see

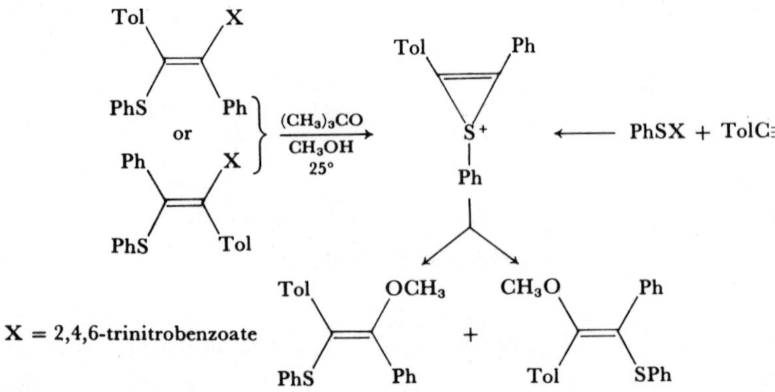

Tol X / PhS Ph
or
Ph X / PhS Tol

$\xrightarrow[\text{CH}_3\text{OH, 25°}]{(\text{CH}_3)_2\text{CO}}$ Tol Ph / S^+ / Ph $\longleftarrow$ PhSX + TolC≡CPh (4-13)

X = 2,4,6-trinitrobenzoate

Tol OCH_3 / PhS Ph + CH_3O Ph / Tol SPh

ratio 1:4.5 (in each case)

Eq. 4-13; Modena *et al.*, 1973). The same thiirenium ion is indicated as an intermediate in the addition of a sulfenyl sulfonate to acetylenes (Modena *et al.*, 1973). Recently the trimethylthiirenium ion has been observed by NMR spectroscopy in liquid sulfur dioxide at $-50°$ (Capozzi *et al.*, 1975).

The high reactivity of β-halosulfides under solvolytic conditions makes them potent biological alkylating agents. Along with its ignoble use in World War I as a powerful blistering agent, β,β'-bis(chloroethyl)sulfide, $S(CH_2CH_2Cl)_2$ ("mustard gas") and more complex relatives such as **4-17** have been used in

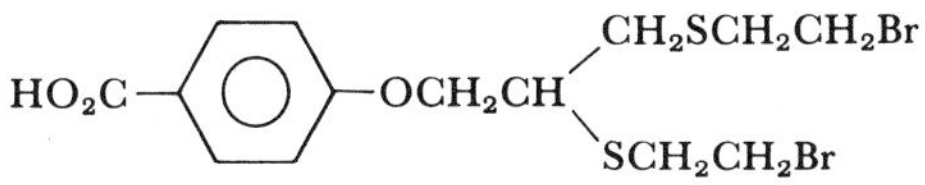

$$HO_2C-C_6H_4-OCH_2CH\begin{cases}CH_2SCH_2CH_2Br\\SCH_2CH_2Br\end{cases}$$

4-17

cancer treatment. The basis for use rests on the fact that rapidly dividing tumor cells are particularly sensitive to destruction by chemical agents (or by ionizing radiation). The halosulfides act *in vivo* as electrophilic reagents alkylating essential cellular macromolecules (presumably cross-linking them, in the case of the bifunctional mustards) and thereby disrupting cellular processes. The kinetic reactivity of the halosulfide toward S_N1 or S_N2 processes is critically important in determining its efficacy as a chemotherapeutic agent. The alkylating agent must survive long enough to reach the tumor bed yet must not be too long lived so that it would pose a threat to sensitive areas of the body such as the bone marrow. Compound **4-17** which has a half-life in the body of only a few seconds is used clinically in the treatment of head and neck cancer through injection directly into blood vessels supplying the tumor (Connors, 1974).

4.5 Reactions of Thiocarbocations

A variety of reaction pathways are open to sulfur-containing carbocations including (among others)

1. trapping by a nucleophile,
2. elimination of a β proton to afford an olefin,
3. α-elimination to give a carbene,
4. addition to an alkene to form a new cation,
5. electrophilic aromatic substitution,
6. rearrangement via a bridged ion,
7. abstraction of a hydride ion to form a new cation, and
8. one-electron reduction to a radical.

The significant intramolecular interaction of a thio group with a cation can provide sufficient stabilization and charge delocalization so that hydride, alkyl, and aryl shifts that are characteristic of less stable carbonium ions are minimized. Examples of the various possible reaction pathways are given in Eqs. 4-14 to 4-43. Other examples appear elsewhere in this chapter.

(1) *Trapping by a nucleophile*

$$CH_3SCH_2Cl \xrightarrow[-80°]{SbCl_5} CH_3\overset{+}{S}{=}CH_2 \xrightarrow{CH_3SCH_3} (CH_3)_2\overset{+}{S}CH_2SCH_3\ (92\%)$$

$$CH_3\overset{+}{S}{=}CH_2 \xrightarrow{CH_3SCH_2CH{=}CH_2} CH_2{=}CHCH_2\overset{+}{S}(CH_3)CH_2SCH_3\ (93\%) \tag{4-14}$$

(Hansen and Olofson, 1971)

$$(CH_3S)_3C^+BF_4^- \xrightarrow{CH_3SH} (CH_3S)_4C\ (97\%) \tag{4-15}$$

(Tucker and Roof, 1967)

$$\text{2-(methylthio)-1,3-dithiolan-2-ylium} + CH_2(CN)_2 \xrightarrow{Py,\ (C_2H_5)_3N} [^-CH(CN)_2]$$

$$\longrightarrow \text{2-(methylthio)-2-[CH(CN)}_2\text{]-1,3-dithiolane} \longrightarrow \text{2-(dicyanomethylene)-1,3-dithiolane}\ (90\%) \tag{4-16}$$

(Gompper and Kutter, 1965)

Hydrolysis of vinyl sulfides to carbonyl compounds involves the formation and trapping of alkylidene sulfonium ions (Eq. 4-17). Generally hydrolysis of enol thioethers requires conditions that are much more vigorous than the conditions needed to hydrolyze enol ethers (Autrey and Scullard, 1968). Thus, for example, thiaoxene undergoes acid catalyzed addition of alcohol

$$CH_3O\text{-}C_6H_3(CN)\text{-}CH_2CH{=}CHSCH_3 \xrightarrow[\text{reflux, 30 min}]{C_2H_5OH/HClO_4} CH_3O\text{-}C_6H_3(CN)\text{-}CH_2CH_2CHO\ \ 64\%$$

$$(\text{via } RCH_2CH{=}\overset{+}{S}CH_3 \xrightarrow{H_2O} RCH_2CH(OH)SCH_3) \tag{4-17}$$

(Autrey and Scullard, 1968)

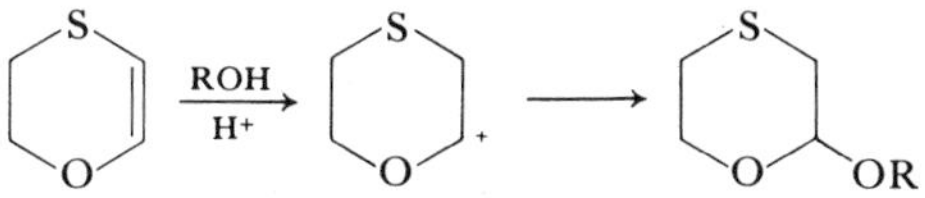

(4-18)

only in the direction that generates a carbocation next to oxygen (Eq. 4-18; Parham *et al.*, 1952). The strained unsaturated sulfide thiete (thiacyclobutene) represents an example of a readily hydrolyzed enol thioether (Eq. 4-19; Dittmer *et al.*, 1967).

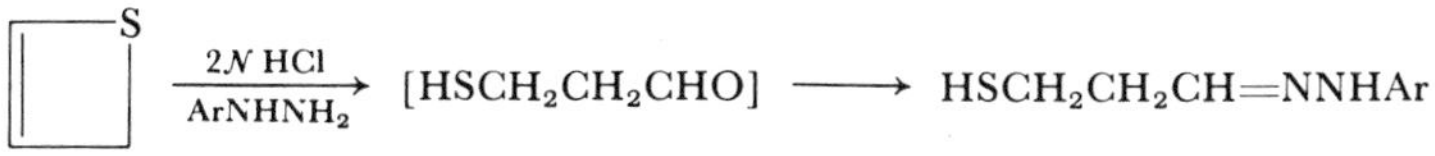

$$\xrightarrow[\text{ArNHNH}_2]{2N\ \text{HCl}} [\text{HSCH}_2\text{CH}_2\text{CHO}] \longrightarrow \text{HSCH}_2\text{CH}_2\text{CH{=}NNHAr} \quad (4\text{-}19)$$

The facile hydrolysis of α-chloro sulfides has been cleverly utilized by Gassman in an ortho-formylation sequence for aromatic amines (Eq. 4-20; Gassman and Drewes, 1974) and in the conversion of aromatic amines into isatins (Gassman *et al.*, 1977).

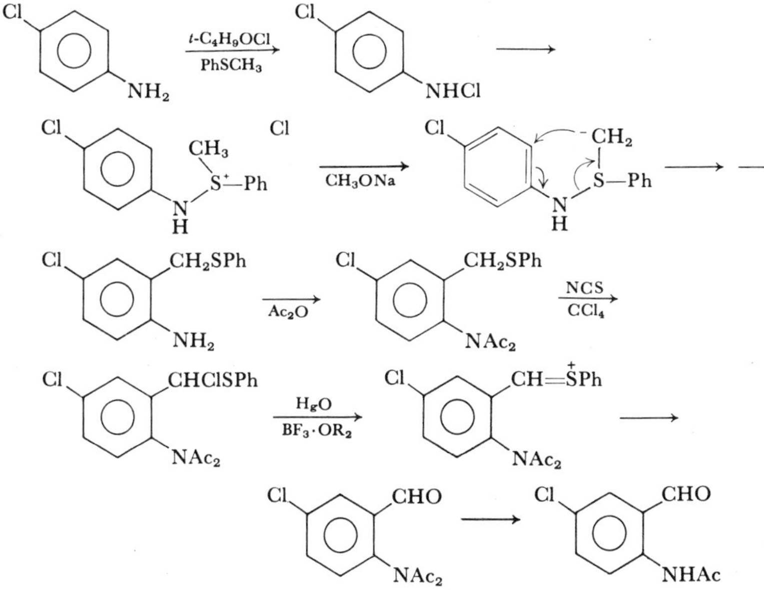

t-C_4H_9OCl, $PhSCH_3$; CH_3ONa; Ac_2O; NCS, CCl_4; HgO, $BF_3 \cdot OR_2$

(4-20)

(50% overall yield)

(2) *β-Elimination*

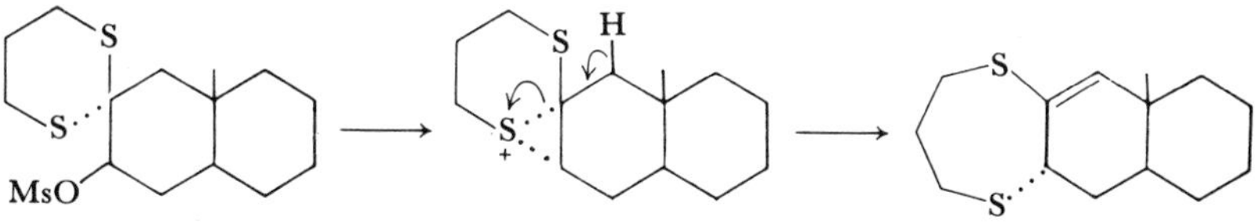

(4-21)

(Marshall and Roebke, 1969)

$$\text{cyclononene} + PhSCl \longrightarrow [\text{bridged episulfonium ion, } \overset{+}{S}\text{–Ph}] \longrightarrow \text{3-(phenylthio)cyclooctene (SPh)} \quad (65\%) \qquad (4\text{-}22)$$

(Burgoine *et al.*, 1974)

$$\text{Ph-thiazolidine(N-Ac)-fused lactone with } CH_2OMs \xrightarrow[\text{NaOAc, } \Delta]{\text{dioxane}} [\text{Ph–episulfonium intermediate}] \longrightarrow \text{Ph-thiazine(N-Ac)-fused unsaturated lactone} \qquad (4\text{-}23)$$

(Stork and Cheng, 1965)

$$(RS)_2C{=}CH{-}\overset{+}{C}(SR)_2 \longrightarrow (RS)_2C{=}C{=}C(SR)_2 \qquad (4\text{-}24)$$

(Gompper and Jersak, 1973)

A general synthesis of vinyl sulfides has been developed by Cohen utilizing cuprous triflate as a Lewis acid to convert thioacetals and thioketals to alkylidene sulfonium ions which then undergo deprotonation or intramolecular capture by a nucleophile (Eqs. 4-25 and 4-26; Cohen *et al.*, 1975a).

$$(PhS)_2CHLi + CH_2{=}CHCH_2Br \longrightarrow CH_2{=}CHCH_2CH(SPh)_2 \xrightarrow[\text{THF, 25°}]{CuOSO_2CF_3\cdot C_6H_6} [CH_2{=}CHCH_2CH{=}\overset{+}{S}Ph] \xrightarrow{-H^+} CH_2{=}CHCH{=}CHSPh \qquad (4\text{-}25)$$

$$[(PhS)_2C(Ph)]_2CuLi + CH_2{=}CHC(O)CH_3 \longrightarrow (PhS)_2C(Ph)CH_2{-}CH_2C(O)CH_3 \xrightarrow[25°]{Cu(I)} \text{2-Ph-5-}CH_3\text{-furan} \ (80\%) \qquad (4\text{-}26)$$

Tucker has reported a reaction which could be considered as a β-elimination of CH_3^+ or, more commonly, as an S_N2 displacement on $CH_3{-}S$ (Eq. 4-27) (Tucker and Roof, 1967).

$$(CH_3S)_3C^+BF_4^- \xrightarrow[\text{18 hr, 80°}]{Py} (CH_3S)_2C{=}S + C_5H_5\overset{+}{N}{-}CH_3\ BF_4^- \qquad (4\text{-}27)$$

(3) *α-Elimination*

$$(CH_3S)_2CH^+BF_4^- \xrightarrow[(i\text{-}C_3H_7)_2N\text{—}C_2H_5]{} (CH_3S)_2C: \xrightarrow[-H^+]{(CH_3S)_2CH^+} (CH_3S)_2C{=}C(SCH_3)_2 \qquad (4\text{-}28)$$

$$(CH_3S)_3C^+BF_4^- \xrightarrow{Ph_3P} (CH_3S)_2C: \xrightarrow{(CH_3S)_3C^+} (CH_3S)_2\overset{+}{C}\text{—}C(SCH_3)_3 \xrightarrow{Ph_3P} (CH_3S)_2C{=}C(SCH_3)_2\ (55\%) \qquad (4\text{-}29)$$

In Eqs. 4-28 and 4-29 the dithioalkyl methylenes are acting as nucleophiles in reacting with the thiocarbocations (Olofson *et al.*, 1968; Tucker and Roof, 1967; cf. Chapter 6).

(4) *Addition to alkenes and 1,3-dienes*

Examples of the addition of α-thiocarbocations to alkenes include Anderson's cation–olefin cyclization (Eq. 4-30; Anderson *et al.*, 1975) and Ichikawa approach to β-thiomethylpropionaldehyde derivatives (Eq. 4-31; Ichikawa *et al.*, 1970). Corey and Walinsky have developed an unusual synthesis of Δ^3-cyclopentenones which incorporates the addition of the 1,3-dithienium ion to 1,3-dienes (Eq. 4-32; Corey and Walinsky, 1972).

CHO; 2-lithio-2-trimethylsilyl-1,3-dithiane (S, S, Li, $Si(CH_3)_3$); CH_3SO_3H, CF_3CH_2OH, C_7H_{16}; OCH_2CF_3, S, S; (1) NBS (2) KOH; H, H, O

(4-30)

$$CH_3SCH_2Cl \xrightarrow{H_2SO_4} CH_3\overset{+}{S}{=}CH_2 \xrightarrow{CH_2{=}CHCl} CH_3SCH_2CH_2\overset{+}{C}HCl$$

CH_3OH / Δ → $CH_3SCH_2CH_2CH(OCH_3)_2$

H_3O^+ \ ArNHNH$_2$ → $CH_3SCH_2CH_2CH{=}NNHAr$ (70%)

(4-31)

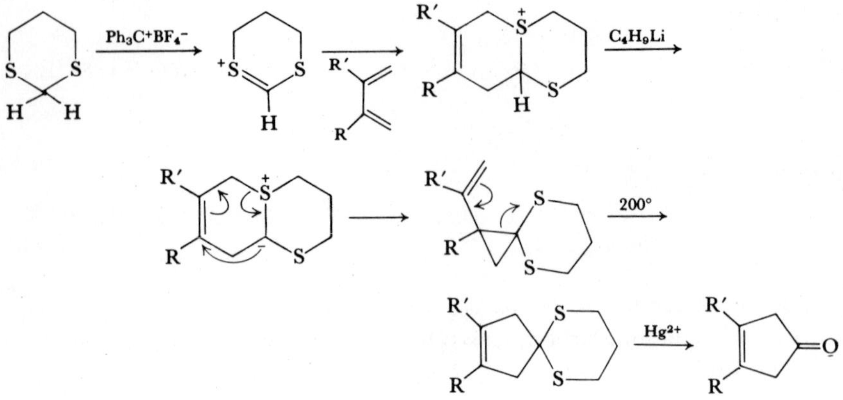

(4-32)

(5) *Electrophilic aromatic substitution*

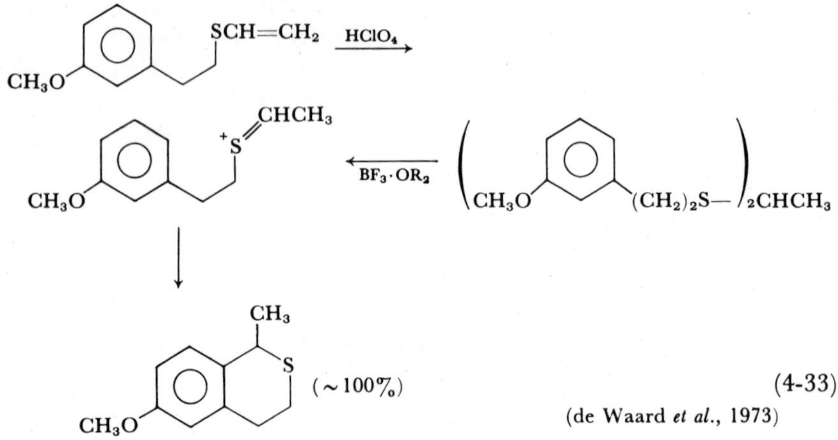

(4-33)

(de Waard *et al.*, 1973)

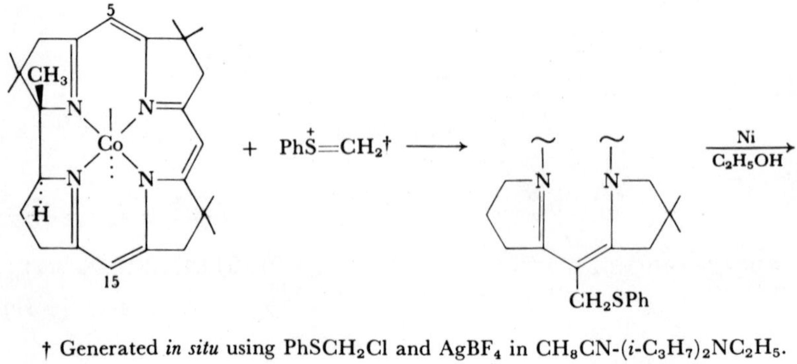

† Generated *in situ* using $PhSCH_2Cl$ and $AgBF_4$ in CH_3CN-$(i\text{-}C_3H_7)_2NC_2H_5$.

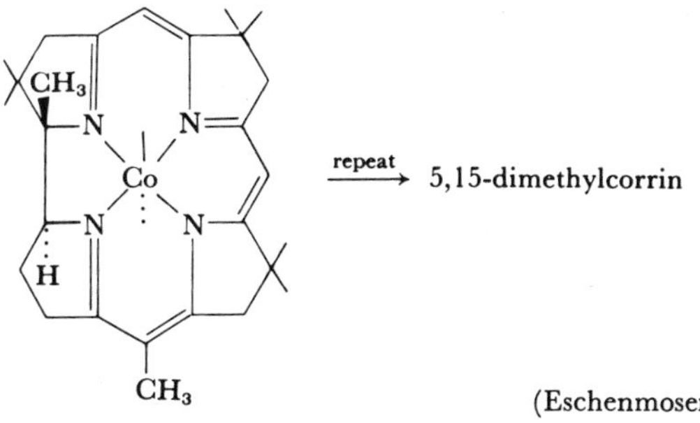

$\xrightarrow{\text{repeat}}$ 5,15-dimethylcorrin (4-34)

(Eschenmoser, 1970)

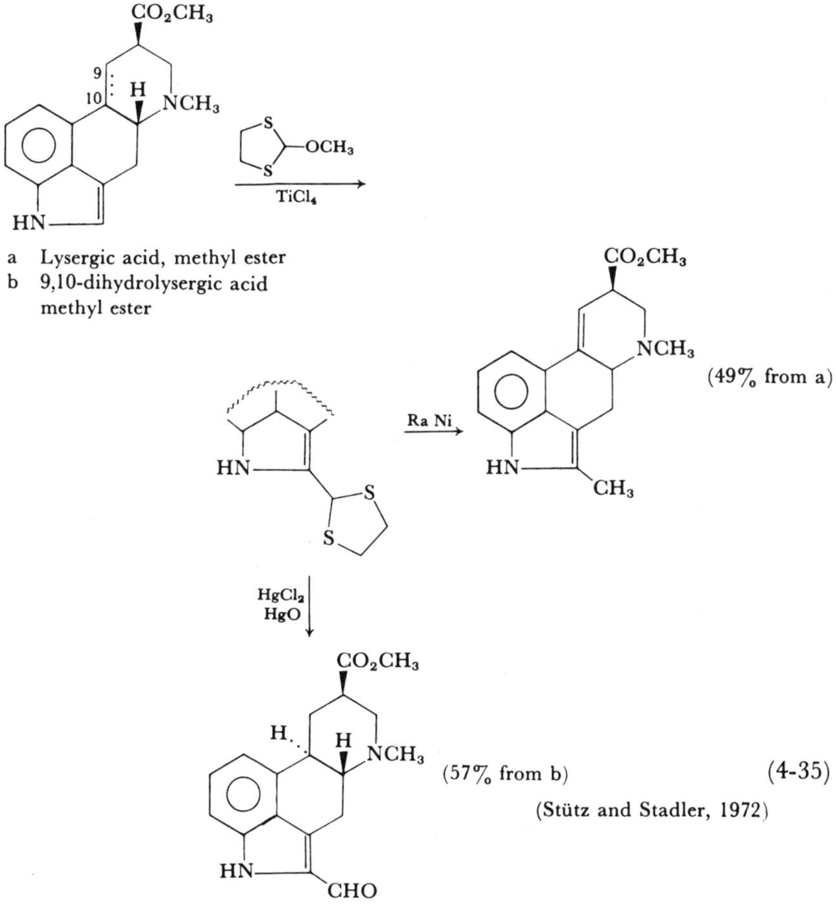

$\xrightarrow[\text{TiCl}_4]{}$

a Lysergic acid, methyl ester
b 9,10-dihydrolysergic acid methyl ester

$\xrightarrow{\text{Ra Ni}}$ (49% from a)

HgCl₂, HgO

(57% from b) (4-35)

(Stütz and Stadler, 1972)

H S S $-H^+$ N H S S + N H S S (38%) (1) C_4H_9Li (2) CH_3I N H S S CH_3 N H (53%)

(4-36)

(Stütz and Stadler, 1972)

(6) *Rearrangement*

S H Br S S S NaOAc HOAc S S S S S S S $-H^+$ CH_2 S S S S (100%)

(4-37)

(Coffen and Lee, 1970)

S H^+ S H $-H^+$ S

(4-38)

(Macnicol and McKendrich, 1973)

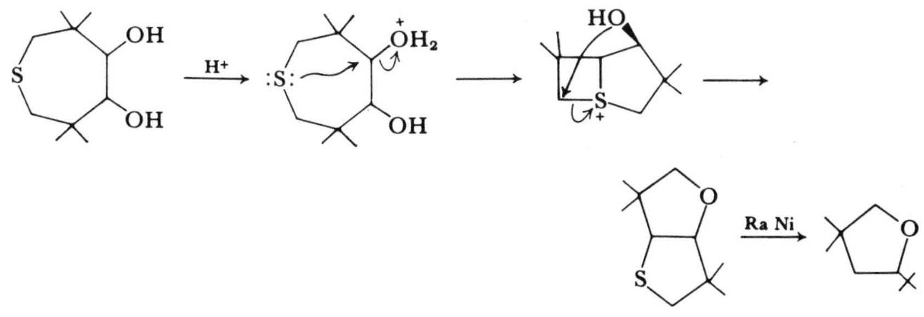

$\xrightarrow{H^+}$ $\xrightarrow{\text{Ra Ni}}$

(66%) (28%)

(4-39)

(de Groot *et al.*, 1968)

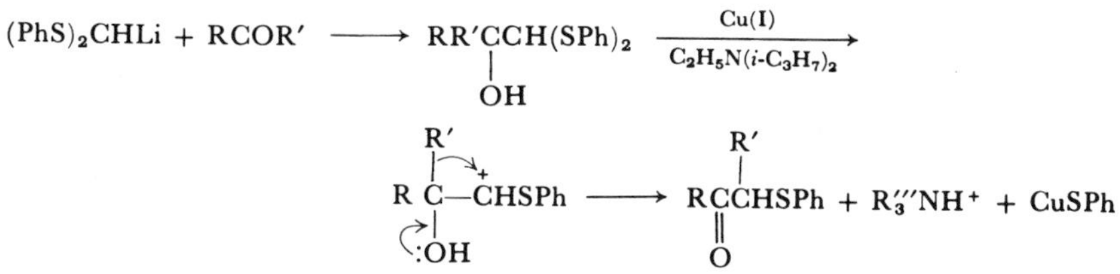

$$(PhS)_2CHLi + RCOR' \longrightarrow RR'C(OH)CH(SPh)_2 \xrightarrow[C_2H_5N(i\text{-}C_3H_7)_2]{Cu(I)}$$

$$RR'C(:OH)\text{—}\overset{+}{C}HSPh \longrightarrow RC(=O)CR'HSPh + R_3''NH^+ + CuSPh$$

(4-40)

(Cohen *et al.*, 1975b)

(7) *Hydride abstraction*

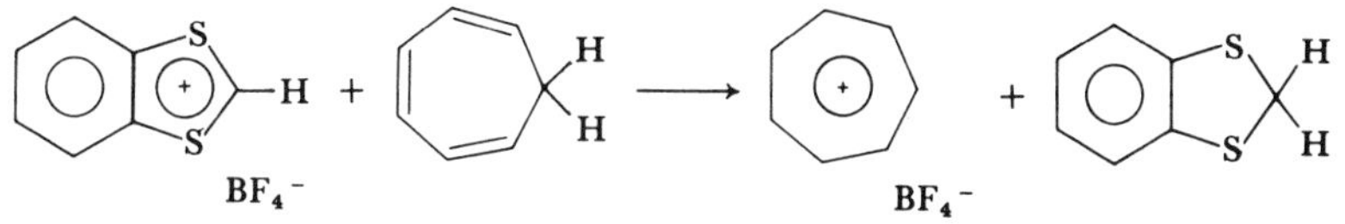

BF_4^- BF_4^-

(4-41)

(Nakayama *et al.*, 1975)

(8) *One-electron reduction*

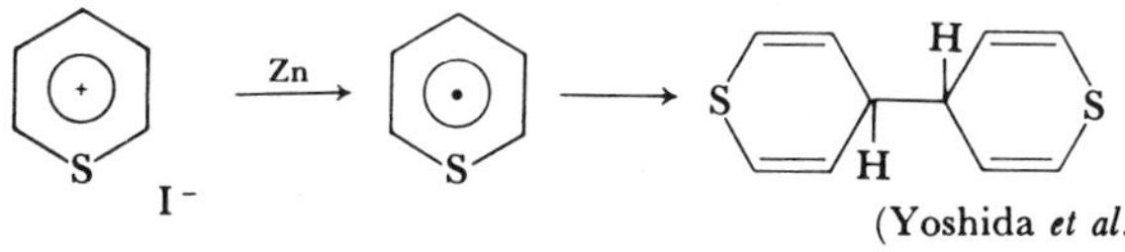

$\xrightarrow{Zn}$ I^- (4-42)

(Yoshida *et al.*, 1971)

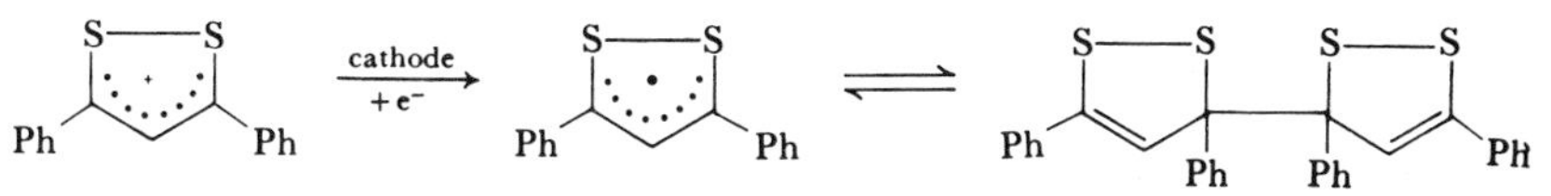

$\xrightarrow[+e^-]{\text{cathode}}$

(4-43)

(Bechgaard *et al.*, 1973)

4.6 The Pummerer Rearrangement

The Pummerer reaction may be generalized as a reaction involving the reduction of a sulfonium sulfur with concomitant oxidation of the α-carbon (Russell *et al.*, 1968; Durst, 1969), e.g., Eq. 4-44. Typical examples of the

$$R\overset{+}{S}(X)CHR'R'' \xrightarrow[+Y^-]{-HX} RSC(Y)R'R'' \qquad (4\text{-}44)$$

Pummerer rearrangement involve the treatment of sulfoxides with an electrophile such as an acid anhydride (Eqs. 4-45 and 4-46; Horner and Kaiser,

$$CH_3S(O)CH_3 + (RCO)_2O \xrightarrow[\text{reflux, 6 hr}]{C_6H_6} \left[CH_3\overset{+}{S}(OC(O)R)CH_3\right] \longrightarrow CH_3SCH_2OC(O)R$$

(R = CH_3, 85%; R = Ph, 79%)

(4-45)

$$\text{tetrahydrothiophene } S{=}O + (CH_3CO)_2O \xrightarrow[25°,\ 3\text{ days}]{CHCl_3} \text{2-}OC(O)CH_3\text{-tetrahydrothiophene} \quad (85\%) \qquad (4\text{-}46)$$

1959). Diphenyl sulfoxide, which is incapable of undergoing the Pummerer reaction, is unreactive toward acetic anhydride even under forcing conditions. A generalized mechanism for the Pummerer reaction is given in Eq. 4-47.

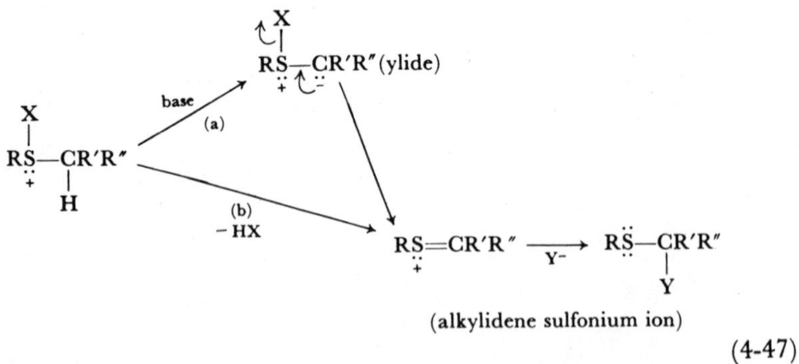

(4-47)

When "X" is acyloxy or alkoxy, the experimental evidence suggests that the reaction involves *ylide* formation in the rate- and product-determining step (path (a)) followed by rapid formation and trapping of an *alkylidene sulfonium ion*, since the nucleophile Y is attached to the α-thio carbon atom where carbanion formation would be most favored as seen, for example, in Eqs. 4-48 to 4-50. In some instances the Pummerer product contains nucleophile

$$CH_3CH_2CH_2S(O)CH_3 \xrightarrow{Ac_2O} CH_3CH_2CH_2SCH_2OAc\ (73\%) \tag{4-48}$$

$$i\text{-}C_3H_7\overset{+}{S}(OCH_3)CH_3 \xrightarrow{NaOAc} \underset{(69\%)}{i\text{-}C_3H_7SCH_2OAc} + \underset{(20\%)}{i\text{-}C_3H_7SCH_2OCH_3} + \underset{(6\%)}{i\text{-}C_3H_7SCH_3} \tag{4-49}$$

(Johnson and Phillips, 1969)

$$PhC(O)CH_2S(O)CH_3 \begin{cases} \xrightarrow{CH_3SOCl} PhC(O)CH(Cl)SCH_3\ (87\%) \\ \xrightarrow{H^+} PhC(O)CH(OH)SCH_3 \end{cases} \tag{4-50}$$

(Russell and Sabourin, 1969a)

"Y" attached at the position where carbocation formation is most preferred suggesting that path (b), Eq. 4-47 has been followed, e.g., as seen in Eq. 4-51 (see Table 4-6 for references). The α-chlorination of 2,2-dideuteriothiacyclo-

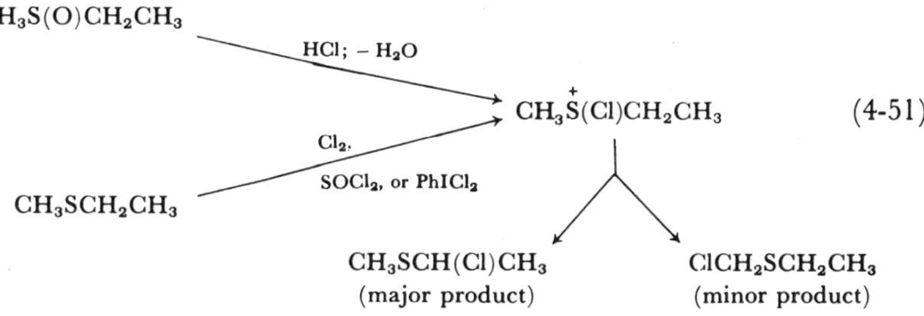

pentane shows an isotope effect (k_H/k_D) of 5.1. Concerted loss of HCl from the chlorosulfonium ion giving the cyclic alkylidenesulfonium ion has been suggested in this case (Wilson and Albert, 1973), an explanation also consistent with the selectivity of chlorination of ethylmethyl sulfide shown in Table 4-6.

Table 4-6. Product Ratios in the Chlorination of Ethylmethyl Sulfide and Sulfoxide

Substrate	Conditions	Product ratio ($CH_3SCH(Cl)CH_3 : ClCH_2SC_2H_5$)	Reference
$CH_3SC_2H_5$	Cl_2, −25°	Single product: $CH_3SCH(Cl)CH_3$	Vilsmaier and Sprugel (1971)
	$SOCl_2$, 4°	5	Tuleen and Stephens (1969)
	$SOCl_2$, 40°	2.8	Vilsmaier and Sprugel (1971)
	$PhICl_2$, −20°	3	Vilsmaier and Sprugel (1971)
$CH_3S(O)C_2H_5$	HCl	3	Rynbrandt (1971)

Convincing evidence for the intermediacy of α-thiocarbocations in Pummerer reactions is provided by the following results (Eqs. 4-52 to 4-54).

Ac_2O → OAc → OAc, $\overset{+}{S}$–OAc → OAc, + → OAc, OAc (27%)

under experimental conditions

OAc, +, H → $CH(OAc)_2$ (27%)

(4-52)

(Parham and Bhausar, 1963)

$ArSC_2H_5(O^{14}CH_3)^+ BF_4^-$ + $Ar'SC_2H_5(OCH_3)^+ BF_4^-$ —acetone, N→ $[Ar\overset{+}{S}{=}CHCH_3 / Ar'\overset{+}{S}{=}CHCH_3]$ →

1389 cpm/μmol

$ArSCH(O^{14}CH_3)CH_3$ + $Ar'SCH(O^{14}CH_3)CH_3$

999 cpm/μmol 423 cpm/μmol

(4-53)

(Johnson and Phillips, 1969)

$CH_3\overset{+}{S}(OC_2H_5)CH_3$ + CH_3O, OCH_3 —$(C_2H_5)_3N$→ OCH_3, OCH_3, CH_2SCH_3 (12%)

(4-54)

(Olofson and Marino, 1971)

Some examples of synthetic utility involving intramolecular trapping of α-thiocarbocations from Pummerer reactions are shown in Eqs. 4-55 to 4-59. In suitable cases, the α-thiocarbocations may undergo proton loss

$\xrightarrow[\text{100°, 2 hr}]{Ac_2O}$ ($>95\%$, R = H or CH_3)

(4-55)

(Numata and Oae, 1972)

$\xrightarrow{CF_3COOH}$ (70%)

(4-56)

(Oikawa and Yonemitsu, 1972)

$\xrightarrow{CCl_3CO_2H}$ (66%)

(4-57)

(Oikawa and Yonemitsu, 1972)

$\xrightarrow{RC(O)CH_2S(O)CH_3}$ $\xrightarrow{HOAc}$

($\sim 15\%$, $R = C_6H_{13}$, etc.)
(a pteridine)

(4-58)

(Rosowsky and Chen, 1973)

$$\text{(2-}CH_2S(O)CH_3\text{-6-}CH_3C_6H_3OH) \xrightarrow[H_3PO_4]{DCC} ArCH_2\overset{+}{S}(CH_3)OC(=NR)NHR \longrightarrow$$

$$\text{(}CH_2\overset{+}{S}{=}CH_2\text{, OH, }CH_3\text{ aryl)} \longrightarrow \text{(benzoxathiin, }CH_3\text{)} \quad (20\%)$$

(4-59)

(Olofson and Marino, 1971)

affording olefin. Thus, a synthetically useful synthesis of α-alkenyl sulfides has been devised based on the reaction of sulfides with benzoic anhydride (Eqs. 4-60 and 4-61). Where the Pummerer sulfonium ion precursor is

$$n\text{-}C_4H_9S(O)C_4H_9\text{-}n \xrightarrow[\text{reflux, 6 hr}]{Bz_2O/C_6H_6} n\text{-}C_4H_9SCH{=}CHC_2H_5 \quad (96\%) \tag{4-60}$$

(Horner, 1959)

$$\text{(thiane S-oxide)} \xrightarrow[\text{reflux, 14 hr}]{Bz_2O/C_6H_6} \text{(dihydrothiopyran)} \quad (54\%) \tag{4-61}$$

(Parham *et al.*, 1964)

particularly sterically crowded, evidence suggests that deprotonation may occur in a direction *opposite* to that normally observed (Jones *et al.*, 1972), e.g., Eq. 4-62. A novel variant of the Pummerer reaction involving α-silylsulfoxides occurs under particularly mild conditions (Eq. 4-63; Brook and Anderson, 1968; Vedejs and Mullins, 1975). Other more complex variants of the Pummerer rearrangement include vinylogous rearrangements and rearrangements involving disulfide *S*-oxides (thiosulfinates) (Eqs. 4-64 to 4-66). A Pummerer-type reaction may also be involved in the α-chlorination of sulfoxides (Eq. 4-67) (Durst *et al.*, 1973).

When combined with carbanionic reactions of sulfoxides, the Pummerer reaction provides a particularly valuable means of introducing a new carbonyl function, as demonstrated, for example, by the following economical synthesis of the amino acid reagent, ninhydrin (Eq. 4-68) (Russell *et al.*, 1969).

Key steps of an oxidative decarboxylation sequence recently described by Trost are closely related to the Pummerer mechanism in a novel way (Eq. 4-69; Trost and Tamaru, 1975).

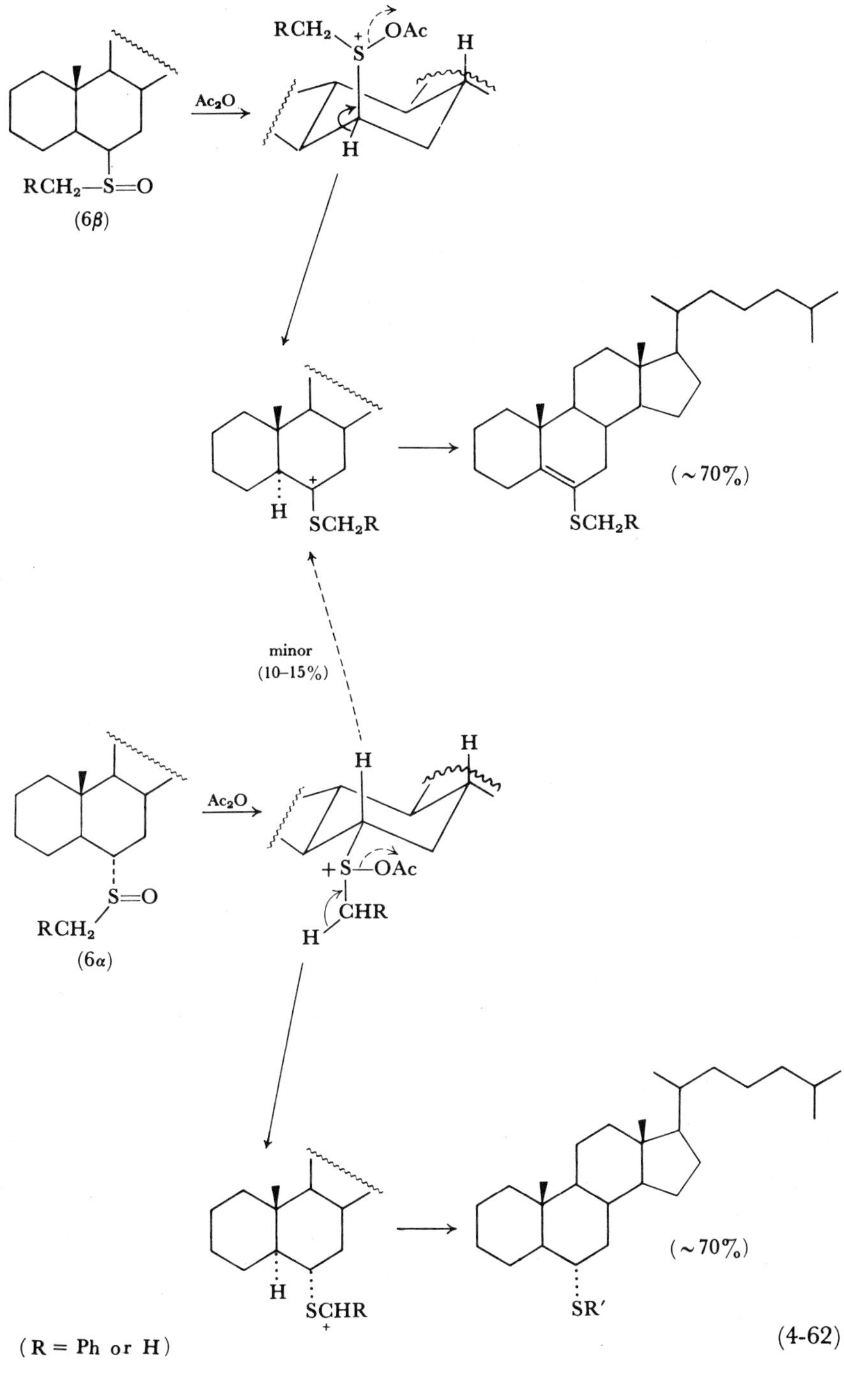

$$\text{(6}\beta\text{)}\ \xrightarrow{Ac_2O}\ \cdots \longrightarrow (\sim 70\%)$$

$RCH_2—S=O$

(6β)

RCH_2 $\overset{+}{S}$ OAc

H

SCH_2R

SCH_2R

(~70%)

minor
(10–15%)

$(6\alpha)\ \xrightarrow{Ac_2O}$

RCH_2 S=O

(6α)

+S—OAc

CHR

H

$\overset{+}{\text{SCHR}}$

SR′

(~70%)

(R = Ph or H) (4-62)

$$PhS(=O)CH_2Si(CH_3)_3 \longrightarrow Ph\overset{+}{S}(OSi(CH_3)_3)-\overset{-}{C}H_2 \longrightarrow Ph\overset{+}{S}{=}CH_2 \;\; {}^{-}OSi(CH_3)_3 \longrightarrow PhSCH_2OSi(CH_3)_3 \quad (79\%) \qquad (4\text{-}63)$$

$$\text{(cyclopentylidene)C(S(O)Ph)}CO_2CH_3 \xrightarrow[\Delta]{H^+} \text{(cyclopentylidene)C(}^{+}S(OH)Ph\text{)}CO_2H \longrightarrow \text{(cyclopentenyl)C(}{=}\overset{+}{S}Ph\text{)C(=O)OH} \xrightarrow{-H^+} \text{SPh-bicyclic lactone} \quad (53\%) \qquad (4\text{-}64)$$

(Kosugi *et al.*, 1975)

$$\text{1,2-dithiolane 1-oxide}-(CH_2)_4COOH \xrightarrow{Ac_2O} \text{3-AcO-1,2-dithiolane}-(CH_2)_4COOH \qquad (4\text{-}65)$$

(Saito and Fukui, 1966)

$$2CH_3S(=O)SCH_3 \xrightarrow[-\,CH_3SOH]{H^+,\ C_6H_6/H_2O} CH_3S(=O)-\overset{+}{S}(SCH_3)-CH_2-H \longrightarrow \longrightarrow$$

$$CH_3S(=O)CH_2SCH_3 \; (84\%) \xrightarrow{Ac_2O} CH_3SCH(OAc)SSCH_3 \; (47\%) \qquad (4\text{-}66)$$

(Block and O'Connor, 1974)

$$CH_3S(=O)CH_2R \xrightarrow{SO_2Cl_2} CH_3S^{+}(Cl)(=O)CH_2R \xrightarrow{-\,HCl} CH_3\overset{+}{S}(=O){=}CHR \xrightarrow{Cl^-} CH_3S(=O)CHClR \qquad (4\text{-}67)$$

CH_3ONa ; CH_3ONa

$CO_2C_2H_5$; $CO_2C_2H_5$; $+ CH_3\overset{O}{\overset{\|}{S}}CH_3$; $\overset{O}{\overset{\|}{C}}CH_2\overset{O}{\overset{\|}{S}}CH_3$; $CO_2C_2H_5$

$S(O)CH_3$; H ; H^+ ; SCH_3 ; OH ; HCl, H_2O ; OH ; OH

(79%)
Ninhydrin

(4-68)

CH_3O ; CO_2H ; $2LiN(i\text{-}C_3H_7)_2$, CH_3SSCH_3 ; CH_3O ; SCH_3 ; CO_2H ; NCS, $NaHCO_3$, C_2H_5OH

Cl ; $S^+—CH_3$; $C—O^-$; O ; $=\overset{+}{S}CH_3$; SCH_3 ; OC_2H_5 ; CH_3O ; O

Corey prostaglandin intermediate; overall yield ~70%.

(4-69)

$CH_3S(O)_nCH_3 \xrightarrow{Cl_2} CH_3\overset{+}{S}(O)_nCH_3$ (with Cl on S) $\xrightarrow{C_6H_{11}CH_2OH}$

$CH_3\overset{+}{S}(O)_nCH_3$ (with $OCH_2C_6H_{11}$ on S) $\xrightarrow{(C_2H_5)_3N}$ $CH_2—\overset{+}{S}(O)_nCH_3$, H, O, CH (cyclohexyl) $\longrightarrow$

$CH_3S(O)_nCH_3 +$ cyclohexyl—CHO (4-70)

(n = 0, 80%
n = 1, 94%)
(Corey and Kim, 1972, 1973)

In addition to undergoing loss of HX, sulfonium or sulfoxonium derivatives of the type $RS^+(X)CHR_2$ and $RS^+(O)(X)CHR_2$ can undergo other reactions such as displacement of group X by some other group Y and, when X = alkoxy, α,β'-elimination to afford carbonyl compounds. These processes, typified by the reaction shown in Eq. 4-70 are of obvious synthetic utility and have already been covered in some detail in Chapter 3 (p. 112 ff). If the sulfoxonium derivative $RS^+(O)(X)CHR_2$ contains a suitably disposed hydroxyl group, intramolecular displacement can occur with some surprising results (Eq. 4-71).

$$PhS(O)(CH_2)_5OH \xrightarrow[SO_2Cl_2]{} PhS(O)CH(Cl)(CH_2)_4OH \quad \text{but} \quad PhS(O)(CH_2)_3OH \xrightarrow{Br_2}$$

$$PhS^+(O)(Br)\text{—}(CH_2)_3OH \longrightarrow PhS^+(=O)\text{—}O\text{(cyclic)} \; Br^- \longrightarrow PhSO_2(CH_2)_3Br \qquad (4\text{-}71)$$

(Durst *et al.*, 1973)

4.7 Beckmann Fragmentation of α-Alkylthioketoximes

While the tosylate of thian-4-one oxime (**4-18**) undergoes a normal Beckmann rearrangement, thian-3-one *anti*-oxime tosylate (**4-19**) undergoes a concerted fragmentation to methylenesulfonium ion **4-20**. Ion **4-20** can be trapped with ethanol, or thionyl chloride, hydrolyzed to formaldehyde or can undergo recyclization eventually giving lactam **4-21** (Grob and Ide, 1974; Hill and Cullison, 1973). In the case of thian-3-one *syn*-oxime tosylate **4-22** Beckmann rearrangement and fragmentation occur in the ratio of 4:1.

It is estimated that in α-alkythio ketoximes of *anti* stereochemistry (e.g., **4-19**) sulfur enhances the rate of fragmentation over rearrangement by a factor of 10^2 to 10^3. A much larger "frangomeric" effect is seen with α-amino functions (rate enhancements as large as 10^8 have been observed) (Grob and Ide, 1974). An unusual reaction occurs in the case of α-methythioisobutyrophenone *anti*oxime tosylate **4-23**. In this case, fragmentation is accompanied by cyclization to the 1,2-thiazetin-1-ium ion, **4-24** (Grob and Ide, 1974).

The Beckmann fragmentation of α-thio ketoximes has been used to synthetic advantage in the synthesis of the alkaloids corynantheine (Eq. 4-72; Autrey and Scullard, 1968) and samandarine (Eq. 4-73; Shimizu, 1972), and in the preparation of 2-cyano aldehydes (Eq. 4-74; Ohno *et al.*, 1966) and δ-lactones (Eq. 4-75; Grieco and Hiroi, 1973).

4-18 —dioxane–water→ (98%)

4-19 → [ ]‡ →

C_2H_5OH → (92%)

4-20 → —H_2O→ NH (6%) **4-21**

$SOCl_2$

H_2O

CH_2O (72)% + ···

(72%; oxime precursor to **4-19** used rather than **4-19** itself in this one reaction)

4-22 → (81%) + —C_2H_5OH→ (18%)

CH_2O (19%) + ···

$CH_3\overset{+}{S}{=}C(CH_3)_2$ + PhCN (53%)

CH_3S N—OTs Ph **4-23** → CH_3—$\overset{+}{S}$—N Ph **4-24** $\xrightarrow{H_2O}$

H_3C S=NH C=O Ph

$\xrightarrow{H_2O}$ $CH_3SC(CH_3)_2CPh$ (O, O) (28%)

$\xrightarrow{-NH}$ $CH_3SC(CH_3)_2CPh$ (O) (13%)

Yohimbone $\xrightarrow{(1)\ HCO_2C_2H_5,\ NaH;\ (2)\ ArSO_2SCH_3,\ CH_3ONa;\ (3)\ NH_2OH}$ [oxime, SCH_3, HO—N] $\xrightarrow{SOCl_2}$

[CH—CH=$\overset{+}{S}CH_3$, H, CN] $\xrightarrow{-H^+}$ [SCH_3, CN] (57%) $\xrightarrow{RaNi}$

[CH=CH_2, CN] $\xrightarrow{\text{several steps}}$ [CH_3O_2C, OCH_3, H, N, H]

Corynantheine

(4-72)

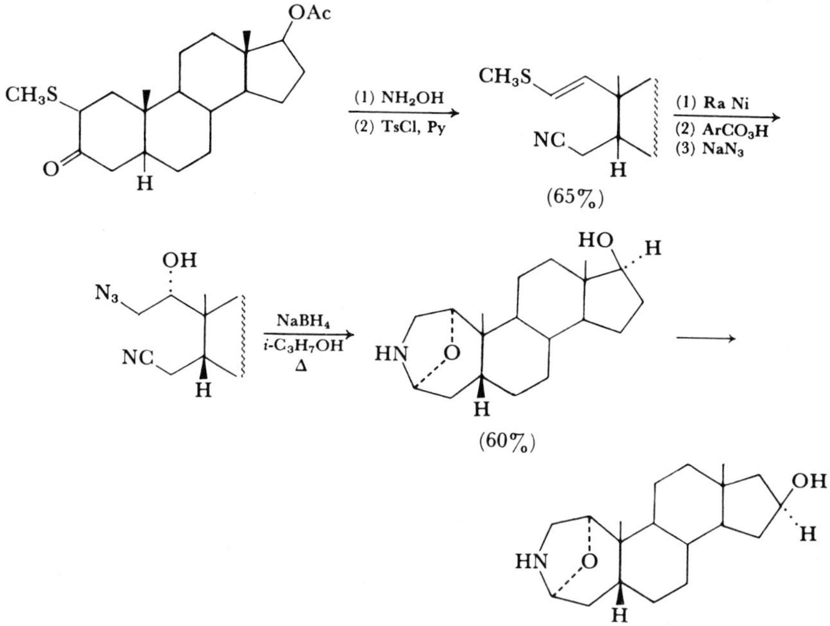

OAc

CH_3S ... (1) NH_2OH (2) TsCl, Py → CH_3S ... NC ... H (65%) (1) Ra Ni (2) $ArCO_3H$ (3) NaN_3 →

OH N_3 ... NC ... H $NaBH_4$ i-C_3H_7OH Δ → HO H HN O H (60%) →

HN O H OH H

Samandarine

(4-73)

$(CH_2)_n$ $\xrightarrow[H^+]{NOCl}$ $(CH_2)_n$ NOH Cl $\xrightarrow{C_2H_5S^-}$ $(CH_2)_n$ NOH SC_2H_5 $\xrightarrow[(2)\ H_2O]{(1)\ PCl_5}$

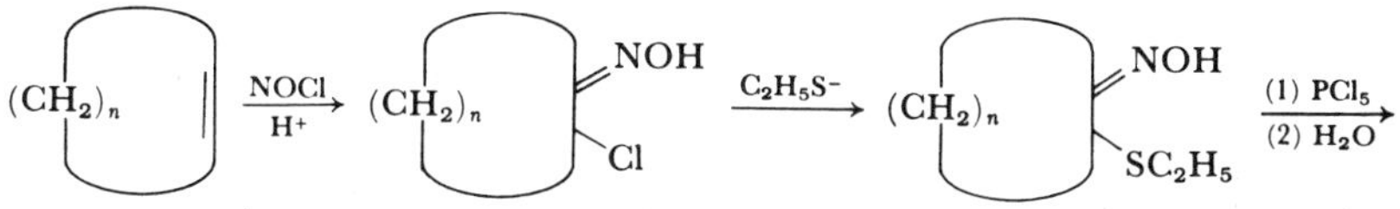

$NC(CH_2)_nCHO$ (n = 4, 6, 10; overall yields 14–44%)

(4-74)

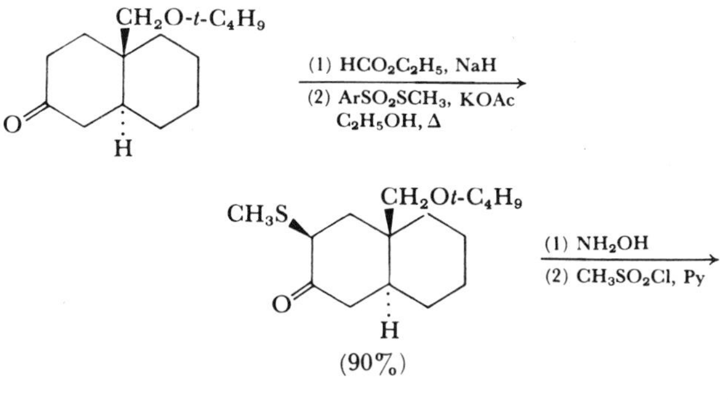

CH_2O-t-C_4H_9 O H (1) $HCO_2C_2H_5$, NaH (2) $ArSO_2SCH_3$, KOAc C_2H_5OH, Δ →

CH_3S CH_2Ot-C_4H_9 O H (90%) (1) NH_2OH (2) CH_3SO_2Cl, Py →

$CH_2Ot\text{-}C_4H_9$, CH_3S, H, CN

(1) Ra Ni
(2) KOH, H_2O
(3) $ArSO_3H—C_6H_6$

O, O, H
(54%)

O, O, H, CH_2
(50%) (4-75)

4.8 Carbocations Containing the Sulfinyl and Sulfonyl Groups

Compounds containing a leaving group α to a sulfinyl or sulfonyl function are resistant to S_N1 processes, at least in the more usual polar solvents, because of the carbocation destabilizing effects of the electropositive sulfur substituents. α-Halosulfoxides show S_N2 reactivity similar to that of simple alkyl halides (see Table 4-7 and Cinquini *et al.*, 1976), whereas α-halosulfones are less reactive and α-halosulfides substantially more reactive.

Conjugative effects involving the divalent sulfur in entry 2 are diminished in entry 3, while inductive rather than conjugative effects seem to be important for the sulfoxides in entries 4 and 5. For all of the compounds in Table 4-7, it is probable that carbocation character is poorly developed in the transition states for displacement (Bordwell and Jarvis, 1964).

Sometimes the diminished S_N2 reactivity of α-halosulfoxides compared to α-halosulfides can be used to advantage. For example, bis(chloromethyl)-sulfoxide reacts smoothly with aqueous sodium sulfide giving 1,3-dithietane 1-oxide which can be reduced to 1,3-dithietane; analogous reaction of

Table 4-7. Relative Rates of Reaction with Potassium Iodide in Acetone at 75°[a]

Entry	Compound	k_{rel}
1	$n\text{-}C_4H_9Cl$	1.0
2	$C_6H_5SCH_2Cl$	540
3	$p\text{-}NO_2\text{-}C_6H_4SCH_2Cl$	104
4	$C_6H_5S(O)CH_2Cl$	0.25
5	$p\text{-}NO_2\text{-}C_6H_4S(O)CH_2Cl$	0.8
6	$C_6H_5SO_2CH_2Cl$	<0.02

[a] Bordwell and Brannen (1964).

bis(chloromethyl)sulfide with sodium sulfide leads only to polymer (Eq. 4-76) (Block *et al.*, 1976). Presumably bis(chloromethyl)sulfide alkylates any 1,3-dithietane formed giving polymer; similar reaction between 1,3-dithietane 1-oxide and bis(chloromethyl)sulfoxide is less favorable.

$$ClCH_2S(=O)CH_2Cl + Na_2S \longrightarrow \text{1,3-dithietane 1-oxide}$$
$$\downarrow$$
$$\text{1,3-dithietane} \not\longleftarrow Na_2S + ClCH_2SCH_2Cl \qquad (4\text{-}76)$$

Steric and field repulsion between the electronegative oxygens and the nucleophile are thought to be responsible for the sluggish S_N2 reactivity of α-halosulfones (see Table 4-7, entry 6; Meyers, 1962; Bordwell and Brannen, 1964). With exceptional leaving groups such as nitrogen (in $RSO_2CHN_2^+$) or "triflate" (Eq. 4-77), or with particularly strong nucleophiles such as mercaptide anions, intermolecular nucleophilic substitution at α-sulfonyl carbon atoms can be readily achieved. In the case of Eq. 4-78, nucleophilic attack at the leaving group has been postulated (Bordwell and Jarvis, 1968; Hovius and Engberts, 1972). Intramolecular nucleophilic substitution at

$$ArSO_2CH_2OSO_2CF_3 \xrightarrow[40°,\ 2\ hr]{Br^-,\ DMF} ArSO_2CH_2Br \quad (92\%,\ Ar = p\text{-tolyl}) \qquad (4\text{-}77)$$

$$PhSO_2CHClPh + PhS^- \xrightarrow[CH_3OH,\ DMF]{70°,\ 0.5\ hr} [PhSO_2\bar{C}HPh] + [PhSCl]$$
$$\swarrow \qquad\qquad \searrow\!\swarrow$$
$$PhSO_2CH_2Ph \qquad\qquad PhSO_2CH(SPh)Ph \qquad (4\text{-}78)$$

α-sulfonyl carbon atoms is quite facile (in contrast to its intermolecular counterpart) forming the basis of the extensively studied Ramberg–Bäcklund reaction (see Section 2.7, p. 75).

Alkylidene sulfoxonium ions (α-sulfinyl carbocations) have been postulated as intermediates in the Pummerer-like rearrangement of chlorosulfoxonium ions (Eq. 4-79; Durst *et al.*, 1973). Efforts to trap the carbonium ions or to extend the rearrangement to include acetoxysulfoxonium ions (Eqs. 4-80 and 4-81) have thus far proven unsuccessful (Durst *et al.*, 1973; Annunziata *et al.*, 1975).

$$CH_3S(O)CH_2R \xrightarrow{SO_2Cl_2} CH_3\overset{+}{S}(Cl)(O)CH_2R \longrightarrow CH_3S(O)\overset{+}{C}HR \longrightarrow CH_3S(O)CHClR \quad (4\text{-}79)$$

$$Ph\overset{+}{S}(Cl)(O)(CH_2)_5OH \longrightarrow PhS(O)\overset{+}{C}H(CH_2)_4OH \longrightarrow PhS(O)CHCl(CH_2)_4OH;\quad PhS(O)\overset{+}{C}H(CH_2)_4OH \not\longrightarrow PhS(O)\text{-(2-tetrahydropyranyl)} \quad (4\text{-}80)$$

$$Ph\overset{+}{S}(Cl)(O)CH_2CH_3 \xrightarrow{PhC^{18}O_2H} PhS(O)(^{18}OC(O)Ph)CH_2CH_3 \xrightarrow{H_2O} PhS(O)(^{18}O)CH_2CH_3;\quad PhS(O)(^{18}OC(O)Ph)CH_2CH_3 \not\longrightarrow PhS(O)\overset{+}{C}HCH_3 \longrightarrow PhS(O)CH(^{18}OC(O)Ph)CH_3 \quad (4\text{-}81)$$

Sulfoxides are readily protonated on oxygen and are quantitatively alkylated on oxygen affording stable alkoxysulfonium salts, e.g., as in Eq. 4-82 (Johnson and McCants, 1965). With methyl iodide–mercuric iodide, various sulfoxides give the thermodynamically more stable *S*-alkylated sulfoxonium salts rather than the product of *O*-alkylation (Eq. 4-83; Kamiyama *et al.*, 1973). Cyclic oxysulfonium salts have been formed through intramolecular trapping of carbocations by the sulfoxide group, for example, as shown below (Eq. 4-84; Hogeveen *et al.*, 1966). As might therefore be expected, the

$$R\text{-}C_5H_9S{=}O + (C_2H_5)_3O^+\ BF_4^- \longrightarrow R\text{-}C_5H_9\overset{+}{S}\text{-}OC_2H_5\ BF_4^- \quad (4\text{-}82)$$

$$PhS(O)CH_3 + CH_3I \xrightarrow[(2)\ AgClO_4]{(1)\ HgI_2} Ph\text{-}\overset{+}{S}(O)(CH_3)\text{-}CH_3\ ClO_4^-\ (73\%) \quad (4\text{-}83)$$

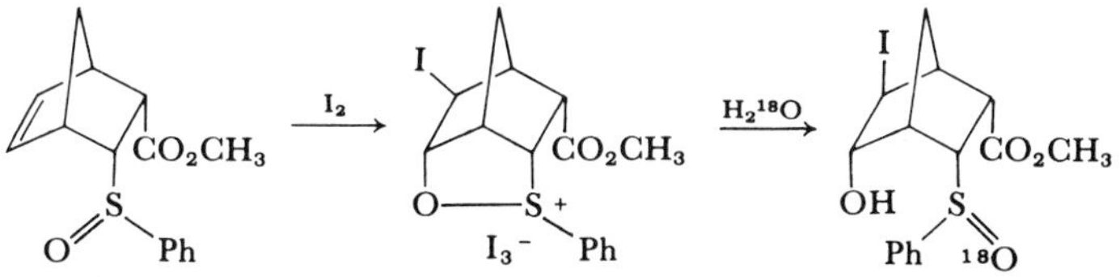

(4-84)

sulfinyl oxygen is sufficiently nucleophilic to provide anchimeric assistance in the solvolysis of various halosulfoxides. Thus, Martin has found *trans*-4-chlorothiane *S*-oxide **4-25** to solvolyze 630 times faster than the *cis* isomer **4-26** (Martin and Vebel, 1964). Generally, neighboring group participation

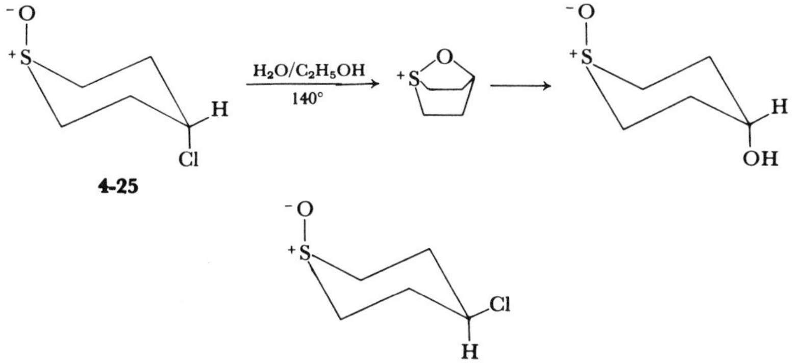

4-25

4-26

by sulfoxide becomes most significant when aprotic polar solvents are used, for the sulfoxide oxygen is strongly hydrogen bonded (and consequently blocked from participating) in protic solvents. Illustrative of participation by the sulfoxide group under different solvent conditions are the data in Table 4-8.

In contrast to sulfoxides, sulfones are protonated and *O*-arylated only with difficulty. Thus concentrated sulfuric acid only partially protonates sulfones; the protonated sulfones have a pK_a estimated to be about -12 (Hall and Robinson, 1964; Arnett and Douty, 1964). Thermal decomposition of aryl diazonium salts in the presence of sulfones has recently been shown to afford aryloxysulfoxonium salts (Eq. 4-85; Chalkley *et al.*, 1970). In view of the

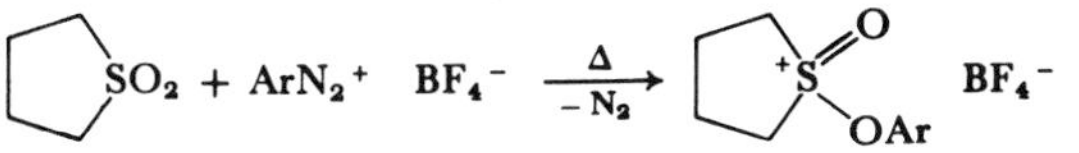

$$SO_2 + ArN_2^+ \ BF_4^- \xrightarrow[-N_2]{\Delta} \ {}^+S(=O)OAr \ BF_4^- \quad (4\text{-}85)$$

extremely low nucleophilicity of the sulfone group, it is not surprising that anchimeric assistance by this group has not been detected (cf. Table 4-8, entries 9 and 10; however also see Braverman and Reisman (1977) for intramolecular trapping of a carbocation by a sulfone group).

Table 4-8. Solvolysis of Chlorosulfoxides and Chlorosulfones

Entry	Compound	Relative rate		Reference
		80% aq. DMF, 35°	30% aq. DMF, 35°	
1	$t\text{-}C_4H_9Cl$	1.0	1.0	Cinquini *et al.* (1966)
2	$PhS(O)(CH_2)_3Cl$	408	0.67	Cinquini *et al.* (1966)
3	$PhS(O)CH_2C(CH_3)_2Cl$	16	—	Cinquini *et al.* (1966)
4	$PhS(O)(CH_2)_2C(CH_3)_2Cl$	100	1.6	Cinquini *et al.* (1966)
5	$PhS(O)(CH_2)_3C(CH_3)_2Cl$	210	—	Cinquini *et al.* (1966)
6	$PhS(O)(CH_2)_4C(CH_3)_2Cl$	0.25	—	Cinquini *et al.* (1966)
		80% aq. sulfolane, 65°; 80% aq. C_2H_5OH 65°		
7	CH_2Cl $CH_2S(O)C_2H_5$	1.0	1.0	Barbieri *et al.* (1968)
8	CH_2Cl $CH_2S(O)C_2H_5$	102	39	Barbieri *et al.* (1968)

9	CH_2Cl (benzene ring, para) $CH_2SO_2C_2H_5$	1.1	0.7	Barbieri *et al.* (1968)
10	CH_2Cl (benzene ring, ortho) $CH_2SO_2C_2H_5$	1.2	0.9	Barbieri *et al.* (1968)
		AcOH, 60°		
11	CH_2Cl (benzene ring, para) 1,3-dithiolan-2-yl (S, S)	1		Hojo *et al.* (1969)
12	CH_2Cl (benzene ring, ortho) 1,3-dithiolan-2-yl (S, S)	220		Hojo *et al.* (1969)

4.9 Carbocations Containing the Disulfide Group

Only recently has the carbocation-stabilizing ability of the disulfide group been examined. Solvolysis studies on chloromethyl methyl disulfide indicate that this compound hydrolyzes 6800 times slower than chloromethyl methyl sulfide, suggesting the poorer carbocation-stabilizing ability of the disulfide group compared to the sulfide group (Block, 1974). This trend is also supported by electron impact data (Block, 1974). It has been suggested (Block, 1974) that the near 90° C—S—S—C dihedral angle adopted by acyclic disulfides is also favored for acyclic α-disulfide carbocations and that this conformation precludes charge delocalization by the second (distant) sulfur while at the same time diminishing by inductive effects the electron availability on the first (nearest) sulfur (see **4-27**). Pummerer-type rearrangements

4-27

of alkyl thiosulfinates are thought to involve α-disulfide carbocations as key intermediates (Eq. 4-86; Block and O'Connor, 1974b). Neighboring group participation by a β-disulfide group is seen in Eq. 4-87 (Karimova *et al.*, 1973).

$$2RS(O)SCH_3 \xrightarrow{-RSOH} RS(O){-}\overset{+}{S}(CH_3){-}SCH_3 \longrightarrow RSOH + CH_2{=}\overset{+}{S}SCH_3 \longrightarrow RS(O)CH_2SSCH_3$$

$$CH_2{=}\overset{+}{S}SCH_3 \xrightarrow{RSO_2^-} RSO_2CH_2SSCH_3 \qquad (4\text{-}86)$$

SSR, Cl, C(O)NH₂ ⇌ ⁺S(SR) episulfonium, C(O)NH₂, Cl⁻ ⇌ SSR, C(O)NH₂, Cl (4-87)

General Reference

Capon, B., and McManus, S. P. (1976). Neighboring Group Participation, Vol. I, Plenum, New York.

References

Anderson, N. H., Yamamoto, Y., and Denniston, A. D. (1975). *Tetrahedron Lett.* 4547.
Annunziata, R., Cinquini, M., and Colonna, S. (1975). *J. Chem. Soc. Perkin Trans.* **I**, 282.
Arnett, E. M., and Douty, C. (1964). *J. Am. Chem. Soc.* **86**, 409.
Autrey, R. L., and Scullard, P. W. (1968). *J. Am. Chem. Soc.* **90**, 4917, 4924.
Bailey, F. P., and Taylor, R. J. (1971). *J. Chem. Soc.* (*B*) 1441.
Barbieri, G., Cinquini, M., and Colonna, S. (1968). *Boll. Sci. Fac. Chim. Ind. Bologna* **26**, 309; see also Montanari, F. (1971). *Int. J. Sulfur Chem. C* **6**, 137.
Bechgaard, K., Parker, V. D., and Pederson, C. Th. (1973). *J. Am. Chem. Soc.* **95**, 4373.
Bernardi, F., Csizmadia, I. G., Schlegel, H. B., and Wolfe, S. (1975). *Can. J. Chem.* **53**, 1144.
Blandamer, M. J., Golinkin, H. S., and Robertson, R. E. (1969). *J. Am. Chem. Soc.* **91**, 2678.
Block, E. (1974). *J. Org. Chem.* **39**, 734.
Block, E., and O'Connor, J. (1974). *J. Am. Chem. Soc.* **96**, 3929.
Block, E., Corey, E. R., Penn, R. E., Renken, T. R., and Sherwin, P. F. (1976). *J. Am. Chem. Soc.* **98**, 5715.
Böhme, H. (1941). *Chem. Ber.* **74**, 248.
Böhme, H., and Sell, K. (1948). *Chem. Ber.* **81**, 123.
Böhme, H., Fischer, H., and Frank, R. (1949). *Justus Liebigs Ann. Chem.* **563**, 54.
Bordwell, F. G., and Jarvis, B. B. (1968). *J. Org. Chem.* **33**, 1182.
Bordwell, F. G., and Brannen, W. T., Jr. (1964). *J. Am. Chem. Soc.* **86**, 4645.
Braverman, S., and Reisman, D. (1977). *Tetrahedron Lett.* 1753.
Brook, A. G., and Anderson, D. G. (1968). *Can. J. Chem.* **46**, 2115.
Brown, H. C., Okamoto, Y., and Inukai, T. (1958). *J. Am. Chem. Soc.* **80**, 4964.
Burgoine, K. T., Davies, S. G., Peagram, M. J., and Whitman, G. H. (1974). *J. Chem. Soc. Perkin Trans.* **I**, 2629.
Burighel, A., Modena, G., and Tonellato, V. (1972). *J. Chem. Soc., Perkin Trans.* **II**, 2026.
Capozzi, G., DeLucchi, O., Lucchini, V., and Modena, G. (1975). *Chem. Commun.* 248.
Carey, F. A., and Court, A. S. (1972). *J. Org. Chem.* **37**, 1926.
Chalkley, G. R., Snodin, D. J., Stevens, G., and Whiting, M. C. (1970). *J. Chem. Soc. C* 682.
Cinquini, M., Colonna, S., and Montanari, F. (1966). *Tetrahedron Lett.* 3181.
Cinquini, M., Colonna, S., Landini, D., and Maia, A. M. (1976). *J. Chem. Soc. Perkin Trans.* **II**, 996.
Coffen, D. L., and Lee, M. L. (1970). *J. Org. Chem.* **35**, 2077.
Cohen, T., Herman, G., Falck, J. R., and Mura, A. J., Jr. (1975a). *J. Org. Chem.* **40**, 812.
Cohen, T., Kuhn, D., and Falck, J. R. (1975b). *J. Am. Chem. Soc.* **97**, 4749.
Connors, T. A. (1974). *Fort. Chem. Forsch.* **52**, 141.
Corey, E. J., and Block, E. (1966). *J. Org. Chem.* **31**, 1663.
Corey, E. J., and Kim, C. V. (1972). *J. Am. Chem. Soc.* **94**, 7586.
Corey, E. J., and Kim, C. V. (1973). *Tetrahedron Lett.* 919.
Corey, E. J., and Walinsky, S. W. (1972). *J. Am. Chem. Soc.* **94**, 8932.
Csizmadia, I. G., Duke, A. J., Lucchini, V., and Modena, G. (1974). *J. Chem. Soc. Perkin Trans.* **II**, 1808.
deGroot, A., Boerma, J. A., and Wynberg, H. (1968). *Tetrahedron Lett.* 2365.
deWaard, E. R., Reus, H. R., Huisman, H. O. (1973). *Tetrahedron Lett.* 4315.
Dittmer, D. C., Takahashi, K., and Davis, F. A. (1967). *Tetrahedron Lett.* p. 4061.
Durst, T. (1969). *In* "Advances in Organic Chemistry" (E. C. Taylor, and H. Wynberg, eds.), Vol. 6. Wiley (Interscience), New York.

Durst, T., Tin, K. C., and Marcil, M. J. V. (1973). *Can. J. Chem.* **51**, 1704; see also Hanessian, S., Yang-Chung, G., Lavallee, P., and Pernet, A. G., *J. Am. Chem. Soc.* **94**, 8929 (1972).
Field, F. H., and Weeks, D. P. (1970). *J. Am. Chem. Soc.* **92**, 6521.
Fletcher, J. H., Dermer, O. C., and Fox, R. B. (eds.) (1974). "Nomenclature of Organic Compounds," Adv. in Chem. Ser. 126. Am. Chem. Soc., Washington, D.C.
Eschenmoser, A. (1970). *Quart. Rev.* **24**, 366.
Gassman, P. G., and Drewes, H. R. (1974). *J. Am. Chem. Soc.* **96**, 3002.
Gassman, P. G., Cue, B. W. Jr., and Luh, T.-Y. (1977). *J. Org. Chem.* **42**, 1344.
Goering, H. L., and Howe, K. L. (1957). *J. Am. Chem. Soc.* **79**, 6542.
Gompper, R., and Jersak, V. (1973). *Tetrahedron Lett.* 3409.
Gompper, R., and Kutter, E. (1965). *Chem. Ber.* **98**, 1365.
Goodman, L., and Taft, R. W. (1965). *J. Am. Chem. Soc.* **87**, 4385.
Gratz, R. F., and Wilder, P., Jr. (1970). *Chem. Commun.* 1449.
Grieco, P. A., and Hiroi, K. (1973). *Tetrahedron Lett.* 1831.
Grob, C. A., and Ide, J. (1974). *Helv. Chim. Acta* **57**, 2562, 2571.
Hall, J. K., and Robinson, E. A. (1964). *Can. J. Chem.* **42**, 113.
Hansen, D. W., Jr., and Olofson, R. A. (1971). *Tetrahedron* **27**, 4221.
Hartzler, H. D. (1973). *J. Am. Chem. Soc.* **95**, 4379.
Hayami, J., Tanaka, N., Kurabayashi, S., Kotani, Y., and Kaji, A. (1971). *Bull. Chem. Soc. Jpn.* **44**, 3091.
Hill, R. K., and Cullison, D. A. (1973). *J. Am. Chem. Soc.* **95**, 2923.
Hogeveen, H., Maccagnani, G., and Montanari, F. (1966). *J. Chem. Soc.* (*C*) 1585.
Hogeveen, H., Kellogg, R. M., and Kuindersma, K. A. (1973). *Tetrahedron Lett.* 3929.
Hojo, M., Ichi, T., Tamaru, Y., and Yoshida, Z. (1969). *J. Am. Chem. Soc.* **91**, 5170.
Horner, L., and Kaiser, P. (1959). *Justus Liebigs Ann. Chem.* **626**, 19.
Hovius, K., and Engberts, J. B. F. N. (1972). *Tetrahedron Lett.* 2477.
Ichikawa, T., Owatari, H., and Kato, T. (1970). *J. Org. Chem.* **35**, 344.
Ikegami, S., Asai, T., Tsuneoka, K., Matsumura, S., and Akaboshi, S. (1974). *Tetrahedron* **30**, 2087.
Ireland, R. E., and Smith, H. A. (1959). *Chem. Ind.* (*London*) 2152.
Johnson, C. R., and McCants, D., Jr. (1965). *J. Am. Chem. Soc.* **87**, 5404.
Johnson, C. R., and Phillips, W. G. (1969). *J. Am. Chem. Soc.* **91**, 682.
Jones, D. N., Helmy, E., and Whitehouse, R. D. (1972). *J. Chem. Soc. Perkin* **I**, 1329.
Kamiyama, K., Minato, H., and Kobayashi, M. (1973). *Bull. Chem. Soc. Jpn*, **46**, 2255.
Karimova, N. M., Lin'kova, M. G., Kil'disheva, O. V., and Knunyants, I. L. (1973). *Khim. Geterotsikl. Soedin.* p. 8; *Chem. Abstr.* **78**, 123970.
Kice, J. L., and Morkved, E. H. (1963). *J. Am. Chem. Soc.* **85**, 3472.
Kosugi, H., Uda, H., and Yamagiwa, S. (1975). *Chem. Commun.* 192.
Krabbenhoft, H. O., Wiseman, J. R., and Quinn, C. B. (1974). *J. Am. Chem. Soc.* **96**, 258.
Krimer, M. Z., Smit, V. A., and Shamshurin, A. A. (1973). *Dokl. Akad. Nauk SSSR* **208**, 867.
Macnicol, D. D., and McKendrich, J. J. (1973). *Tetrahedron Lett.* 2593.
Marshall, J. A., and Roebke, H. (1969). *J. Org. Chem.* **34**, 4188.
Martin, J. C., and Vebel, J. J. (1964). *J. Am. Chem. Soc.* **86**, 2936.
McManus, S. P., and Pittman, C. U., Jr. (1973). *In* "Organic Reactive Intermediates" (S. P. McManus, ed.), Chapter 4. Academic Press, New York.
Meerwein, H., Zenner, K.-F., and Gipp, R. (1965). *Justus Liebigs Ann. Chem.* **688**, 67.
Meyers, C. Y. (1962). *Tetrahedron Lett.* 1125.
Modena, G., Scorrano, G., and Tonellato, V. (1973). *J. Chem. Soc. Perkin Trans.* **II**, 493.

Nakayama, J., Fujiwara, K., and Hoshino, M. (1975). *Chem. Lett.* (*Jpn.*). 1099.
Noe, E. A. (1977). *J. Am. Chem. Soc.* **99**, 2803.
Numata, T., and Oae, S. (1972). *Chem. Ind.* (*London*) 726.
Ohno, M., Naruse, N., Torimitsu, S., and Teresawa, I. (1966). *J. Am. Chem. Soc.* **88**, 3168.
Oikawa, Y., and Yonemitsu, O. (1972). *Tetrahedron Lett.* 3393; see also Oikawa, Y., and Yonemitsu, O. (1976). *J. Chem. Soc. Perkin Trans.* **I**, 1479.
Olofson, R. A., and Marino, J. P. (1971). *Tetrahedron* **27**, 4195.
Olofson, R. A., Walinsky, S. W., Marino, J. P., and Jernow, J. L. (1968). *J. Am. Chem. Soc.* **90**, 6554.
Parham, W. E., Gordon, I., and Swalen, J. D. (1952). *J. Am. Chem. Soc.* **74**, 1824.
Parham, W. E., and Bhausar, M. D. (1963). *J. Org. Chem.* **28**, 2686.
Parham, W. E., Christensen, L., Groen, S. H., and Dodson, R. M. (1964). *J. Org. Chem.* **29**, 2211.
Paquette, L. A., Meehan, G. V., and Wise, L. D. (1969). *J. Am. Chem. Soc.* **91**, 3231.
Roof, G. L., and Tucker, W. P. (1968). *J. Org. Chem.* **33**, 3333.
Rosowsky, A., and Chen, K. K. N. (1973). *J. Org. Chem.* **38**, 2073.
Russell, G. A., and Mikol, G. J. (1968). *In* "Mechanisms of Molecular Migrations" (B. S. Thyagarajan, ed.), Vol. I, p. 157ff. Wiley (Interscience), New York.
Russell, G. A., and Sabourin, E. T. (1969). *J. Org. Chem.* **34**, 2336.
Russell, G. A., Sabourin, E. T., and Hamprecht, G. (1969). *J. Org. Chem.* **34**, 2339.
Rynbrandt, R. H. (1971). *Tetrahedron Lett.* 3553.
Saito, I., and Fukui, S. (1966). *J. Vitaminol.* (*Kyoto*) **12**, 244.
Shimizu, Y. (1972). *Tetrahedron Lett.* 2919.
Stork, G., and Cheng, H. T. (1965). *J. Am. Chem. Soc.* **87**, 3784.
Stütz, P., and Stadler, P. A. (1972). *Helv. Chim. Acta* **55**, 75.
Tabushi, I., Tomaru, Y., Yoshida, Z., and Sugimoto, T. (1975). *J. Am. Chem. Soc.* **97**, 2886; also see Tsuji, T., Komeno, T., Itani, H., and Tanida, H. (1971). *J. Org. Chem.* **36**, 1648.
Taft, R. W., and Rakshys, J. W., Jr. (1965). *J. Am. Chem. Soc.* **87**, 4387.
Taft, R. W., Martin, R. H., and Lampe, F. W. (1965). *J. Am. Chem. Soc.* **87**, 2490.
Trost, B. M., and Tamaru, Y. (1975). *J. Am. Chem. Soc.* **97**, 3528.
Tucker, W. P., and Roof, G. L. (1967). *Tetrahedron Lett.* 2747.
Tuleen, D. L., and Stephens, T. B. (1969). *J. Org. Chem.* **34**, 31.
Uneyama, K., Torii, S., and Oae, S. (1971). *Bull. Chem. Soc. Jpn.* **44**, 815.
Vedejs, E., and Mullins, M. (1975). *Tetrahedron Lett.* 2017.
Vilsmaier, E., and Sprügel, W. (1971). *Justus Liebigs Ann. Chem.* **749**, 62.
Wilson, G. E., Jr., and Albert, R. (1973). *J. Org. Chem.* **38**, 2156.
Winstein, S., Allred, E., Heck, R., and Glick, R. (1958). *Tetrahedron* **3**, 1.
Yoshida, Z., Yoneda, S., Sugimoto, T., and Kikukawa, O. (1971). *Tetrahedron Lett.* 3999.

Chapter 5 | Sulfur-Containing Radicals

5.1 Introduction

Various sulfur-stabilized carbanions and carbocations can be preserved indefinitely under appropriately inert conditions at room temperature. Radicals (species having one or more unpaired electrons) centered either at sulfur (i.e., $RS\cdot$) or on carbon adjacent to sulfur (i.e., $RSCH_2\cdot$) are most often highly reactive and are encountered only as fleeting intermediates in chemical reactions. Even if radical–molecule reactions can be suppressed, radicals cannot be obtained in any appreciable concentration at ambient temperatures because of very rapid self-annihilation by dimerization. Radical dimerizations generally proceed with diffusion-controlled rate constants as in the example of Eq. 5-1 (Graham *et al.*, 1964). This reaction is favored

$$2CH_3S\cdot \longrightarrow CH_3SSCH_3 \qquad (5\text{-}1)$$
$$k = 3 \times 10^{10}\ M^{-1}\ \text{sec}^{-1}$$

thermodynamically since the heat of formation of the new bond (73 kcal/mol) is liberated.

Among the most stable sulfur-containing radicals are resonance stabilized **5-1** (Gardner and Fraenkel, 1956; found in liquid sulfur) and **5-2** (Baldock *et al.*, 1974), and radicals **5-3** (Rundle and Scheffler, 1965) and **5-4** (Seebach *et al.*, 1972b) for which dimerization is sterically unfavorable. In radical **5-3**,

$$\cdot S{-}S_n{-}\ddot{\underset{..}{S}}{-}\ddot{\underset{..}{S}}\cdot \longleftrightarrow \cdot S{-}S_n{-}\overset{+}{\underset{..}{\dot{S}}}{-}\overset{-}{\underset{..}{\ddot{S}}}: \longleftrightarrow \text{etc.}$$

5-1

which has a half-life of only 30 min at room temperature, the large sulfur atom is less effectively shielded by the bulky *tert*-butyl groups than in the case of the very stable 2,4,6-tri-*tert*-butylphenoxyl radical. Highly sensitive methods such as ESR and CIDNP are available for the detection of free radicals with lifetimes as short as 10^{-7} sec and at concentrations as low as 10^{-10} M. Clever procedures have been devised for the photochemical (or pyrolytic) generation of steady state concentrations of radicals of high reactivity. In most instances, however, evidence for the involvement of radicals is of a more indirect and circumstantial character.

5-2

5-3 5-4

Despite the short lifetimes and elusive nature of sulfur-containing radicals, these ephemeral species figure prominently in a wide variety of organosulfur reactions, many of which are of great synthetic utility. We shall find that organosulfur radicals have important roles in polymer chemistry, in air pollution, and in the action of antioxidants and radioprotective agents.

5.2 Generation and Reactions of Sulfur-Containing Radicals

Sulfur-containing radicals may be generated from nonradical precursors by thermal or photolytic bond homolysis, by radiolysis, or by electron transfer (oxidation or reduction; in some cases this may be accomplished by electrolysis). The reaction of radicals with sulfur-containing substrates can lead to sulfur-containing radicals through abstraction, displacement, addition, or fragmentation processes. Scheme 5-1 presents some representative examples.

A wide range of reactions are available to sulfur-containing radicals

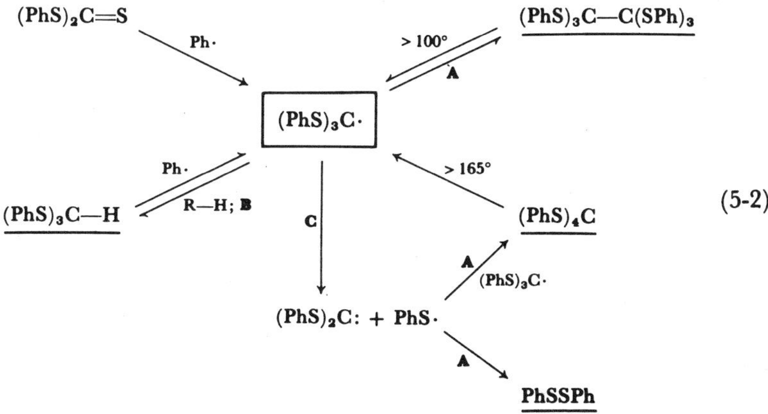

(5-2)

(Seebach *et al.*, 1972a; Uneyama and Torii, 1969)

including combination (**A**), abstraction (**B**), fragmentation (**C**), addition to multiple bonds (**D**), homolytic aromatic substitution (**E**), displacement (**F**), disproportionation (**G**), electron transfer (**H**), and rearrangement (**I**). Often several of these processes may occur together in the same reaction as may formation of several different types of sulfur-containing radicals. The examples shown in Eqs. 5-2 to 5-15, involving carbon centered α- or β-thio

$$C_2H_5SCH_2CH_3 \xrightarrow{t\text{-}C_4H_9O\cdot} C_2H_5S\dot{C}HCH_3 + C_2H_5SCH_2CH_2\cdot$$

$C_2H_5S\dot{C}HCH_3 + C_2H_5SCH_2CH_2\cdot \xrightarrow{A} \underline{C_2H_5SCH_2CH_2CH(CH_3)SC_2H_5}$ (3.6%)

$C_2H_5S\dot{C}HCH_3 \xrightarrow{A} \underline{(C_2H_5SCH(CH_3)-)_2}$ (32%)

$C_2H_5SCH_2CH_2\cdot \xrightarrow[C]{-CH_2{=}CH_2} C_2H_5S\cdot$

$C_2H_5S\cdot + C_2H_5S\dot{C}HCH_3 \xrightarrow{A} \underline{C_2H_5SCH(CH_3)SC_2H_5}$ (2.9%)

$C_2H_5S\cdot + C_2H_5SCH_2CH_2\cdot \xrightarrow{A} \underline{(C_2H_5SCH_2)_2}$ (2.2%) (5-3)

(Migita *et al.*, 1973)

$$C_2H_5SCH_2CH_2Br \xrightarrow[\text{from } R_3SnH]{R_3Sn\cdot} C_2H_5SCH_2CH_2\cdot \xrightarrow[B]{R_3SnH} \underline{C_2H_5SCH_2CH_3}$$

$$C_2H_5SCH_2CH_2\cdot \xrightarrow{G} C_2H_5S\cdot + CH_2{=}CH_2 \quad (5\text{-}4)$$

$$\underline{C_2H_5SH} \xleftarrow[B]{R_3SnH} C_2H_5S\cdot + CH_2{=}CH_2$$

(Kuivala, 1968)

$$Ph\cdot + CH_2{=}CHCH_2SC_2H_5 \longrightarrow PhCH_2\dot{C}HCH_2SC_2H_5 \xrightarrow{C}$$

$$\underset{(45\%)}{\underline{PhCH_2CH{=}CH_2}} + C_2H_5S\cdot \longrightarrow \underset{(13\%)}{\underline{(C_2H_5S)_2}} \quad (5\text{-}5)$$

(Migita *et al.*, 1973)

$$PhSC(CH_3)_2N{=}NC(CH_3)_2SPh \longrightarrow 2PhS\dot{C}(CH_3)_2 + N_2$$

$2PhS\dot{C}(CH_3)_2 \xrightarrow{A} \underline{PhSC(CH_3)_2C(CH_3)_2SPh}$

$2PhS\dot{C}(CH_3)_2 \xrightarrow{G} \underline{PhSC(CH_3){=}CH_2} + \underline{PhSCH(CH_3)_2}$ (5-6)

(Ohno *et al.*, 1971a)

$$PhSCH(CH_3)_2 \xrightarrow{t\text{-}C_4H_9O\cdot} PhSCH(CH_3)CH_2\cdot \xrightarrow{C}$$

$$CH_3CH{=}CH_2 + PhS\cdot \xrightarrow{A} \underline{PhSSPh}$$

$$\xrightarrow{B,\,D} \underline{PhSCH_2CH_2CH_3} \quad (5\text{-}7)$$

(Ohno *et al.*, 1971a)

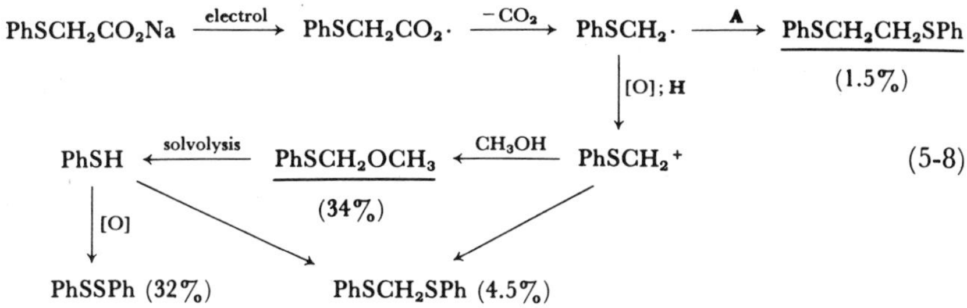

$$PhSCH_2CO_2Na \xrightarrow{\text{electrol}} PhSCH_2CO_2\cdot \xrightarrow{-CO_2} PhSCH_2\cdot \xrightarrow{\mathbf{A}} \underline{PhSCH_2CH_2SPh}\ (1.5\%)$$

$$PhSCH_2\cdot \xrightarrow{[O];\ \mathbf{H}} PhSCH_2^+ \xrightarrow{CH_3OH} \underline{PhSCH_2OCH_3}\ (34\%) \xrightarrow{\text{solvolysis}} PhSH$$

$$PhSH \xrightarrow{[O]} PhSSPh\ (32\%); \quad PhSH + PhSCH_2^+ \longrightarrow PhSCH_2SPh\ (4.5\%) \qquad (5\text{-}8)$$

(Uneyama *et al.*, 1971)

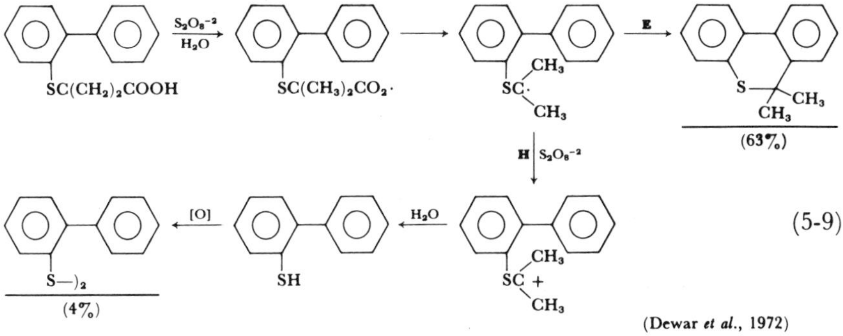

SC(CH₂)₂COOH → ($S_2O_8^{-2}$, H_2O) → SC(CH₃)₂CO₂· → SC·(CH₃)₂ → (E) → S–C(CH₃)₂ cyclized (63%)

SC·(CH₃)₂ → (H, $S_2O_8^{-2}$) → SC⁺(CH₃)₂ → (H_2O) → SH → ([O]) → (S—)₂ (4%) (5-9)

(Dewar *et al.*, 1972)

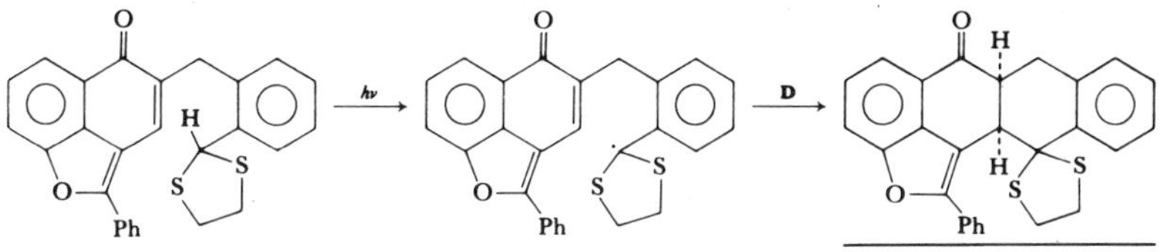

$h\nu$ → D (5-10)

(Barton *et al.*, 1971)

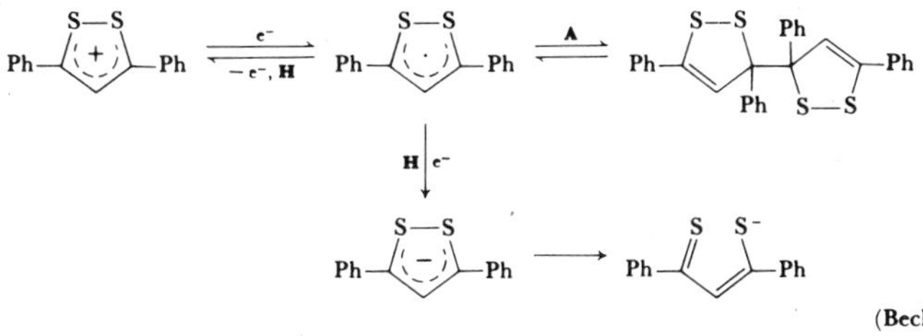

Ph–(S–S, +)–Ph ⇌ (e^- / $-e^-$, H) Ph–(S–S, ·)–Ph ⇌ (A) dimer (Ph, S—S); H, e^- → Ph–(S–S, −)–Ph → Ph–C(=S)–CH=C(S⁻)–Ph (5-11)

(Bechgaard *et al.*, 1973)

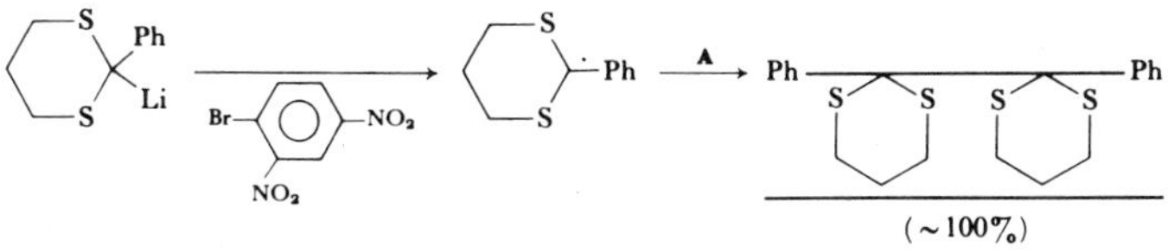

2-Ph-2-Li-1,3-dithiane → (Br–C₆H₃(NO₂)₂) → 2-Ph-1,3-dithian-2-yl· → (A) Ph–dithiane–dithiane–Ph (~100%) (5-12)

(Baarschers and Loh, 1971)

Scheme 5-1. Generation of Radicals

$(PhS)_3C{-}C(SPh)_3 \overset{130°}{\rightleftharpoons} 2(PhS)_3C\cdot$ $PhS{-}SPh \xrightarrow{h\nu} 2PhS\cdot$ $(PhS)_3N \overset{80°}{\rightleftharpoons} (PhS)_2N\cdot + PhS\cdot$	bond homolysis Seebach and Beck (1972); Barton *et al.* (1973)
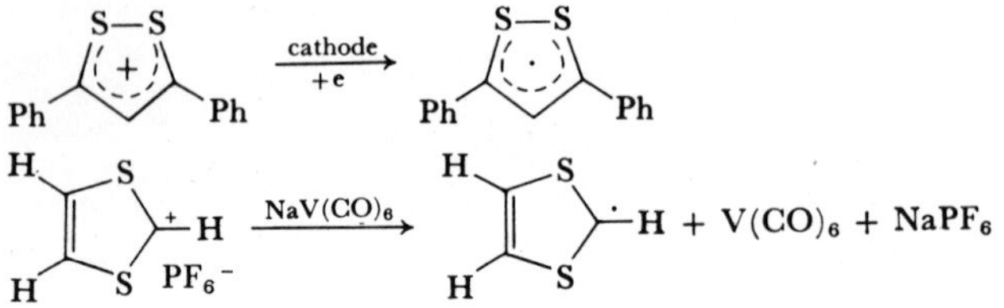 3,5-diphenyl-1,2-dithiolium cation $\xrightarrow[+e]{cathode}$ 3,5-diphenyl-1,2-dithiolyl radical 1,3-dithiolium PF_6^- $\xrightarrow{NaV(CO)_6}$ 1,3-dithiol-2-yl radical $+ V(CO)_6 + NaPF_6$	electron transfer Bechgaard *et al.* (1973) Siedle and Johannesen (1975)
$PhSCH_3 \xrightarrow{Ph\cdot} PhSCH_2\cdot$ $t\text{-}C_4H_9SOH \xrightarrow{t\text{-}C_4H_9O\cdot} t\text{-}C_4H_9SO\cdot$	abstraction Russell (1973); Howard and Furimsky (1974)
$R\cdot + CH_2{=}C(SC_2H_5)_2 \longrightarrow RCH_2\dot{C}(SC_2H_5)_2$ $Ph\cdot + (PhS)_2C{=}S \longrightarrow (PhS)_3C\cdot$	addition Tagaki *et al.* (1968); Uneyama *et al.* (1969)
$PhSCH_2CO_3t\text{-}C_4H_9 \xrightarrow[-t\text{-}C_4H_9O\cdot]{70°} PhSCH_2CO_2\cdot \xrightarrow{-CO_2} PhSCH_2\cdot$ $RSC(CH_3)_2N{=}NC(CH_3)_2SR \xrightarrow{\Delta} 2RSC(CH_3)_2\cdot + N_2$	fragmentation Uneyama *et al.* (1971); Ohno *et al.* (1971); Timberlake *et al.* (1973)
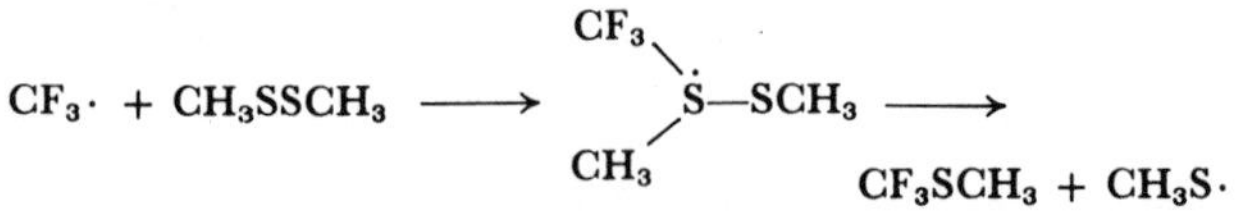 $CF_3\cdot + CH_3SSCH_3 \longrightarrow (CF_3)(CH_3)\dot{S}{-}SCH_3 \longrightarrow CF_3SCH_3 + CH_3S\cdot$	substitution Haszeldine *et al.* (1972)

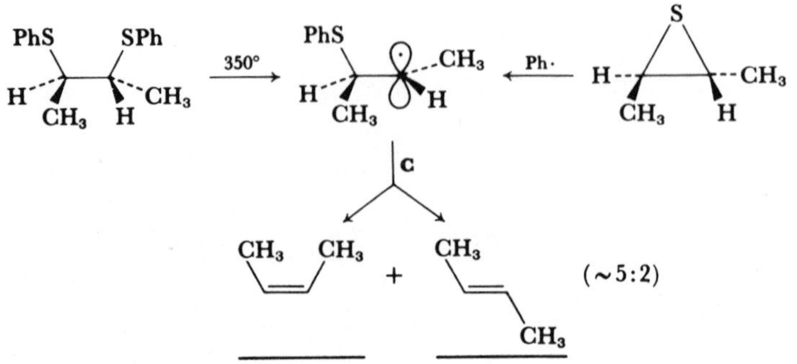

$$\text{PhS(H)(CH}_3\text{)C–C(CH}_3\text{)(H)SPh} \xrightarrow{350°} \text{PhS(H)(CH}_3\text{)C–}\dot{\text{C}}\text{H(CH}_3\text{)} \xleftarrow{Ph\cdot} \text{trans-2,3-dimethylthiirane}$$

$$\xrightarrow{c} \text{cis-2-butene} + \text{trans-2-butene} \quad (\sim 5:2) \tag{5-13}$$

(Shevlin and Greene, 1972)

CH_3S MgX, CH_3S MgX; $\xrightarrow{CH_3MgX}$; $\rightleftharpoons$; CH_3S MgX, CH_3S, MgX ·; $\xrightarrow{c}$ CH_3S, CH_3S (65%)

(1) ~ H
(2) I

SCH_3, CH_3S (30%) ⟵ CH_3S MgX, CH_3S

(5-14)

(Dagonneau *et al.*, 1973)

$$\equiv R{-}OH \longrightarrow ROC(S)SCH_3 \xrightarrow{(C_4H_9)_3Sn\cdot}$$

$$ROC(SCH_3)SSn\,(C_4H_9)_3 \xrightarrow{c} (C_4H_9)_3SnSC(O)SCH_3 + R\cdot$$

$$R\cdot \xrightarrow{(C_4H_9)_3SnH} (C_4H_9)_3Sn\cdot + \underline{R{-}H}\ (85\%) \qquad (5\text{-}15)$$

(Barton and McCombie, 1975)

substituted radicals, are illustrative (the types of reactions involved are indicated by the appropriate letter; isolated products are underscored).

5.3 Kinetic Assessment of Stability of Sulfur-Substituted Radicals

A variety of kinetic methods have been used to assess the radical-stabilizing ability of a substituent X, such as rate studies of radical additions to substituted olefins (5-16), hydrogen abstractions (5-17), and thermolysis of azoalkanes (5-18). Relative rate data on these three types of studies as they pertain to sulfur substituents are collected in Tables 5-1, 5-2, and 5-3,

$$R\cdot + CH_2{=}CHX \longrightarrow RCH_2{-}\dot{C}HX \longrightarrow \text{products} \qquad (5\text{-}16)$$

$$R\cdot + H{-}\overset{|}{\underset{|}{C}}{-}X \longrightarrow RH + \cdot\overset{|}{\underset{|}{C}}X \longrightarrow \text{products} \qquad (5\text{-}17)$$

$$XC(CH_3)_2N{=}NC(CH_3)_2X \longrightarrow N_2 + 2(CH_3)_2\dot{C}{-}X \longrightarrow \text{products} \qquad (5\text{-}18)$$

respectively. In these studies it is assumed that factors which stabilize the substituted radical also stabilize the transition state leading to the radical and that the transition states for the reactions compared resemble the radical intermediate to the same extent. Since ESR evidence indicates only modest spin delocalization by substituents (see below), these relatively small stabilizing effects should be felt only when the transition state occurs late on the reaction coordinate, and so already has a strong radical character (Rüchardt, 1970).

Reaction 5-16 occurs as the propagation step in radical polymerization. The reactivity of a monomer, e.g., $CH_2{=}CHX$, in copolymerization with some other standard monomer such as styrene can be expressed by the ratio of the two monomers found in the copolymer using two constants, one of which (Q) is related to the reactivity of the double bond toward a neutral free radical and is therefore claimed to be governed chiefly by the stability of the free radical in the adduct (Alfrey and Price, 1947). Table 5-1 summarizes Q values determined for a variety of monomers. A current view of the relative reactivity sequence indicated by Q values in copolymerization is that it reflects the extent of transition state polarization rather than stability of the incipient radical. The common tendency for a growing polymer chain to alternate when two monomers are present is thought to be a reflection of the reduction in activation energy by transition state polarization, often with complete electron transfer and charge separation as bond formation takes place (Abell, 1973). Apparently, the Q treatment emphasizes the carbanionic character of the substituents, for ethers show abnormally low Q values, while

Table 5-1. Q Values for the Copolymerization of Various Monomers with Styrene

Monomer	Q	Reference
$(C_2H_5O)_2C{=}CH_2$	0	Tagaki *et al.* (1968)
$C_2H_5OCH{=}CH_2$	0.02–0.03	Young (1961)
$CH_3C(O)OCH{=}CH_2$	0.03	Young (1961)
$CH_3S(O)CH{=}CH_2$	0.06–0.10	Young (1961)
$PhSO_2CH{=}CH_2$	0.07	Young (1961)
$CH_3SO_2CH{=}CH_2$	0.11	Young (1961)
$C_2H_5S(O)CH{=}CH_2$	0.13	Inoue *et al.* (1971)
$C_2H_5S(O)CH{=}CH_2$[a]	0.15	Inoue *et al.* (1971)
$C_6Cl_5SCH{=}CH_2$	0.22	Young (1961)
$CH_3C(O)SCH{=}CH_2$	0.31	Young (1961)
$CH_3SCH{=}CH_2$	0.32–0.34	Young (1961)
$PhSCH{=}CH_2$	0.34	Young (1961)
$C_2H_5SCH{=}CH_2$	0.37	Young (1961)
$PhCH{=}CH_2$	1.00	Young (1961)
$(C_2H_5S)_2C{=}CH_2$	2.70	Tagaki *et al.* (1968)

[a] With added $ZnCl_2$ in benzene.

Table 5-2. Relative Reactivity toward *t*-Butoxy and Phenyl Radicals per Hydrogen Atom[e]

Compound	Ph·[a]	t-C_4H_9O·[b]	Compound	Ph·[a]	t-C_4H_9O·[b]
$Ph{-}C\underline{H}_3$	1.00	1.00	$CH_3SSC\underline{H}_3$	1.2[c]	—
$PhC\underline{H}_2CH_3$	4.56	3.2	$((CH_3)_2C\underline{H})_2S_2$	7.1(9.1)[c]	—
$PhC\underline{H}(CH_3)_2$	9.7	6.8[d]	$CH_3S(O)C\underline{H}_3$	0.08	—
$CH_3{-}C\underline{H}_3$	0.11	0.1[d]	$((CH_3)_2C\underline{H})_2SO$	0.78	—
$CH_3C\underline{H}_2CH_3$	1.0	1.2[d]	$PhSO_2C\underline{H}_3$	—	0.05
$(CH_3)_3C\underline{H}$	5.22	4.3[d]	$(CH_3C\underline{H}_2)_2SO_2$	1.2	—
$CH_3SC\underline{H}_3$	1.89	—	Thiacyclopentane		
$PhSC\underline{H}_3$	1.44	2.12	*S,S*-dioxide (α-H)	0	—
$(CH_3C\underline{H}_2)_2S$	6.0	7.10	$CH_3OC\underline{H}_3$	0.5	—
$CH_3C\underline{H}_2SPh$	2.9	—	$PhOC\underline{H}_3$	0.34	1.44
$((CH_3)_2C\underline{H})_2S$	12.0	—	$(CH_3C\underline{H}_2)_2O$	3.7	—
$(CH_3)_2C\underline{H}SPh$	9.8	—	$((CH_3)_2C\underline{H})_2O$	9.9	—
Thiacyclopentane (α-H)	7.22	—	Tetrahydrofuran		
Thiacyclohexane (α-H)	4.2	—	(α-H)	3.44	—
1,3-Dithiolane (2-H)	—	49.2	$(PhO)_2C\underline{H}_2$	0.5	4.14
$(C_2H_5S)_2C\underline{H}_2$	—	16.3	$(C\underline{H}_3)_3N$	3.67	—
$(PhS)_2C\underline{H}_2$	—	10.0	$(CH_3C\underline{H}_2)_3N$	16.1	—
$(PhS)_3C\underline{H}$	—	13.9	$PhN(C\underline{H}_3)_2$	—	81.9

[a] 60° (Russell, 1973).
[b] 130° in chlorobenzene (Uneyama *et al.*, 1968).
[c] Pryor and Smith (1970).
[d] Russell (1973).
[e] Hydrogen atom abstracted is underlined.

sulfones and sulfoxides display abnormally high Q values compared to other indicators of radical stability. It is noteworthy that the reactivity of ethyl vinyl sulfoxide in copolymerization is enhanced by the addition of a Lewis acid such as zinc chloride (cf. Table 5-1), an effect attributable to the increased electron-accepting ability of the complexed monomer, perhaps via a charge separated transition state (Eq. 5-19).

$$\sim\!CH_2{-}\overset{\overset{\displaystyle Ph}{|}}{C}H\cdot \xrightarrow{CH_2{=}CH\overset{\overset{O\cdots ZnCl_2}{\|}}{S}C_2H_5} \sim\!CH_2\underset{\delta+}{\overset{\overset{\displaystyle Ph}{|}}{C}H}{-}{-}{-}CH_2{=}{=}\underset{\delta-}{CH\overset{\overset{O\cdots ZnCl_2}{\|}}{S}C_2H_5} \longrightarrow$$

$$\sim\!CH_2CH(Ph)CH_2\dot{C}HS(O)C_2H_5 \xrightarrow{CH_2{=}CHPh} \text{etc.} \qquad (5\text{-}19)$$

The ease of hydrogen atom abstraction (reaction 5-17) by a variety of radicals such as Cl·, Ph·, and *t*-BuO· has been used as a measure of radical stability. From the data in Table 5-2, it is seen that substituents facilitate hydrogen abstraction by the phenyl radical in the order

$$RSO_2 \sim RS(O) < CH_3 < PhO < CH_3O < Ph < RSS < PhS < CH_3S < (CH_3)_2N$$

The transition state for hydrogen abstraction by Ph· is felt to possess significant bond cleavage (and therefore show substantial radical character) based on the size of the deuterium isotope effect observed ($k_H/k_D = 4.5$ with toluene as substrate; by way of comparison Cl·, *t*-BuO·, and CH_3· give respective isotope effects of 1.3, 5.5, and 7.9) (Russell, 1973). In contrast to Cl· and *t*-BuO·, it is argued that Ph· should have little propensity to enter into polar structures of the type illustrated in Eq. 5-20 which, with substituents capable of charge delocalization, could result in a lowering of activation energy (Russell, 1973). However a modest polar effect might be expected with Ph·

$$\gt C{-}H + R\cdot \longrightarrow \left[\gt \overset{\delta+}{C}\cdots H \cdots \overset{\delta-}{R} \right] \longrightarrow \gt C\cdot + H{-}R \qquad (5\text{-}20)$$

in view of the location of the lone electron in the somewhat electronegative sp^2 orbital (Hay, 1974). This proposal finds support in the observation of an unusually small effect of a phenyl group compared with methyl groups in activating adjacent C—H bonds toward abstraction by Ph·. For example, Table 5-2 indicates that the methyl hydrogens of toluene are only *nine* times as reactive toward Ph· as the hydrogens of ethane and identical in reactivity with the methylene hydrogen of propane, even though a phenyl group is thought to stabilize a radical 10 kcal/mol more than a methyl substituent (Russell, 1973). A very similar situation exists with *t*-C_4H_9O·. Apparently the electron withdrawal effects by the phenyl group in toluene are felt to some extent in the transition state for H· abstraction. By way of contrast, the more discriminating Br· radical favors H· abstraction from the methyl group of toluene over H· abstraction from ethane and the methylene position of propane by factors of 64,000 and 291, respectively (Russell, 1973). The point to be made here is that polar effects associated with H· abstraction by Ph· (or *t*-C_4H_9O·) may be sufficiently large to invalidate the use of relative reactivity tabulations as precise indicators of radical stability.

Evidence of a more qualitative nature is available on the ease of hydrogen abstraction from organosulfur compounds. ESR indicates that H· abstraction from 1,4-thioxane by *t*-C_4H_9O· occurs alpha to sulfur (Biddles *et al.*, 1972). Formation of CD_3H to the virtual exclusion of CD_4 in the photolysis of $CD_3S(O)CD_3$ in various nondeuterated solvents suggests that hydrogen abstraction from dimethyl sulfoxide is energetically unfavorable (Gollnick and Schade, 1973). Because of its unreactivity toward free radicals, dimethyl sulfoxide is widely used as a solvent for radical polymerizations. Some interesting chemistry has been described on free-radical halogenation of sulfoxides and sulfones. Radical bromination of sulfoxide **5-5a** leads to clean functionalization of the C_3 methyl group (**5-6a**). Under similar conditions, sulfide **5-5b** is extensively decomposed, presumably through formation of the highly reactive 2-bromo derivative by functionalization adjacent to sulfur

$$\text{5-5a} \; (n = 1),\ \text{5-5b} \; (n = 0) \xrightarrow[h\nu]{\text{NBS}} \text{5-6a} \; (n = 1)$$

5-5a $n = 1$
5-5b $n = 0$

5-6a $n = 1$

(Cooper, 1972). Early work indicated that simple benzyl and alkyl sulfones do not undergo free-radical bromination (Backer *et al.*, 1948). Only with sulfones containing hydrogen atoms which are both benzylic and tertiary can free-radical bromination (*N*-bromosuccinimide/benzoyl peroxide) be achieved (Eq. 5-21a; Bordwell and Doomes, 1974).† With the more reactive

$$\text{PhCH(CH}_3\text{)SO}_2\text{CH}_2\text{Ph} \xrightarrow[(\text{BzO})_2]{\text{NBS}} \text{PhCBr(CH}_3\text{)SO}_2\text{CH}_2\text{Ph} \quad (75\%) \tag{5-21a}$$

chlorine atoms, benzyl sulfones can be α-chlorinated; however with simple aliphatic sulfones and alkanesulfonyl chlorides, very little α-substituted product is obtained (Eq. 5-21b; Tabushi *et al.*, 1974; Horner and Schlafer, 1960). In these particular reactions, polar effects are apparently directing the electrophilic Cl· to attack positions remote from the sulfone group.

$$\text{sulfolane} \xrightarrow[\text{SO}_2\text{Cl}_2]{(t\text{-C}_4\text{H}_9\text{O})_2} \text{2-chlorosulfolane} \; (2.6\%) + \text{3-chlorosulfolane} \; (97.4\%) \tag{5-21b}$$

A radical abstraction reaction of some synthetic utility, the peroxyester reaction, has been developed by Sosnovsky (Rawlinson and Sosnovsky, 1972). This copper ion catalyzed reaction involves a three-step chain process with an alternative to the third step being oxidation of R'· to a carbonium ion by copper(II) (Eq. 5-22). This reaction works particularly well with

$$\begin{aligned} \text{RC(O)OO}t\text{-C}_4\text{H}_9 + \text{Cu(I)} &\longrightarrow t\text{-C}_4\text{H}_9\text{O}\cdot + \text{RC(O)OCu(II)} \\ \text{R}'\text{—H} + \cdot\text{O}t\text{-C}_4\text{H}_9 &\longrightarrow \text{R}'\cdot + t\text{-C}_4\text{H}_9\text{OH} \\ \text{R}'\cdot + \text{RC(O)OCu(II)} &\longrightarrow \text{R}'\text{—OC(O)R} + \text{Cu(I)} + \cdots \end{aligned} \tag{5-22}$$

sulfides, as indicated by the examples in Eqs. 5-23 to 5-25. The product ratio in reaction 5-25 (Sugawara *et al.*, 1974) reflects the relative reactivities of the three kinds of methylene hydrogens in 1,3-dithiane (cf. Table 5-2).

$$\text{1,4-oxathiane} \xrightarrow[\text{Cu(I)}]{\text{PhC(O)OO-}t\text{-C}_4\text{H}_9} \text{3-OC(O)Ph-1,4-oxathiane} \quad (56\%) \tag{5-23}$$

† Apparently, radical α-bromination of benzyl sulfonamides occurs in reasonable yield (Sheehan *et al.*, 1974).

$$C_3H_7SC_3H_7 \xrightarrow[Cu(I)]{PhC(O)OO\text{-}t\text{-}C_4H_9} C_3H_7SCH(OC(O)Ph)C_2H_5 \ (78\%) \qquad (5\text{-}24)$$

$$\text{1,3-dithiane} \xrightarrow[Cu(I)]{PhC(O)OO\text{-}t\text{-}C_4H_9} \text{2-(OC(O)Ph)-1,3-dithiane} \ (56\%) + \text{4-(OC(O)Ph)-1,3-dithiane} \ (14\%) \qquad (5\text{-}25)$$

The decomposition of symmetric azo compounds (reaction 5-18) should be a particularly useful reaction for the assessment of radical-stabilizing ability, since in this process the C—N bonds should be extensively broken and the radicals strongly anticipated at the transition state, with minimal contributions from polarization effects (Rüchardt, 1970). The data in Table 5-3 indicate that stabilization by an oxygen atom α to the (tertiary) radical center is small, while the effect of an α-sulfur is more substantial. Similar conclusions have been drawn from a comparison of rates for reaction 5-18 with X = CH_3CO_2, $CH_3C(O)S$, CH_3O, and CH_3S (Timberlake *et al.*, 1973). The one possible drawback of the 2-substituted azobis(2-propane) system to demonstrate radical stabilization is that the resonance contribution of the substituent X in $(CH_3)_2\dot{C}$—XR is less than in the primary radical $RXCH_2\cdot$ (see below for ESR evidence on this point), presumably because greater steric compression is required in the tertiary radical to orient the RX group so that the lone pair orbital of X is parallel to the unpaired electron orbital on carbon. Such steric effects could vary inversely with C—X bond length.

The "leveling" effect of methyl substitution may be noted from data in Table 5-2 which show that the ratios of hydrogen abstraction rates of the pairs $CH_3SCH_3/CH_3—CH_3$, $(C_2H_5)_2S/(CH_3)_2CH_2$, and $(i\text{-}C_3H_7)_2S/$

Table 5-3. Rate Data for Thermolysis of X—$C(CH_3)_2N{=}NC(CH_3)_2$—X at 150° in Tetralin[a]

X	Relative rate	$\Delta G^‡$ (kcal/mole)	$\Delta H^‡$ (kcal/mole; ±0.5)	$\Delta S^‡$ (e.u.; ±2)
$C_2H_5OCH_2$	0.13	35.5	43.3	18.4
$PhCH_2$	0.63	33.9	35.6	4.0
$n\text{-}C_3H_7$	1.0	33.8	37.9	9.6
PhO	1.7	33.4	32.3	−2.6
$C_2H_5SCH_2$	2.4	32.8	38.6	13.6
C_2H_5O	3.7	32.6	36.9	10.2
PhS	56	30.1	24.9	−12.4
C_2H_5S	175	29.3	29.9	1.4

[a] Ohno *et al.* (1971b), Ohno and Ohnishi (1971).

$(CH_3)_3C—H$ for the primary, secondary, and tertiary hydrogens, respectively, decrease in the order 17, 6, 2. One intriguing aspect of the data in Table 5-3 is that the 2-thio substituted azo compounds show much more negative values for $\Delta S^{\ddagger}$ than do the 2-oxy analogs. ESR evidence suggests that α-thio radicals show a greater degree of planarity than α-oxy radicals (see below). The more negative values for $\Delta S^{\ddagger}$ for the decompositions leading to formation of α-thio radicals may possibly be a kinetic expression of the greater need for coplanarity for the developing α-thio radicals than the α-oxy radicals. That the PhX substituents give rise to more negative values for $\Delta S^{\ddagger}$ than the alkyl-X substituents may be rationalized in terms of the resonance structure **5-7**.

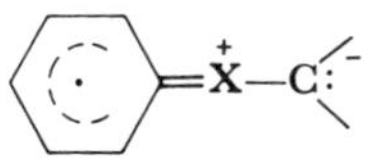

5-7

5.4 The Study of Sulfur-Containing Radicals by Electron Spin Resonance Spectroscopy

The ESR spectrum of a sulfur-containing radical can provide valuable information about the structure of the radical, especially with respect to the delocalization of the unpaired electron and the geometry of the species. Much can be learned through consideration of the hyperfine splitting constants and g-value (spectroscopic splitting factor).

The g value is equal to 2.0023 for a free electron and has slightly larger or smaller values for radicals because of coupling of the electron spin with orbital angular momentum (spin–orbit coupling). Since it is possible to measure g values with great precision, they can be used to identify radicals or give information about the orbital occupied by the unpaired electron. In particular, high g values are obtained for radicals in which there is some localization of spin density on sulfur because of the large spin–orbit coupling associated with a heavy atom such as sulfur. Table 5-4 indicates the range of g values seen for sulfur-containing radicals.

The variation in g values in the series of α-thioradicals $RSCH_2\cdot$, $(PhS)_3C\cdot$, $RS\dot{C}HCH_3$, $RS\dot{C}(CH_3)_2$, and $RS\dot{C}[C(CH_3)_3]_2$ is significant. Increased substitution of sulfur (in $(PhS)_3C\cdot$) might be expected to substantially raise the g value. That this is not observed is due to increased nonplanarity of the radical (the estimated deviation from coplanarity is 11–12°) which reduces delocalization to the same level as in $RSCH_2\cdot$ (Stegmann *et al.*, 1975). For the same reason, the successive introduction of methyl or *tert*-butyl groups on the tervalent carbon should also lead to less effective conjugation with sulfur, and therefore lower *g* values (Gilbert *et al.*, 1973a; Scaiano and Ingold, 1976).

In addition to the *g* values, the hyperfine splitting constants are also a valuable source of information on radical properties. The hyperfine splitting *a* reflects the interaction of the unpaired electron with nuclei possessing spin (^1H, ^{13}C, ^{14}N, etc.; in solution, the value of *a* depends on the probability of the unpaired electron finding itself at the nucleus in question). On the basis of α-H hyperfine coupling constants ($a_{\alpha\text{-H}}$) for radicals

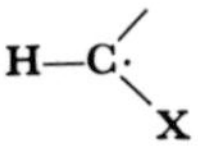

$$\mathrm{H{-}\dot{C}{<}^{\diagup}_{X}}$$

Table 5-4. *g* Values for Some Sulfur Radicals

Radical[a]	*g* Values	Reference
$R_2N{-}S{-}S{-}S\cdot$	2.030	Hodgson *et al.* (1963)
$\cdot S{-}S_n{-}S\cdot$	2.024	Gardner and Fraenkel (1956)
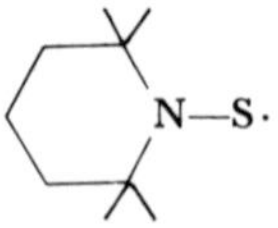 N—S·	2.0173	Bennett *et al.* (1967)
$t\text{-}C_4H_9SO\cdot$	2.0106	Howard and Furimsky (1974)
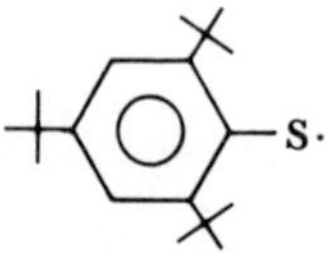 S·	2.0104	Rundle and Scheffler (1965)
$[(CH_3)_2SS(CH_3)_2]^{\dot{+}}$	2.0103	Gilbert *et al.* (1973b)
$CH_3SO\cdot$	2.0100	Gilbert *et al.* (1977b)
$CH_3\dot{S}(OC_2H_5)_2$	2.0096	Gara *et al.* (1977)
$(p\text{-}CH_3C_6H_4)_2S^{\dot{+}}$	2.0091	Schmidt *et al.* (1970)
$(PhS)_2N\cdot$	2.0082	Miura *et al.* (1975)
$(CH_3)_2\dot{S}OSi(CH_3)_3$	2.0076	Gara and Roberts (1977)
$(C_2H_5O)_3S\cdot$	2.0068	Chapman *et al.* (1976)
$CH_3SO_2\cdot$	2.0050	Kochi (1972)
$(PhS)_3C\cdot$	2.0049	Seebach, 1972b
$RSCH_2\cdot$	2.0048–2.0049	Gilbert *et al.* (1973a)
$RS\dot{C}HCH_3$	2.0044	Gilbert *et al.* (1973a)
$RS\dot{C}(CH_3)_2$	2.0036–2.0038	Gilbert *et al.* (1973a)
$RSCPh_2\cdot$	2.0030–2.0032	Dagonneau *et al.* (1973)
$HOCH_2\cdot$	2.0033	Norman and Pritchett (1965)
$RS\dot{C}[C(CH_3)_3]_2$	2.0026–2.0027	Scaiano (1976)
$CH_3CH_2\cdot$	2.0025	Carton *et al.* (1975)
$CH_3SO_2CH_2\cdot$	2.0025	Carton *et al.* (1975)
$CH_3S(O)CH_2\cdot$	2.0025	Carton *et al.* (1975)
$^-OSO_2CH_2\cdot$	2.0025	Carton *et al.* (1975)

[a] R = simple alkyl group.

Table 5-5. Decrease in the Spin Density in the α-Carbon Atom in $CH_3\dot{C}(X)^-$ by Substituents X as Determined by ESR Measurements[a]

X	Relative spin density (%)	*X*	Relative spin density (%)
H	100	$C(O)C_2H_5$	84
$SO_2C_2H_5$	100	OPh	83
$S(O_2)O^-$	96	OC_2H_5	81
$S(O)C_2H_5$	93	SPh	77
COOR	93	SSC_2H_5	77
CH_3	92	SC_2H_5	74
CN	85		

[a] Carton *et al.* (1975); Rüchardt (1970); Biddles *et al.* (1972); Adams (1970).

it has been argued that the radicals with X = RO have a less planar structure than in the case of sulfur conjugated radicals, X = RS (Biddles *et al.*, 1972). The β-H hyperfine coupling constant (a_{CH_3}) of a radical of the type

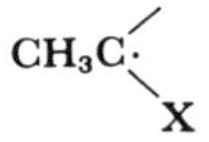

$$CH_3\dot{C}{<}^{}_{X}$$

is proportional to the spin density on the α-carbon atom (assuming planarity at this tervalent carbon) and is therefore felt to be a good measure of the ability of substituent X to reduce the α-carbon spin density by resonance (Rüchardt, 1970). Table 5-5 gives the percent decrease in the spin density at the tervalent carbon as a function of substituent X as calculated from a_{CH_3}. The general agreement of these data with the data in Table 5-3 is gratifying.

The larger ring proton splittings in the α-phenylthiylmethyl radical **5-9** compared to the α-phenoxymethyl radical **5-8** is also cited as evidence of increased delocalization of the unpaired electron by sulfur (Biddles *et al.*, 1972).

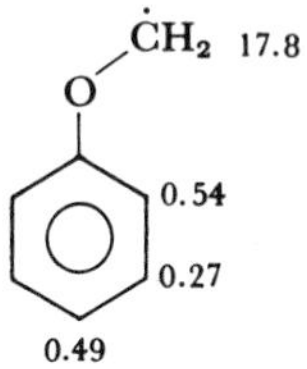

$PhO\dot{C}H_2$: $\dot{C}H_2$ 17.8; ring 0.54, 0.27, 0.49
5-8

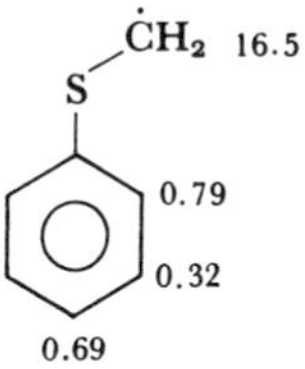

$PhS\dot{C}H_2$: $\dot{C}H_2$ 16.5; ring 0.79, 0.32, 0.69
5-9

5.5 The Nature of Radical Stabilization by Sulfur

From the data previously presented, the following trends appear with regard to α-substituents:

1. The approximate order of stabilization offered by an α-substituent is $R_2N > RS > RO > C > RS(O) > RSO_2$.

2. The greater ease of formation of α-thioradicals over α-oxyradicals is in general substantially smaller than the corresponding preference for *carbanion* formation (i.e. $RSCH_2^-$ versus $ROCH_2^-$).

3. The greater ease of formation of α-thioradicals over α-sulfinyl and α-sulfonyl radicals, and α-alkylthiyl over α-phenylthiyl radicals is clearly opposite to the trend in *carbanion* formation and parallels the trend seen in *carbonium ion* formation.

4. The extent of spin delocalization onto sulfur as indicated by ESR amounts to 25% at most.

5. A near planar geometry is indicated for α-thio- and α-oxyradicals for maximum heteroatom conjugation, deviations from planarity being smaller for the former radicals.

The resonance structures **5-10a** to **5-11c** can be considered for α-oxy- and α-thioradicals. It has already been indicated that the major contributors are

$$\mathrm{R{-}\ddot{O}{-}\dot{C}\langle} \longleftrightarrow \mathrm{R\overset{+}{\dot{O}}{-}\overset{-}{\ddot{C}}\langle} \qquad \mathrm{R\ddot{S}{-}\dot{C}\langle} \longleftrightarrow \mathrm{R\overset{+}{\dot{S}}{-}\overset{-}{\ddot{C}}\langle} \longleftrightarrow \mathrm{R\dot{S}{=}C\langle}$$

5-10a **5-10b** **5-11a** **5-11b** **5-11c**

likely to be structures **5-10a** and **5-11a**. Structure **5-11c** has been widely used in accounting for the greater stability of α-thioradicals compared to α-oxyradicals (e.g., Ohno *et al.*, 1971b; Uneyama *et al.*, 1968). This contributor requires the use of sulfur 3d orbitals either in the formation of a 3d–2p π bond or to accommodate the unpaired electron allowing sulfur to participate in 3p–2p π bonding. However, recent theoretical calculations by Bernardi, Wolfe, and co-workers suggest that 3d orbital contributions in this system are negligibly small, as might be expected for divalent sulfur (Bernardi *et al.*, 1976). These calculations suggest the use of an antibonding π-type orbital to accommodate the additional electron, predict a deviation from planarity for an α-thioradical of 10°, and indicate greater stabilization for an α-thioradical than for an α-oxyradical. Hybrid contributor **5-11b** for sulfur should be more important than contributor **5-10b** for oxygen because of the lower electronegativity and greater polarizability of sulfur compared to oxygen (cf. Dobbs *et al.*, 1971, for a related comparison of α-oxy- and α-fluororadicals). Solvation, which is apparently critically important in the stabilization of α-oxycarbonium ions in solution, should be relatively unimportant in these radical systems.

In light of the well-documented ability of sulfur to stabilize carbonium ions at a β-location with respect to sulfur, evidence for a similar effect in β-thio radicals was sought. These radicals could be generated either by addition or abstraction processes (Eq. 5-26). The stereospecificity of thiol addition

$$RS\cdot + \text{olefin} \rightleftharpoons [RS\text{—C—C}\cdot] \xleftarrow{-Y\cdot} RS\text{—C—C—}Y; \quad [RS\text{—C—C}\cdot] \xrightarrow{X\text{—}R' \text{ or } X\cdot} RS\text{—C—C—}X \tag{5-26}$$

to olefins under photochemical conditions in the presence of D—Br as an excellent transfer agent for D· led Skell to postulate a sulfur-bridged radical (Eq. 5-27; Readio and Skell, 1966).

$$\text{MeSD} \xrightarrow[\textit{cis}\text{-}CH_3CH{=}CHCH_3]{h\nu,\ DBr} \text{(SMe; H, Me; H, Me; D)} \qquad \textit{via} \ \text{(S-bridged radical, S–Me)} \tag{5-27}$$

$$\text{MeSD} \xrightarrow[\textit{trans}\text{-}CH_3CH{=}CHCH_3]{h\nu,\ DBr} \text{(SMe; H, Me; Me, H; D)}$$

However, the data on $C_2H_5SCH_2C(CH_3)_2$—N=N—$C(CH_3)_2CH_2SC_2H_5$ decomposition in Table 5-3 indicate that, to the extent that the transition state for this reaction resembles the radical intermediate, no particular stabilization can be attributed to the β-ethylthiyl group (Ohno and Ohnishi, 1971c). Furthermore, Kampmeier has shown that the hydrogens of di-*tert*-butyl sulfide, which are β to sulfur, show no special reactivity in abstraction

Table 5-6. Rate Data for Thermolysis of Some Azobiscyclohexanes [a]

Compound	K_{rel}	$\Delta H^{\ddagger}$, kcal/mole	$\Delta S^{\ddagger}$, e.u. (80°)
1,1'-azobis(1-cyano-3-thiacyclohexane) (S adjacent)	27.6	26.9	0.8
1,1'-azobis(1-cyano-4-thiacyclohexane)	3.3	28.5	1.0
1,1'-azobis(1-cyanocyclohexane)	1.0	32.3	9.5

[a] Ohno and Ohnishi (1972).

by the phenyl radical (Hepinstall and Kampmeier, 1973). A slight rate enhancement is observed for the formation of 3- and 4-thiocyclohexyl radicals (Table 5-6). It is argued that the observed rate enhancement together with the diminished $\Delta S^{\ddagger}$ supports the existence of transannular participation by sulfur (Ohno and Ohnishi, 1972).

ESR evidence indicates that the conformation of β-thioethyl radicals in which the C—S bond nearly eclipses the p orbital containing the unpaired electron (conformer **5-12a**) is preferred by 3.3 kcal over the conformer in which the C—S bond lies in the p orbital nodal plane (conformer **5-12b**; this particular conformer is favored for β-alkoxylethyl radicals) (Krusic and Kochi, 1971). Furthermore these studies reveal that the β-sulfur atom is

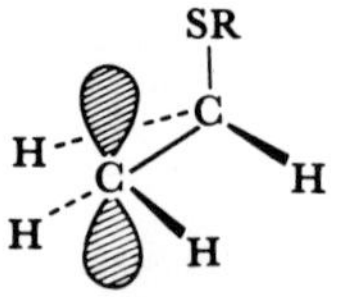

5-12a 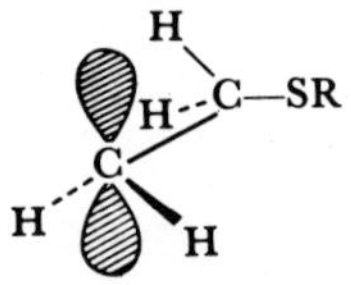 **5-12b**

displaced from its tetrahedral position toward the p orbital. However, Krusic and Kochi conclude that "these radicals are not truly bridged structures but rather exist in preferred conformational orientations which may be sufficient to control stereochemistry."

The reversible formation of β-thioradicals from olefins and thiyl radicals is the basis of a very useful method of olefin isomerization (Eq. 5-28; Moussebois and Dale, 1966). The most satisfactory conditions involve irradiation (at wavelengths >300 nm) of a solution containing the olefin(s) and diphenyl disulfide or diphenyl sulfide (the latter forms diphenyl disulfide on irradiation). The isomerization is facilitated by the weakness of the C—SPh bond and the relatively slow transfer rate of the initially formed adduct radical with disulfide. Double bond migration or other side reactions are absent even with sensitive substrates (see Table 5-7). It might be noted that

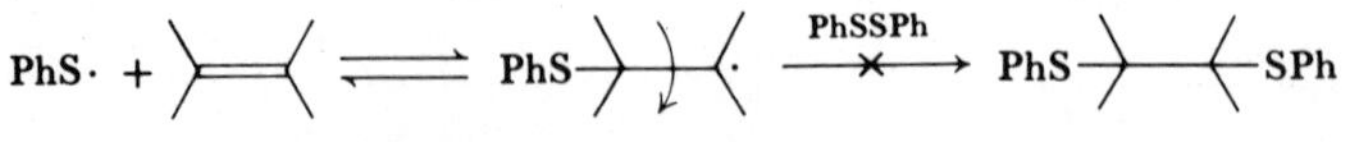

PhS· + olefin ⇌ PhS–C–C· —PhSSPh (✕)→ PhS–C–C–SPh (5-28)

while radical addition of disulfides to olefins is generally not observed because of the facile fragmentation of β-thioalkyl radicals, radical addition of disulfides to terminal acetylenes can be realized in good yields, e.g., as shown in Eq. 5-29 (Heiba and Dessau, 1967). Apparently, fragmentation of the

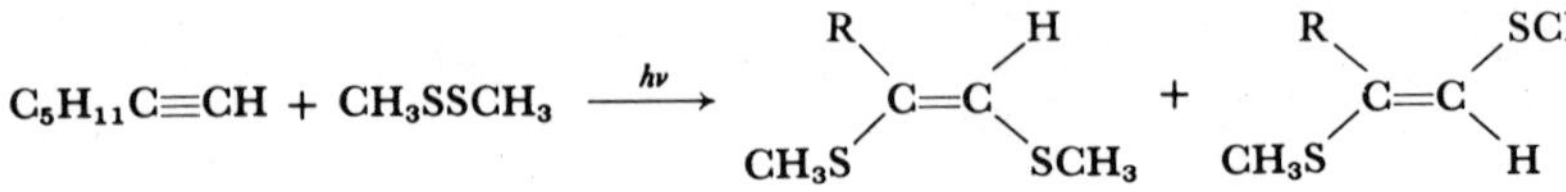

$$C_5H_{11}C{\equiv}CH + CH_3SSCH_3 \xrightarrow{h\nu} (R)(CH_3S)C{=}C(H)(SCH_3) + (R)(CH_3S)C{=}C(SCH_3)(H) \quad (1:3) \qquad (5\text{-}29)$$

Table 5-7. Olefin Isomerization by the Moussebois–Dale Procedure

Starting olefin	Product	Reference
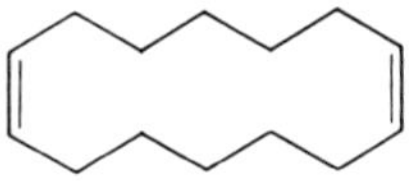	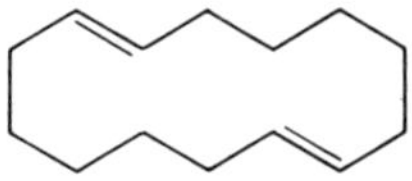	Moussebois and Dale (1966)
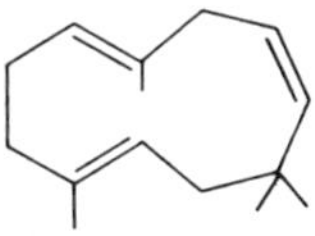	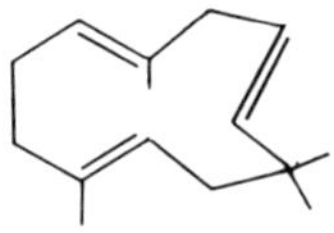 (Humulene)	Corey and Hamanaka (1967)
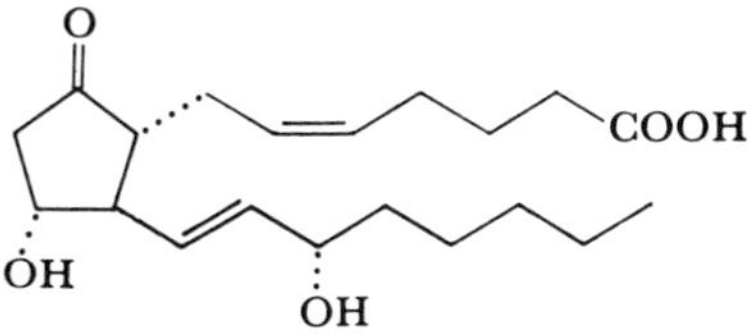	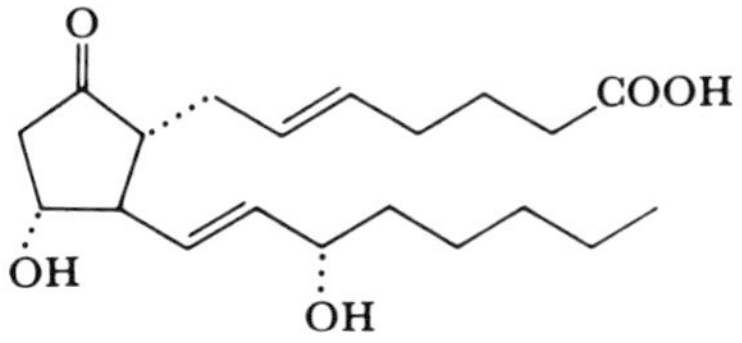	Bundy *et al.* (1972)
$(PhCH{=}CH)_2CH_2$ (isomers)		Block and Orf (1972)
Caryophyllene	Isocaryophyllene	Gollnick and Schade (1968)

intermediate β-thiovinyl radical is slow relative to radical displacement at sulfur.

5.6 α-Disulfide Radicals

Data presented above suggest that the spin delocalizing ability of an α-dithiyl group is less than that of an α-thiyl group (a trend that has its parallel in carbonium ion chemistry). Thus, Table 5-5 indicates a 23% decrease in spin density by attachment of the C_2H_5SS group to a carbon radical compared to decreases of 26% and 23% seen with the C_2H_5S and PhS groups, respectively.

From a study of the temperature dependence of ESR linewidths, it is concluded that the conformations of the methylthiomethyl and methyldithiomethyl radicals are best represented as **5-13a** and **5-13b**, respectively

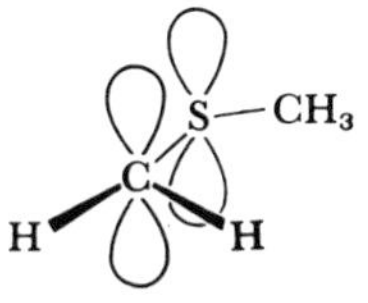

5-13a 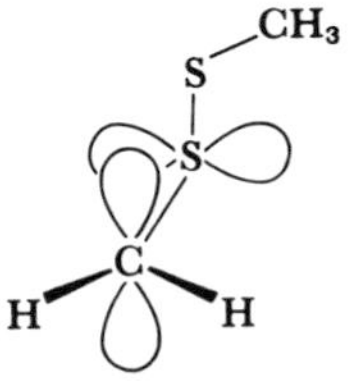 **5-13b** 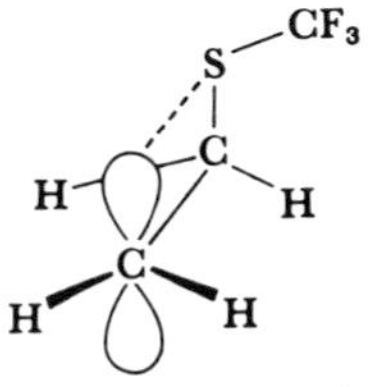 **5-13c**

(Krusic and Kochi, 1971). In radical **5-13b** there is the possibility for 1,3-interaction between the more distant sulfur and the tervalent carbon, analogous to that suggested for β-thioalkyl radicals (e.g., **5-13c**). In radical **5-13c**, significant hyperfine splitting due to the trifluoromethyl is seen ($CF_3OCH_2CH_2\cdot$ also shows fluorine coupling but of less than half the value obtained for **5-13c**) (Krusic, 1971). There is no indication of hyperfine splitting across a disulfide bond (i.e., by the CH_2 protons in $CH_3\dot{C}HSS\underline{CH_2}CH_3$) (Adams, 1970). Because of the larger splitting associated with fluorine, it would be of interest to search for fluorine splitting in $\cdot CH_2SSCF_3$.

The hydrogen atom abstraction relative rate data for dimethyl and diisopropyl disulfide in Table 5-2 are also consistent with the diminished stability of α-dithiyl radicals relative to α-thiyl radicals.

Virtually nothing is known about the chemistry of α-dithiyl radicals, an unfortunate situation due in part to the lack of simple methods for generating these radicals without simultaneously producing thiyl radicals by direct S—S homolysis or by homolytic displacement at sulfur.

5.7 α-Sulfinyl Radicals

Relatively little is known about radicals adjacent to a sulfinyl function. The pyrolysis of dimethyl sulfoxide is suggested to involve a chain decomposition with $CH_3S(O)CH_2\cdot$ and sulfine CH_2SO as intermediates (Eq. 5-30; Thyrion and Debecker, 1973). Block and co-workers (1976) have confirmed

$$\begin{aligned} CH_3S(O)CH_3 &\xrightarrow{650^\circ} CH_3SO\cdot + CH_3\cdot \\ CH_3S(O)CH_3 + CH_3\cdot &\longrightarrow CH_4 + CH_3S(O)CH_2\cdot \\ CH_3S(O)CH_2\cdot &\longrightarrow CH_3\cdot + CH_2SO \end{aligned} \tag{5-30}$$

this sequence, using microwave spectroscopy to identify and determine the structure of sulfine. These same authors have also provided another pyrolytic route to $CH_3S(O)CH_2\cdot$ and sulfine, as shown in Eq. 5-31. Gollnick has

$$\begin{aligned} CH_3S(O)CH_2I &\xrightarrow{350^\circ} CH_3S(O)CH_2\cdot + I\cdot \\ CH_3S(O)CH_2\cdot &\longrightarrow CH_2SO + CH_3\cdot \\ CH_3S(O)CH_2I + CH_3\cdot &\longrightarrow CH_3I + CH_3S(O)CH_2\cdot \end{aligned} \tag{5-31}$$

studied the photochemistry of dimethyl sulfoxide and offers Eq. 5-32 to explain the formation of methane, formaldehyde, and dimethyl disulfide (Gollnick and Schade, 1973). Supporting this photochemical sequence is the

$$\begin{aligned} CH_3S(O)CH_3 &\xrightarrow{h\nu} CH_3\cdot + CH_3SO\cdot \\ CH_3S(O)CH_3 + CH_3\cdot &\longrightarrow CH_4 + CH_3S(O)CH_2\cdot \\ CH_3S(O)CH_2\cdot &\longrightarrow CH_3\dot{S}\underset{O}{\diagdown\!\diagup}CH_2 \longrightarrow CH_2{=}O + CH_3S\cdot \\ 2CH_3S\cdot &\longrightarrow CH_3SSCH_3 \end{aligned} \tag{5-32}$$

report (Petrova and Friedlina, 1966) that diisopropyl sulfoxide, on treatment with the excellent radical initiator dicyclohexyl peroxydicarbonate in chlorobenzene at 60°, affords acetone and diisopropyl disulfide. While it is known that formaldehyde is a decomposition product of sulfine (Block *et al.*, 1976), it is not known at present whether the fragmentation reaction of $CH_3S(O)CH_2\cdot$ (Eq. 5-30) occurs together with or to the exclusion of the rearrangement reaction of this same radical as proposed in Eq. 5-32.

Norman and colleagues have generated α-sulfinyl radicals in solution by a variety of routes (Eqs. 5-33 to 5-35) and have studied them by ESR using flow-system techniques (Carton *et al.*, 1975). Reaction 5-35 takes advantage of the fact that the reaction of hydroxyl radicals with dialkyl sulfoxides leads to attack at sulfur rather than hydrogen abstraction (Eq. 5-36, Lagercrantz and Forshult, 1969). Contrary to the pyrolysis (Block *et al.*, 1976) and photolysis (Gollnick and Schade, 1973) studies on neat dimethyl sulfoxide, Norman

$$CH_3S(O)CH_3 + PhN_2^+\ BF_4^- + Ti(III) \xrightarrow{pH\ 8} [Ph\cdot] \xrightarrow{CH_3S(O)CH_3} CH_3S(O)CH_2\cdot \quad (5\text{-}33)$$

$$CH_3S(O)CH_3 + HPO_2^- \longrightarrow CH_3S(O)CH_2\cdot \quad (5\text{-}34)$$

$$\text{1,3-dithiane 1,3-dioxide} \xrightarrow{HO\cdot} \text{adduct (S(O}\cdot\text{)(OH))} \longrightarrow HO_2S(CH_2)_3S(O)CH_2\cdot \quad (5\text{-}35)$$

$$CH_3S(O)CH_3 \xrightarrow{HO\cdot} CH_3S(O\cdot)(OH)CH_3 \longrightarrow CH_3\cdot + CH_3SO_2H \quad (5\text{-}36)$$

$$CH_3S(O)CH_3 + CH_3\cdot \not\longrightarrow CH_3S(O)CH_2\cdot \quad (5\text{-}37)$$

indicates that methyl radicals do not abstract hydrogen from dimethyl sulfoxide in solution (Eq. 5-37; Carton *et al.*, 1975). Norman also presents evidence that α-sulfinyl radicals are coplanar at the tervalent carbon atom, that the sulfinyl group withdraws 6% of the spin from that carbon, which is far less than the effect of a thiyl group (see Table 5.5), and that the *g* value for the radical is indistinguishable from that for a simple alkyl radical, again in contrast to α-thio radicals (see Table 5.4).

While there is no evidence from any of these studies for the formation of the dimer of α-sulfinyl radicals, e.g., $RS(O)CH_2CH_2S(O)R$, it is possible to produce these bis-sulfoxides in respectable yields by oxidation of α-sulfinyl carbanions with copper(II) as reported by Mislow (Eq. 5-38; Maryanoff *et al.*, 1973).

$$ArS(O)CH_2^- \xrightarrow{Cu(II)} ArS(O)CH_2CH_2S(O)Ar \quad (5\text{-}38)$$

(45% for Ar = *p*-tolyl)

5.8 α-Sulfonyl Radicals

Employing the same ESR flow-system techniques described in Section 5.7, Norman and co-workers have recently generated α-sulfonyl radicals (Eqs. 5-39 and 5-40; Carton *et al.*, 1975). Interestingly, they find that while sulfolane undergoes only α-H atom abstraction with the phenyl radical, β-H atom abstraction is the dominant process with the more electrophilic hydroxyl radical (Eq. 5-40; Carton *et al.*, 1975). These results are nicely in

$$CH_3S(=O)_2CH_3 \xrightarrow{Ph\cdot} CH_3S(=O)_2CH_2\cdot \qquad (5\text{-}39)$$

$$\text{sulfolane} \xrightarrow{Ph\cdot} \alpha\text{-sulfolanyl radical}; \quad \text{sulfolane} \xrightarrow{HO\cdot} \alpha\text{-radical (weak)} + \beta\text{-radical (strong)} \qquad (5\text{-}40)$$

(weak) (strong)

accord with the observations on chlorination of sulfones described previously in Section 5.3. Norman concludes that α-sulfonyl radicals are coplanar at the tervalent carbon atom and that *the sulfonyl group is wholly ineffective at removing spin* (see the data in Tables 5.4 and 5.5). Thus canonical form **5-14b** is irrelevant (Carton *et al.*, 1975). Kaiser has determined by the reactions

$$CH_3\dot{C}H{-}S^{+}(O^-)(=O){-}C_2H_5 \;\longleftrightarrow\!\!\!\!\!\!\times\;\; CH_3CH{=}S^{+}(O^-)(O\cdot){-}C_2H_5$$

5-14a **5-14b**

$$PhSO_2{-}\overset{*}{C}(CO_3t\text{-Bu})(CH_3){-}R \xrightarrow[\text{hydrogen atom donor}]{\Delta \text{ or } h\nu} PhSO_2{-}CH(CH_3){-}R \text{ (racemic)}$$

$$PhSO_2{-}\overset{*}{C}H(CH_3){-}R + PhSO_2{-}C(CO_3t\text{-Bu})(CH_3){-}R \xrightarrow{h\nu} PhSO_2{-}CH(CH_3){-}R \qquad (5\text{-}41)$$

optically active, *n* parts; racemic, *m* parts; *n* parts still optically active

shown in Eq. 5-41 that α-sulfonyl radicals generated from chiral precursors are not asymmetric (Kaiser and Mayers, 1965). Irradiation of α-bromosulfones is a novel method of generating α-sulfonyl radicals as indicated by reactions 5-42 (Koshar and Mitsch, 1973) and 5-43 (Oku and Philips, 1973). The heavy preponderance of tricyclic products in Eq. 5-43 presumably reflects the instability of radical **5-15a** in equilibrium with radical **5-15b**.

$$(CF_3SO_2)_2CHBr \xrightarrow[\text{AIBN}]{h\nu \text{ or}} (CF_3SO_2)_2CH\cdot \xrightarrow{CH_2{=}CHR} (CF_3SO_2)_2CHCH_2CHRBr \qquad (5\text{-}42)$$

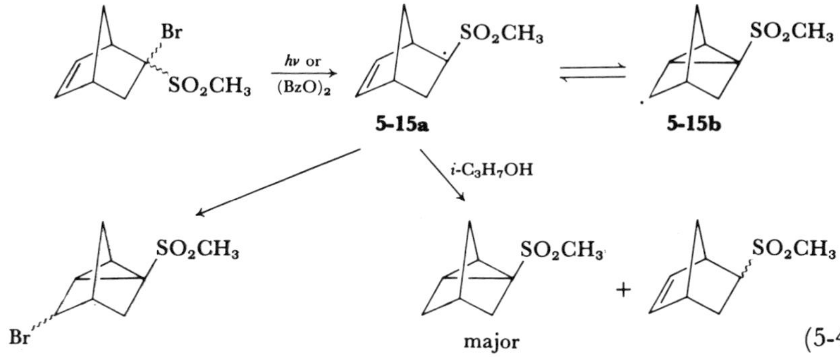

Br, SO_2CH_3; $h\nu$ or $(BzO)_2$; SO_2CH_3 **5-15a**; SO_2CH_3 **5-15b**; $i\text{-}C_3H_7OH$; SO_2CH_3, Br; SO_2CH_3 major; + SO_2CH_3 (5-43)

An intramolecular homolytic aromatic substitution reaction has been described for an α-sulfonyl radical generated by treatment of a sulfonylacetic acid with persulfate (Eq. 5-44; Dewar *et al.*, 1972). A minor product in this

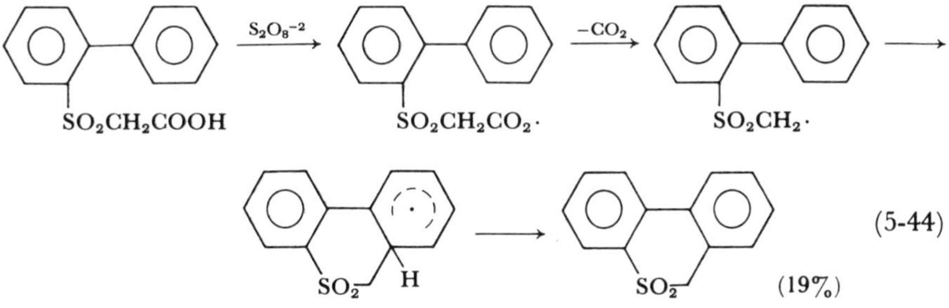

$SO_2CH_2COOH \xrightarrow{S_2O_8^{-2}} SO_2CH_2CO_2\cdot \xrightarrow{-CO_2} SO_2CH_2\cdot \longrightarrow$ SO_2, H $\longrightarrow$ SO_2 (19%) (5-44)

SO_2, O, C=O, CH_2 $\longrightarrow$ SO_2, O, C=O $\longrightarrow$

$OCCH_2SO_2\cdot$ (O) $\xrightarrow[+H\cdot]{-SO_2}$ OAc (5-45)

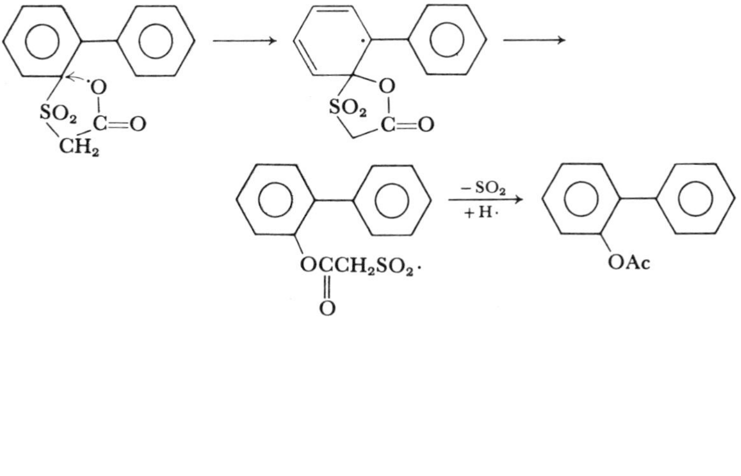

reaction is 2-acetoxybiphenyl. While the authors offer no explanation for the formation of this by-product, a novel radical rearrangement of the type indicated in Eq. 5-45 might be competitive with loss of carbon dioxide because of the high energy of the α-sulfonyl radical.

5.9 The Dibenzenesulfenamidyl Radical

When tribenzenesulfenamide $(PhS)_3N$ is heated at 78°, nitrogen and diphenyl disulfide are formed in virtually quantitative yield. A purple coloration is seen and an ESR signal detected on decomposition in solution. A plausible mechanism for this reaction has been provided by Barton who also reports a number of other intriguing transformations (shown in Scheme 5-2), all involving the dibenzenesulfenamidyl radical as the key intermediate (Barton *et al.*, 1973; the reader is urged to work out mechanisms before consulting the original references).

Scheme 5-2. Reactions of the Dibenzenesulfenamidyl Radical

$$2(PhS)_3N \xrightarrow{\Delta} 3PhSSPh + N_2$$

$$2(PhS)_3N \longrightarrow 2PhS\cdot + 2(PhS)_2N\cdot \quad \text{(purple)}$$

$$2(PhS)_2N\cdot \rightleftharpoons (PhS)_2N{-}N(SPh)_2$$

$$(PhS)_2N{-}N(SPh)_2 \longrightarrow PhSN{=}NSPh + 2PhS\cdot$$

$$PhSN{=}NSPh \longrightarrow N_2 + 2PhS\cdot$$

$$6PhS\cdot \longrightarrow 3PhSSPh$$

2-Naphthol + $(PhS)_3N$ $\xrightarrow{\Delta}$ 1-(NSPh)-naphthalen-2(1H)-one (85%) + PhSSPh

2,3,4,5-Tetraphenylpyrrole + $(PhS)_3N$ $\xrightarrow{\Delta}$ 2,4,5,6-tetraphenylpyrimidine (82%) + PhSSPh

5.10 The Stevens Rearrangement

We have previously considered the Stevens rearrangement, a four-electron 1,2 rearrangement, as a synthetically useful transformation of sulfur ylides (Eq. 5-46). Mechanistically this reaction poses a problem since concerted S_N2-like displacement of R by the ylide electron pair would involve a geometrically difficult inversion at R. In terms of the Woodward–Hoffmann

$$\overset{+}{\ddot{S}}(R)\text{—}\overset{-}{\ddot{C}} \longrightarrow \ddot{S}\text{—}C(R) \qquad (5\text{-}46)$$

rules a concerted anionic [1,2] sigmatropic shift, as the Stevens rearrangement might be classified, may occur either as a suprafacial migration with inversion at the migrating atom or as an antarafacial migration with retention at the migrating atom (Woodward and Hoffmann, 1970). In fact the Stevens rearrangement occurs with predominant retention of configuration with chiral R. An alternative possibility would be a stepwise radical dissociation–recombination sequence involving an intermediate radical pair which could collapse to product fast enough that when R is chiral, net retention of stereochemistry is observed (Eq. 5-47). Work by Baldwin,

$$\overset{+}{\ddot{S}}(R)\text{—}\overset{-}{\ddot{C}} \longrightarrow \left[\overset{+}{\dot{S}}\text{—}\overset{-}{\ddot{C}}\ \ R\cdot \longleftrightarrow \ddot{S}\text{—}\dot{C}\ \ R\cdot\right] \longrightarrow \ddot{S}\text{—}C(R) \qquad (5\text{-}47)$$

Schöllkopf, and others has provided support for the stepwise sequence. Thus, brief heating of the S-methyl-S-benzylphenacylsulfoniumylide, **5-16**, in refluxing toluene affords products derived from radical dimerization in addition to the major product **5-17** which is the result of a Stevens rearrangement (Baldwin *et al.*, 1970; Schöllkopf *et al.*, 1969).

$$\underset{\textbf{5-16}}{CH_3\overset{+}{S}(CH_2Ph)\text{—}\overset{-}{C}H\text{—}C(=O)\text{—}Ph} \xrightarrow[110^\circ]{\Delta} CH_3S\text{—}\dot{C}HC(O)Ph \quad PhCH_2\cdot \longrightarrow$$

$$\underset{\textbf{5-17}}{CH_3S\text{—}CH(CH_2Ph)C(O)Ph}\ (45\%) + CH_3S\text{—}CHC(O)Ph\text{—}CHC(O)Ph\text{—}SCH_3\ (22\%) + (PhCH_2\text{—})_2\ (14\%) + \cdots$$

At 40°, the sulfonium ylide **5-16** disappeared with first-order kinetics for three half lives. Large positive entropies of activation were obtained (+19 e.u. in benzene and +38 e.u. in deuteriochloroform) which is consistent with a dissociative mechanism (Schöllkopf *et al.*, 1970). Deuterium labeling experiments proved informative in these studies. Thus, rearrangement of an intimate mixture of ylides labeled in different positions indicated that approximately 18% (±6%) of the Stevens product was formed by intermolecular combination processes (Eq. 5-48; Baldwin *et al.*, 1970).

Rearrangement of a chiral monodeuterated sample of this same ylide gave diastereomeric Stevens products from which it could be determined that 44 ± 20% retention of configuration had occurred for product formed within the solvent cage (i.e., correcting for the 18 ± 6% intermolecular Stevens product) (Eq. 5-49; Baldwin *et al.*, 1970). A CIDNP effect, illustrated in

$$CH_3\overset{+}{S}(CD_2Ph)-\overset{-}{C}HC(O)Ph + CH_3\overset{+}{S}(CH_2Ph)-\overset{-}{C}HC(O)C_6D_5 \xrightarrow{130^\circ} \left.\begin{array}{l} CH_3SCH(CD_2Ph)C(O)Ph \\ CH_3SCH(CH_2Ph)C(O)C_6D_5 \end{array}\right\} \text{major} \quad \left.\begin{array}{l} CH_3SCH(CH_2Ph)C(O)Ph \\ CH_3SCH(CD_2Ph)C(O)C_6D_5 \end{array}\right\} \text{minor} \tag{5-48}$$

$$CH_3-\overset{+}{S}(CHDPh)-\overset{-}{C}HC(O)Ph \xrightarrow{110^\circ} CH_3-S-CH(CHDPh)C(O)Ph \quad (44 \pm 20\% \text{ retention}) \tag{5-49}$$

Fig. 5-1, was observed from the thermal rearrangement of ylide **5-16**. The emission seen for the methylene multiplet of **5-17** at δ3.2, the enhanced absorption associated with the methyne triplet of **5-17** at δ4.4, and the methylene singlet of bibenzyl at δ2.87 reflect the spin population perturbations associated with products formed by radical combination (Lepley, 1973; Ward, 1973). Other groups have reported CIDNP effects with other Stevens rearrangements involving sulfur ylides (Ando *et al.*, 1972; Iwamura *et al.*, 1971). Recent detailed study of the mechanism of the Stevens rearrangement of nitrogen ylides (Ollis *et al.*, 1975; Dolling *et al.*, 1975) affords results that are compatible with two possible mechanisms:

1. a radical pair mechanism with an average geminate recombination rate which is exceptionally fast, or
2. dual pathways involving concurrent radical pair and concerted processes.

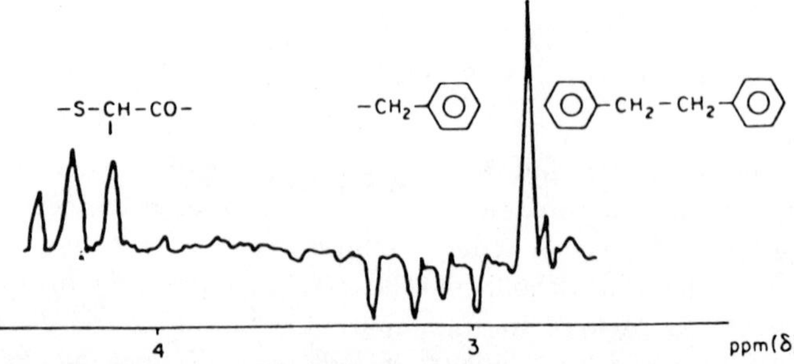

Figure 5-1. CIDNP NMR spectrum from thermal rearrangement of sulfonium ylide **5-16** at 90° in chloroform (Schöllkopf *et al.*, 1969). Reprinted with permission from *Tetrahedron Letters* **31**, 2619 (1969), Pergamon Press, Ltd.

Furthermore, Dewar has suggested that despite the apparent orbital-symmetry restrictions, Stevens rearrangements may nonetheless occur in a concerted rather than stepwise manner (Dewar and Ramsden, 1974).

5.11 The Wittig Rearrangement

Dibenzylsulfide on treatment with *n*-butyllithium can undergo either a Sommelet-type rearrangement (a [2,3] sigmatropic shift; Eq. 5-50, path a) or a Wittig rearrangement (Eq. 5-50, path b). The latter reaction is an anionic [1,2] sigmatropic shift which would not be expected to occur in a concerned manner (see above discussion of the Stevens rearrangement). The isolation of dimeric products lends support to a dissociative mechanism for the Wittig rearrangement of dibenzylsulfide (Biellmann and Schmitt, 1973; also see Lepley, 1973, for evidence supporting the dissociative character of the Wittig rearrangement of benzyl ethers). It is suggested that path a is

$$(PhCH_2)_2S \xrightarrow{BuLi} Ph\bar{C}HSCH_2Ph \xrightarrow{a} \text{[methylenecyclohexadiene}-CH(S^-)Ph]$$

$$Ph\bar{C}HSCH_2Ph \xrightarrow{b} [PhC(H)\cdot{-}S^-\cdot CH_2Ph] \longrightarrow PhCH(S^-)CH_2Ph + PhCH(S^-){-}CH(S^-)Ph + PhCH_2CH_2Ph \quad (5\text{-}50)$$

favored over path b at lower temperatures and under conditions where there is a high degree of dissociation of ion pairs; intimate ion pairs are said to lead exclusively to the dissociative Wittig pathway (Biellmann and Schmitt, 1973).

5.12 Sulfur-Centered Radicals

General

Sulfur-centered radicals, such as the thiyl ($RS\cdot$), polysulfide ($RS_n\cdot$), sulfinyl ($RSO\cdot$), and sulfonyl ($RSO_2\cdot$) radicals, have figured prominently in the chemical literature in the development of rules regarding the regioselectivity of olefin addition reactions, in studies of industrially important antioxidants and radical initiators and chain-transfer agents in polymerization reactions, in investigations of radiation-induced reactions both *in vitro* and *in vivo* (including radiation protection mechanisms), in the study of homolytic displacement and pyrolysis reactions, and in ESR studies. Because of the enormity of the literature in this area and the fact that a number of extensive review articles have recently appeared on these subjects (Block, 1969; Kellogg, 1969; Kice, 1973; Strausz *et al.*, 1972), this section will necessarily be a selective résumé.

The most obvious means of generating a sulfur-centered radical is the homolysis of a bond to sulfur. A number of organosulfur compounds undergo facile homolysis at temperatures below 150°, in most cases a consequence of the special stability of radicals of the type $R{-}\ddot{\underset{..}{X}}{-}\ddot{\underset{..}{Y}}\cdot$ (see Eq. 5-51) which is reflected in the bond dissociation energies, $\Delta H^{\ddagger}$ (Kice, 1973):

$$\begin{aligned}
CH_3S{-}SCH_3 &\longrightarrow CH_3S\cdot + CH_3SS\cdot \quad && \Delta H^{\ddagger} = 45 \text{ kcal/mole} \\
ArS(O){-}CH_2Ph &\longrightarrow ArSO\cdot + PhCH_2\cdot && \Delta H^{\ddagger} = 44 \\
ArSO_2{-}SO_2Ar &\longrightarrow 2ArSO_2\cdot && \Delta H^{\ddagger} = 41 \\
ArS(O){-}SAr &\longrightarrow ArSO\cdot + ArS\cdot && \Delta H^{\ddagger} = 35 \\
CH_3S_2{-}S_2CH_3 &\longrightarrow 2CH_3SS\cdot && \Delta H^{\ddagger} = 37 \\
ArS(O){-}SO_2Ar &\longrightarrow ArSO\cdot + ArSO_2\cdot && \Delta H^{\ddagger} = 28
\end{aligned} \tag{5-51}$$

However, most sulfur compounds (particularly the simple alkyl derivatives of sulfides, disulfides, and sulfones) show considerable thermal stability reflecting the substantial bond dissociation energies (~74 kcal for C—S or S—S bonds).† Indeed because of their great stability, certain polysulfones have been used in the face shields worn by the astronauts. Benzylic, allylic, or strained or otherwise activated carbon–sulfur bonds can be cleaved pyrolytically at 300–600°C (the C—S bond dissociation energies for benzyl sulfides and sulfones are in the range of 50–53 kcal/mole (Mackle, 1963)) or by irradiation with ultraviolet light.

As we shall see, pyrolytic and photolytic C—S homolysis has formed the basis for synthetically valuable methods of carbocyclic ring synthesis based on the introduction and subsequent extrusion of a sulfur-containing temporary bridging element, e.g., as shown in Eq. 5-52. Combining the photolytic

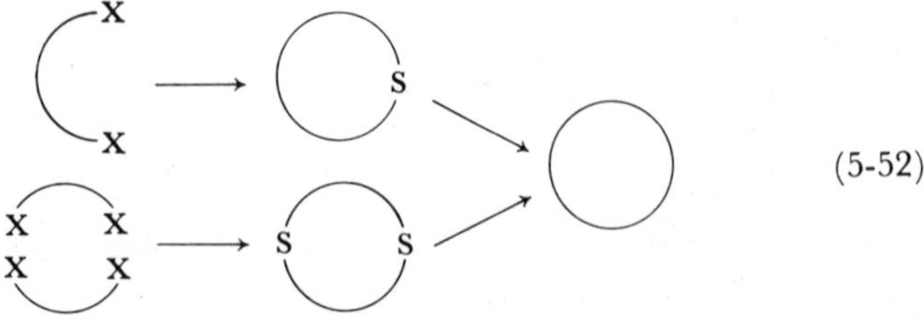

(5-52)

scission of benzylic or allylic C—S bonds with the desulfurization of the resultant thiyl radicals by tervalent phosphorus solvents (a reaction discovered by Walling), Corey and Block developed a method for desulfurization with concomitant C—C bond formation (Corey and Block, 1969). The

† See Table B-6, page 302, for typical bond dissociation energies.

Table 5-8. Ring Synthesis via Cophotolysis of Sulfides and Phosphites

Entry	Sulfide	Solvent	Product(s) (Yield)	Reference
1	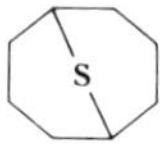 S	$(C_8H_{17}O)_3P$	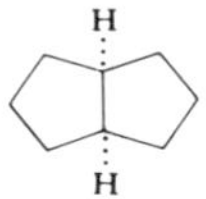 H, H (22%); cyclooctene (22%); cyclooctane (8%)	Corey and Block (1969)
2	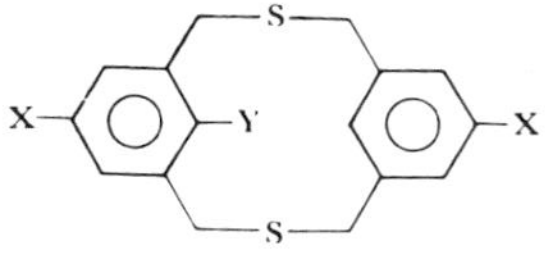 X, Y, S, S, X	$(CH_3O)_3P$	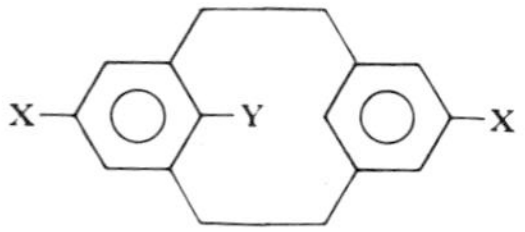 X, Y, X ($X = CH_3$, $Y = H$, 85%; $X = H$, $Y = CH_3$, 49%)	Boekelheide *et al.* (1973)
3	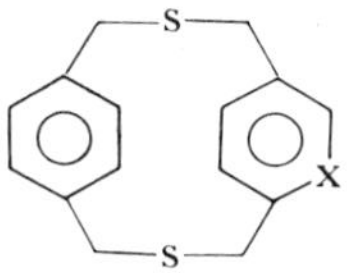 S, S, X	$(C_2H_5O)_3P$	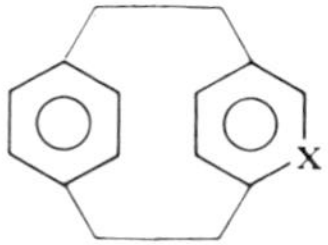 X ($X = N$, 60%; $X = CH$, 75%)	Bruhin *et al.* (1973)
4	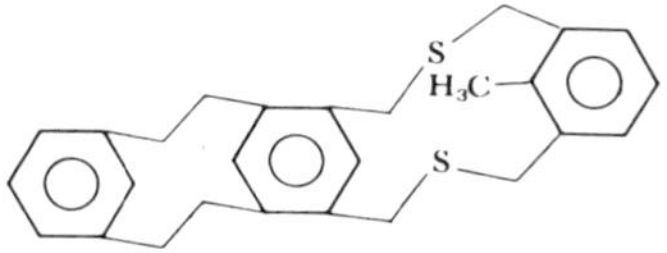 S, H_3C, S	$(C_2H_6O)_3P$	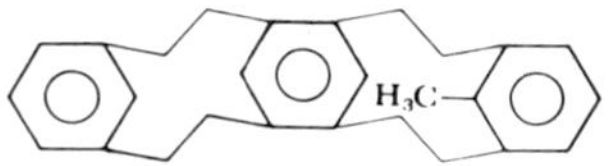 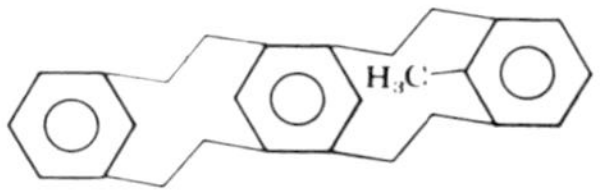 H_3C **a** (18%); H_3C **b** (43%)	Umemoto *et al.* (1974)
5	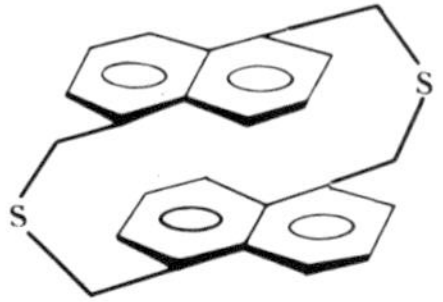 S, S	$(C_2H_5O)_3P$/THF	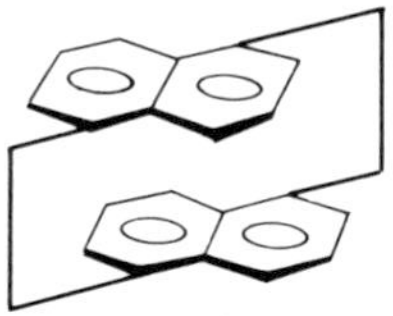 (37%)	Haenel (1974)

Table 5-9. Ring Synthesis via Sulfone Pyrolysis

Entry	Sulfone	Temperature (°C)	Product(s) (Yield)	Reference
1	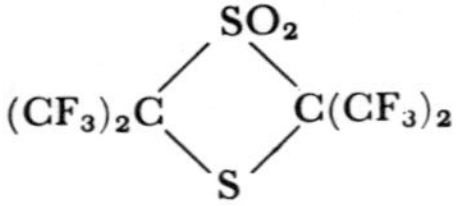 SO_2 $(CF_3)_2C$ $C(CF_3)_2$ S	550°	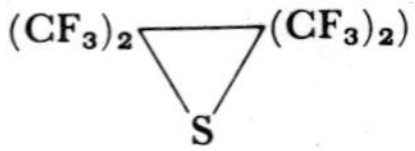 $(CF_3)_2$ $(CF_3)_2$ S (95%)	Middleton (1964)
2	 SO_2	710°	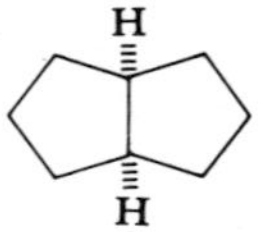 H H (50%); cyclooctene (10%)	Corey and Block (1969)
3	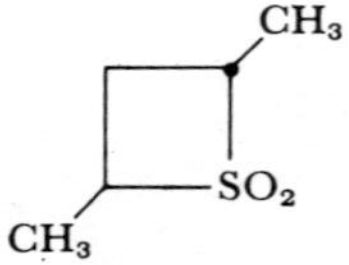 CH_3 SO_2 CH_3	350°	*cis*-1,2-Dimethylcyclopropane (23%); *trans*-1,2-dimethylcyclopropane (20%); 2-pentenes (17%)	Trost *et al.* (1971)
4	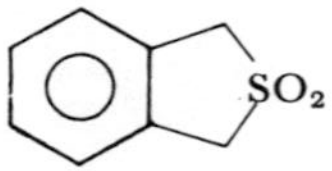 SO_2	500°	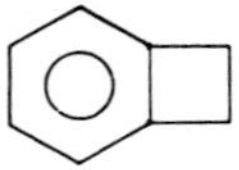 (60%)	Cava and Kuczkowski (1970)
5	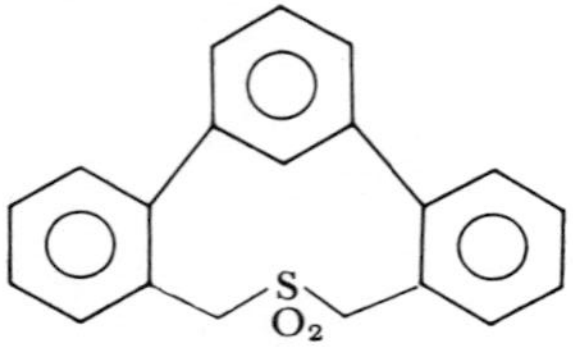 S O_2	300°	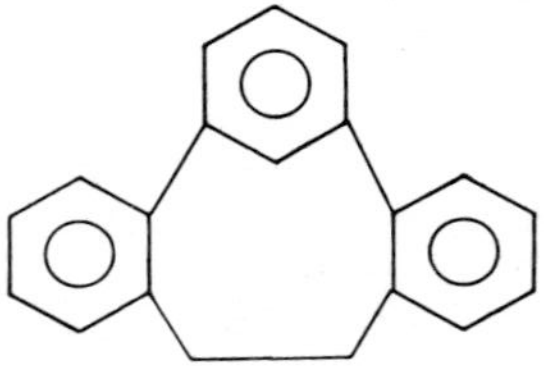 (100%)	Vögtle (1969)

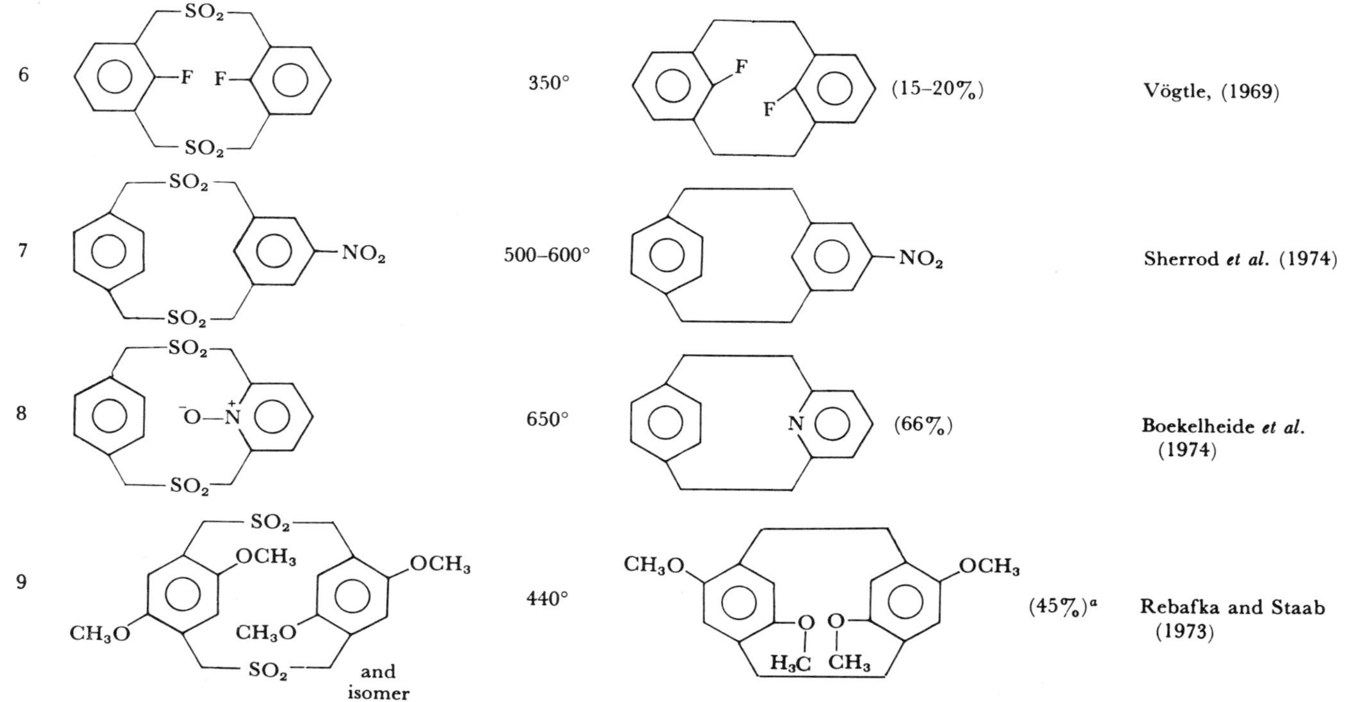

6	350°		(15–20%)	Vögtle, (1969)
7	500–600°			Sherrod *et al.* (1974)
8	650°		(66%)	Boekelheide *et al.* (1974)
9	440°		(45%)[a]	Rebafka and Staab (1973)

[a] This cyclophane could also be prepared in 50% yield by irradiation of the bissulfone.

homolytic sequence is illustrated for dibenzyl sulfide in Eq. 5-53. This procedure has proven particularly useful in cyclophane synthesis (cf. Table 5-8), complementing the previously discussed Stevens rearrangement

$$\begin{aligned} &PhCH_2SCH_2Ph \xrightarrow{h\nu} PhCH_2\cdot + PhCH_2S\cdot \\ &PhCH_2S\cdot + R_3P \longrightarrow [PhCH_2S\text{—}\dot{P}R_3] \longrightarrow PhCH_2\cdot + R_3P{=}S \\ &2PhCH_2\cdot \longrightarrow PhCH_2CH_2Ph\ (59\%) \end{aligned} \quad (5\text{-}53)$$

method. For example products **a** and **b** in entry 4 of Table 5-8 were obtained in 50 and 2.3% yield, respectively, using the Stevens route. In these cyclophane syntheses from bis-sulfides, the extrusion of the two sulfur atoms occur in a stepwise fashion, as indicated by the isolation of the monosulfides in some cases. The formation from unsymmetrical bis-sulfides of unsymmetrical cyclophanes uncontaminated by the symmetrical coupling products (cf. entries 2 and 3 in Table 5-8) also argues against the generation of a pair of diradicals by double extrusion of sulfur.

Pyrolysis of a sulfone affords sulfonyl radicals which, at the elevated temperatures employed, undergo facile loss of sulfur dioxide (Eq. 5-54).

$$R\text{—}SO_2\cdot \longrightarrow R\cdot + SO_2 \quad (5\text{-}54)$$

Despite the rather high temperatures required to effect C—S homolysis,† sulfone pyrolysis represents a particularly valuable and versatile synthetic approach to cyclophanes and benzocyclobutenes as indicated in Table 5-9.‡ As in the case of the R_2S/R'_3P photochemical method of cyclophane synthesis, separate desulfonylation steps are probably involved in pyrolysis of the disulfones. Boekelheide has suggested that loss of the first SO_2 from the

SO_2 ... $\xrightarrow[2\ mm]{500°}$ ... $\xrightarrow{-SO_2}$ ... $\longrightarrow$ [... SO_2]* $\longrightarrow$ ... $SO_2\cdot$ $\xrightarrow{-SO_2}$ $\longrightarrow$ ...

(5-55)

† Cyclic sulfones which on desulfonylation afford π systems (e.g., episulfones, sulfolene, etc.) lose sulfur dioxide very readily. These "chelotropic" processes will be considered, together with other pericyclic processes, in Chapter 7.

‡ It is *not* essential that the C—S bonds be benzylic (Grütze and Vögtle, 1977; Haenel *et al.*, 1977; Vögtle and Rossa, 1977).

disulfone gives a diradical, closure of which is highly exothermic so that at the low pressure employed (usually ~2 mm) monosulfone might be in a highly excited vibrational state which might lose SO_2 competitively with collisional deactivation (Eq. 5-55; Sherrod *et al.*, 1974).

Extrusion of SO_2 from bis-sulfones to give cyclophanes may also be effected photochemically with yields very similar (if not slightly higher) than those obtained by pyrolysis (Rebafka and Staab, 1973).

Thiyl Radicals

We have considered a variety of reactions involving thiyl radicals elsewhere in this chapter. These radicals, most usually generated by thermal or photochemical homolysis of S—N, S—S, or C—S bonds undergo such characteristic reactions as addition to unsaturated systems, H-abstraction, radical combination, and attack on neutral atoms (such as sulfur or phosphorus). In contrast to oxy-radicals (such as $t\text{-}C_4H_9O\cdot$ or $CH_3CO_2\cdot$), thiyl radicals show no tendency to undergo fragmentation (the processes indicated should not be favored thermodynamically) (Eq. 5-56).

$$\begin{aligned} R_3C\text{—}S\cdot &\not\longrightarrow R_2C{=}S + R\cdot \\ RC(O)S\cdot &\not\longrightarrow OCS + R\cdot \end{aligned} \qquad (5\text{-}56)$$

Free radical addition of thiols to olefins and acetylenes is generally a rapid process giving high yields of the adducts. Intramolecular as well as intermolecular addition may be realized as shown by the reactions of Eqs. 5-57 and 5-58. Several recent detailed surveys are available on radical addition

$$\text{(cyclohexenyl)}CH_2SH \xrightarrow{h\nu} \text{(bicyclic sulfide)} \quad (70\%) \qquad \text{(Surzur } et\ al.\text{, 1971)} \qquad (5\text{-}57)$$

$$PhC{\equiv}CCH_2CH_2CH_2SH \xrightarrow{h\nu} PhCH{=}\text{(tetrahydrothiophen-2-ylidene)} \quad (75\%) + \cdots \qquad \text{(Surzur } et\ al.\text{, 1969)} \qquad (5\text{-}58)$$

of thiols to unsaturated compounds (Griesbaum, 1970; Kellogg, 1969). The addition reactions are chain processes with relatively long chain lengths (Eq. 5-59). Thiols are excellent hydrogen atom transfer agents and rate

$$\begin{aligned} &(1)\quad RSH \xrightarrow[\text{or } \Delta]{h\nu} RS\cdot + H\cdot \\ &(2)\quad RS\cdot + {>}C{=}C{<} \rightleftharpoons RS\text{—}\overset{|}{C}\text{—}\overset{|}{C}\cdot \\ &(3)\quad RS\text{—}\overset{|}{C}\text{—}\overset{|}{C}\cdot + RSH \longrightarrow RS\text{—}\overset{|}{C}\text{—}\overset{|}{C}\text{—}H + RS\cdot \\ &(4)\quad 2RS\cdot \longrightarrow RSSR \end{aligned} \qquad (5\text{-}59)$$

constants of the order of $10^6\ M^{-1}\ \text{sec}^{-1}$ have been reported for step (3).† Indeed, the effectiveness of thiols in counteracting the harmful action of ionizing radiation reflects the ability of thiols to supply hydrogen atoms to radiation-produced radicals, thus "repairing" the damage. There are a number of synthetic applications which exploit the hydrogen atom transfer capabilities of thiols, e.g., in Cr(II) deiodinations (Bachi *et al.*, 1968) and aldehyde decarbonylations (Harris and Waters, 1952). Thiyl radicals can also function as H-atom abstractors as in the example in Eq. 5-60 (Morizur, 1964; elemental sulfur may also be used here).

PhSSPh, Δ or $h\nu$ (5-60)

Guaiazulene

Sulfinyl Radicals

Sulfinyl radicals have been produced by homolysis of S—X or O—X bonds in sulfinyl and sulfenyl derivatives RS(O)—X and RSO—X, respectively, by abstraction of X· from these same derivatives, and by fragmentation processes in Eqs. 5-61.

The ESR spectra of alkanesulfinyl radicals have been reported (Gilbert *et al.*, 1977b; Howard and Furimsky, 1974) and it is concluded that the unpaired electron is localized in a S—O π-orbital (actually π-antibonding) as might be represented with resonance structures **5-18**. The principal fate

$$\text{ArS(O)CH}_2\text{Ph} \xrightarrow{\Delta} \text{ArSO}\cdot + \text{PhCH}_2\cdot \quad \text{(Miller } et\ al., \text{ 1968)}$$
$$\text{PhS(O)SPh} \xrightarrow{\Delta} \text{PhSO}\cdot + \text{PhS}\cdot \quad \text{(Koch } et\ al., \text{ 1970)}$$
$$\text{ArS(O)SO}_2\text{Ar} \xrightarrow{\Delta} \text{ArSO}\cdot + \text{ArSO}_2\cdot \quad \text{(Kice, 1973)}$$
$$\text{CH}_3\text{S(O)CH}_2\text{C(O)Ph} \xrightarrow{h\nu} \text{CH}_3\text{SO}\cdot + \text{PhC(O)CH}_2\cdot \quad \text{(Majeti, 1971)}$$
$$\text{ArSO—NO}_2 \xrightarrow{\Delta} \text{ArSO}\cdot + \text{NO}_2 \quad \text{(Topping and Kharasch, 1962)}$$
$$\text{ArS(O)Cl} \xrightarrow{\text{Zn or } h\nu} \text{ArSO}\cdot \quad \text{(Bernard, 1957; Gilbert } et\ al., \text{ 1977b)}$$

(5-61)

$$\text{R—}\ddot{\underset{\cdot\cdot}{\text{S}}}\text{—}\dot{\underset{\cdot\cdot}{\text{O}}}\text{:} \longleftrightarrow \text{R—}\overset{+}{\dot{\underset{\cdot\cdot}{\text{S}}}}\text{—}\overset{-}{\ddot{\underset{\cdot\cdot}{\text{O}}}}\text{:} \quad \text{or} \quad \text{R—S}\dot{\cdots}\text{O:}$$

5-18

of sulfinyl radicals is radical combination. Dimerization of sulfinyl radicals is thought to occur by head-to-tail combination to give a sulfinyl sulfinate which rearranges to a thiosulfonate. At $-100°\text{C}$, *t*-butylsulfinyl radicals decay with a bimolecular rate constant of $6 \times 10^7\ M^{-1}\ \text{sec}^{-1}$ (Howard and

† A rate constant of $10^3\ 1\ M^{-1}\ \text{sec}^{-1}$ has recently been determined for the reverse of step 3, namely hydrogen abstraction by RS· (Miyashita *et al.*, 1977).

Furimsky, 1974). Interestingly, the half life for this process is nine orders of magnitude smaller than the half life for dimerization of the di-*t*-butylperoxy radical under the same conditions. Howard indicates that if a solution in which a significant concentration of $t\text{-}C_4H_9SO\cdot$ had decayed was heated in the cavity of the ESR spectrometer from $-140°$ to $-30°$, there was no evidence for radical regeneration (Howard and Furimsky, 1974). The significance of this observation is to provide support for head-to-tail combination over oxygen-to-oxygen or sulfur-to-sulfur coupling which would be expected to be reversible (Eq. 5-62).

$$t\text{-}C_4H_9SO\cdot \longrightarrow [t\text{-}C_4H_9S\text{—}O\text{—}S(O)t\text{-}C_4H_9] \longrightarrow t\text{-}C_4H_9SSO_2t\text{-}C_4H_9$$
$$t\text{-}C_4H_9SO\cdot \;\not\rightleftarrows\; t\text{-}C_4H_9S\text{—}O\text{—}St\text{-}C_4H_9 \quad \text{or} \quad t\text{-}C_4H_9S(O)S(O)t\text{-}C_4H_9 \tag{5-62}$$

A number of dialkyl sulfoxides and thiosulfinates (RS(O)SR) act as inhibitors in the autoxidation of hydrocarbons. It has been suggested that this inhibitory action derives from the ability of these compounds to form sulfenic acids or related compounds on thermal decomposition (Eq. 5-63; Block and O'Connor, 1974; Koelewijn and Berger, 1972). The sulfenic acids

$$\begin{aligned} t\text{-}C_4H_9S(O)t\text{-}C_4H_9 &\xrightarrow{\Delta} t\text{-}C_4H_9SOH + C_4H_8 \\ t\text{-}C_4H_9SS(O)t\text{-}C_4H_9 &\xrightarrow{\Delta} t\text{-}C_4H_9SSOH + C_4H_8 \\ C_2H_5S(O)SC_2H_5 &\xrightarrow{\Delta} C_2H_5SOH + CH_3CHS \end{aligned} \tag{5-63}$$

have been found to be extremely active radical scavengers showing rate constants of at least $10^7\ M^{-1}\ sec^{-1}$ for reactions with peroxy radicals (Koelewijn and Berger, 1972). The efficacy of sulfenic acids as hydrogen atom transfer agents is undoubtedly a consequence of the appreciable stability of the sulfinyl radical (due to its delocalized structure).

From the elegant studies of Mislow and co-workers, it is known that sulfinyl radicals are involved in the thermal racemization of benzyl sulfoxides. Thus racemization of benzyl *p*-tolyl sulfoxide is accompanied by the appearance of bibenzyl and *p*-tolyl-*p*-toluenethiosulfonate as decomposition products (Eq. 5-64; Miller *et al.*, 1968). Furthermore, it was shown that with PhCHDS(O)Ar the rate of racemization at the chiral methylene carbon adjacent to sulfur is the same as the rate of racemization at sulfur. The

$$Ar\ddot{S}(=O)CH_2Ph \underset{\Delta}{\rightleftharpoons} [ArSO\cdot \ \cdot CH_2Ph] \underset{\Delta}{\rightleftharpoons} PhCH_2\ddot{S}(=O)Ar$$
$$[ArSO\cdot \ \cdot CH_2Ph] \longrightarrow ArSO_2SAr + PhCH_2CH_2Ph \tag{5-64}$$

appearance of CIDNP emission in the rearrangement of sulfenates to sulfoxides (Jacobus, 1970; a four-electron [1,2] sigmatropic shift) has been cited as evidence that a radical pair is involved in this process as well (Eq. 5-65; Ward, 1973; Lepley, 1973). Sulfinyl radicals are also thought to be involved in various reactions of thiirane oxides (Eqs. 5-65 to 5-68; Baldwin *et al.*, 1971; Lemal and Chao, 1973; Kondo *et al.*, 1972). In the

$$ArSO{-}CH_2Ph \xrightarrow{\Delta} [ArSO\cdot \cdot CH_2Ph] \longrightarrow ArS(O)CH_2Ph \qquad (5\text{-}65)$$

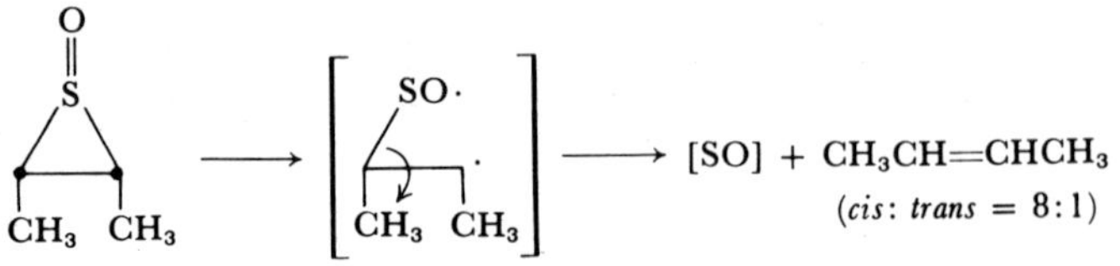

$$\text{cis-2,3-dimethylthiirane oxide} \longrightarrow [\cdot CH(CH_3)CH(CH_3)SO\cdot] \longrightarrow [SO] + CH_3CH{=}CHCH_3 \qquad (5\text{-}66)$$

(*cis*: *trans* = 8:1)

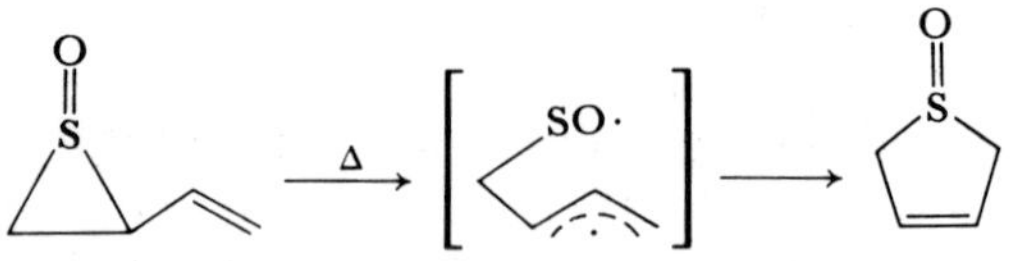

$$\text{vinylthiirane oxide} \xrightarrow{\Delta} [\text{allyl radical}\cdots CH_2SO\cdot] \longrightarrow \text{thiolene oxide} \qquad (5\text{-}67)$$

$$\text{2,3-diphenylthiirane oxide} + Ar_2C{=}S \xrightarrow[\Delta]{PhCH_3} [\cdot CH(Ph)CH(Ph)SO\cdot] \xrightarrow{(81\%)} \text{4,5-diphenyl-2,2-diaryl-1,3-dithiolane 1-oxide} \qquad (5\text{-}68)$$

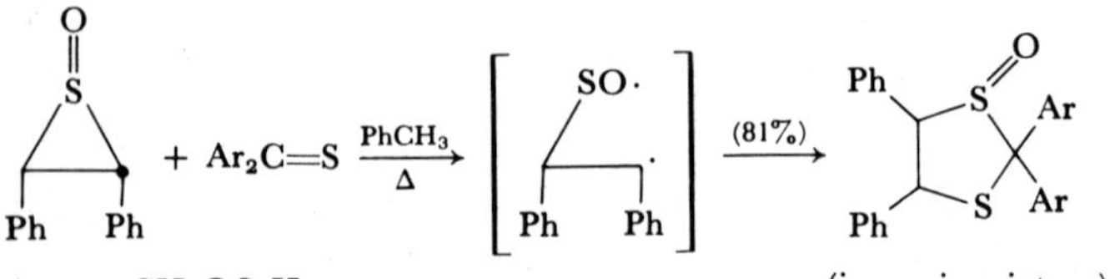

$Ar = p\text{-}CH_3OC_6H_4$ (isomeric mixture)

first case, a concerted loss of SO is unfavorable since SO is a *triplet* in its ground state (Lemal and Chao, 1973). Recent studies of the pyrolysis of *cis*- and *trans*-2,3-dideuteriothiirane oxide indicate that deuterated ethylenes are formed with 95% retention of stereochemistry (Aalbersberg and Vollhardt, 1977). The biradical intermediate [$\cdot$CHDCHDSO$\cdot$] is suggested to be stereochemically rather rigid, akin to the π-thiacyclopropane radical proposed in thiirane decomposition (Lown *et al.*, 1968). Alternatively there may be a significant contribution of a concerted process to thiirane oxide decomposition (Aalbersberg and Vollhardt, 1977).

Sulfonyl Radicals

Sulfonyl radicals, produced by chlorine atom abstraction from alkane- or arenesulfonyl chlorides with triethylsilyl radicals (reaction 5-69) or by hydrogen atom abstraction from alkanesulfinic acids with *t*-butoxy radicals (reaction 5-70), have been studied by ESR spectroscopy (Davies *et al.*, 1973; Kawamura *et al.*, 1972). The data suggest that the unpaired electron is centered mainly on sulfur in an orbital partaking predominantly of 3p

$$t\text{-}C_4H_9O\cdot + (C_2H_5)_3SiH \longrightarrow (C_2H_5)_3Si\cdot + t\text{-}C_4H_9OH$$

$$(C_2H_5)_3Si\cdot + RSO_2Cl \longrightarrow RSO_2\cdot + (C_2H_5)_3SiCl \quad (5\text{-}69)$$

$$RSO_2H + t\text{-}C_4H_9O\cdot \longrightarrow RSO_2\cdot + t\text{-}C_4H_9OH \quad (5\text{-}70)$$

character. The sulfonyl radicals are indicated to be pyramidal with respect to the sulfur atom. Rotation about the C—S bond in arenesulfonyl radicals is hindered as indicated by nonequivalence of the ortho-hydrogens in *p*-chloro- and *p*-fluorobenzenesulfonyl radicals. A possible equilibrium conformation is shown in **5-19**.

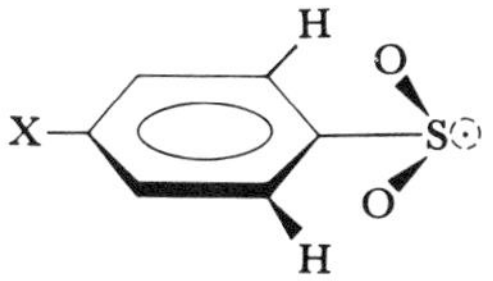

5-19

A reaction which is particularly germane to the study of air pollution is the reversal of the previously discussed desulfonylation reaction of sulfonyl radicals, namely, radical addition to sulfur dioxide, e.g., reactions 5-71 and 5-72. Sulfur dioxide is an excellent free radical trap, for the rate constant

$$CH_3\cdot + SO_2 \xrightarrow{k_1} CH_3SO_2\cdot \quad (5\text{-}71)$$

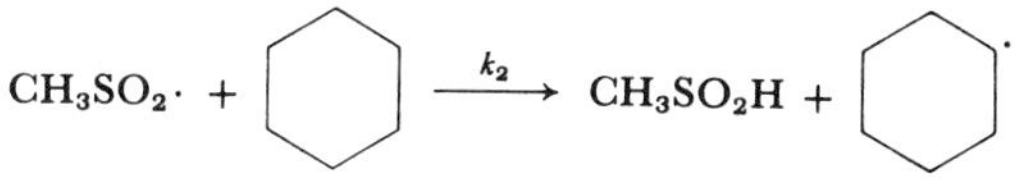

$$CH_3SO_2\cdot + C_6H_{12} \xrightarrow{k_2} CH_3SO_2H + C_6H_{11}\cdot \quad (5\text{-}72)$$

k_1 has the same value as the rate constant for the very fast reaction of methyl radical with oxygen, $2 \times 10^8\ M^{-1}\ \text{sec}^{-1}$ (James *et al.*, 1973). Alkanesulfonyl radicals in turn readily abstract hydrogen atoms from hydrocarbons ($k_2 = 3 \times 10^2\ M^{-1}\ \text{sec}^{-1}$ at 120°) forming alkanesulfinic acids which may contribute to aerosol formation in polluted atmospheres (Badcock *et al.*, 1971; Horowitz and Rajbenbach, 1975). Sulfur dioxide can even serve as an efficient biradical trap (Eq. 5-73; Wilson, 1974).

Under conditions where desulfonylation and hydrogen abstraction are precluded, sulfonyl radicals may undergo other reactions such as coupling or addition to multiple bonds. Arenesulfonyl radicals, generated photochemically from arenesulfonyl iodides apparently prefer to couple in a head-to-tail fashion (Eq. 5-74; da Silva Correa and Waters, 1968). On the other hand, oxidation of alkanesulfinic acids by Co(III), a process which may also involve alkanesulfonyl radicals (but more likely involves ligand transfer processes at cobalt), affords an α-disulfone in good yield (Eq. 5-75; Denzer *et al.*, 1966). The free radical addition of alkanesulfonyl bromides to acetylenes provides a useful synthesis of α,β-unsaturated sulfones (Eq. 5-76; Amiel, 1974).

$$\text{p-benzoquinone} + CH_2{=}CH{-}t\text{-Bu} \xrightarrow[SO_2,\ -11^\circ]{h\nu} \text{(radical adduct)} \xrightarrow{SO_2} \text{(spiro sulfone dienone)} \longrightarrow \text{(benzoxathiin sulfone, OH)}\ (81\%) \tag{5-73}$$

$$ArSO_2{-}I \xrightarrow[Ph{-}H]{h\nu} ArS(=O)_2\cdot \longleftrightarrow ArS(=O){-}O\cdot + I\cdot$$

$$ArSO_2SO_2Ar\ (5\%) \qquad [ArSO_2{-}O{-}S(=O)Ar] \longrightarrow ArSO_3\cdot + ArSO\cdot \tag{5-74}$$

$$ArSO_3\cdot \xrightarrow{ArSO_2\cdot} (ArSO_2)_2O\ (23\%) \qquad ArSO\cdot \longrightarrow ArSO_2SAr$$

$$RSO_2H \xrightarrow{Co(III)} RSO_2SO_2R \tag{5-75}$$

$$CH_3SO_2{-}Br \xrightarrow[PhC{\equiv}CH]{100^\circ,\ 9\ hr} CH_3SO_2\cdot + Br\cdot$$

$$CH_3SO_2\cdot \xrightarrow{PhC{\equiv}CH} Ph\dot{C}{=}CH(SO_2CH_3) \xrightarrow[(-CH_3SO_2\cdot)]{CH_3SO_2Br} Ph(Br)C{=}CH(SO_2CH_3)\ (90\%) \tag{5-76}$$

A novel and potentially useful sulfone synthesis is based on the reaction of the electrochemically generated sulfur dioxide radical anion with alkyl halides (Knittel and Kastening, 1973). This reaction (Eq. 5-77) involves generation of sulfonyl radicals in the slow step. The sulfur dioxide anion

$$\begin{aligned} SO_2 + e^- &\rightleftharpoons SO_2^{\dot{-}} \\ SO_2^{\dot{-}} + RX &\longrightarrow RSO_2\cdot + X^- \\ SO_2^{\dot{-}} + RSO_2\cdot &\longrightarrow RSO_2^- + SO_2 \\ RSO_2^- + RX &\longrightarrow R_2SO_2 + X^- \end{aligned} \qquad (5\text{-}77)$$

radical has a blue color and in fact the reaction with alkyl halide can be conducted as a titration using the blue color as an indicator. Some representative yields may be cited (given as starting material, product and percentage yield in acetonitrile as solvent): allyl bromide, diallyl sulfone, 88%; *n*-butyl bromide, di-*n*-butylsulfone, 74%; phenacyl bromide, diphenacylsulfone, 57%; *o*-xylylenedibromide, 1-oxa-2-thiatetrahydronaphthalene *S*-oxide, 67%.

Sulfur Radical Cations

Divalent organosulfur compounds can be converted to radical cations by one-electron oxidation. Sulfuric acid has been shown to convert diaryl sulfides and sulfoxides to the corresponding cation radicals which may be

$$\text{Ar}\ddot{\text{S}}\text{Ar} \xrightarrow{H_2SO_4} \left[\text{Ar}\dot{\text{S}}^{+}\text{Ar}\right] \xleftarrow{-OH\cdot} \text{Ar}\overset{+}{\ddot{\text{S}}}(\text{OH})\text{Ar} \xleftarrow{H_2SO_4} \text{Ar}\ddot{\text{S}}(=\text{O})\text{Ar} \qquad (5\text{-}78)$$

$(g = 2.00737 \quad \text{for} \quad Ar = p\text{-tolyl})$

characterized by ESR (Eq. 5-78; Shine *et al.*, 1967; Schmidt, 1964; Oae and Kunieda, 1968) spectroscopy. Treatment of sulfides with Ti(III)–H_2O_2 leads to a variety of reactions said to involve sulfur-centered cation radicals

$$\begin{aligned} HOCH_2CH_2SR &\xrightarrow{Ti(III)-H_2O_2} HOCH_2CH_2\dot{S}^{+}R \\ H\!-\!O\!-\!CH_2\!-\!CH_2\dot{S}^{+}R &\longrightarrow H^+ + CH_2O + RSCH_2\cdot \\ HOCH_2CH(H)\dot{S}^{+}R &\longrightarrow H^+ + HOCH_2\!-\!\dot{C}HSR \\ HOCH(H)\!-\!CH_2\!-\!\dot{S}^{+}R &\longrightarrow HOCH\!=\!CH_2 + H^+ + RS\cdot \\ R_2S^{\dot{+}} + R_2S &\longrightarrow [R_2S\!-\!SR_2]^{\dot{+}} \end{aligned} \qquad (5\text{-}79)$$

(Eq. 5-79; Gilbert *et al.*, 1973a, b). Thallium(III) nitrate has also been employed as a one-electron oxidant (Nagao *et al.*, 1977).

Very recently, bicyclic cation radical **5-20** has been generated by one-electron oxidation of 1,5-dithiacyclooctane (Musker and Wolford, 1976).

$$\text{(1,5-dithiacyclooctane)} \xrightarrow[\text{NOBF}_4]{\text{Cu(II) or}} \text{(S}\dot{-}\text{S}^+\text{ bicyclic cation radical)}$$

5-20
(g = 2.012)

Species **5-20** is said to possess the longest lifetime of any nonaromatic cation radical known.

Electrolytically produced sulfide cation radicals have been found to participate in some novel reactions of potential synthetic utility (Eqs. 5-80 and 5-81; Kim and Shine, 1974; Uneyama and Torii, 1972). Disulfide cation

$$\text{thianthrene} \xrightarrow{-e^-} \text{thianthrene}^{\cdot +} \xrightarrow{CH_3C(O)CH_3} \text{thianthrene-}S^+\text{-}CH_2C(O)CH_3 \qquad (5\text{-}80)$$

$$PhSPh \longrightarrow Ph\ddot{S}^{\cdot +}Ph \longrightarrow Ph_2\overset{+}{S}\text{-}C_6H_4\text{-}SPh\ (71\%) \qquad (5\text{-}81)$$

$$Ph\ddot{S}^{\cdot +}Ph \xrightarrow{-e,\ H_2O} Ph_2S{=}O\ (1\%) \qquad Ph_2\overset{+}{S}\text{-}C_6H_4\text{-}SPh \xrightarrow{+e,\ (-Ph\cdot)} PhS\text{-}C_6H_4\text{-}SPh\ (4\%)$$

radicals RSSR$^{\cdot +}$ can be generated through electron transfer processes on interaction of hydroxyl radicals with simple dialkyl disulfides (Bonifacic *et al.*, 1975; also see Zweig and Hoffmann, 1965).

Sulfuranyl Radicals

In reactions involving radical substitution at sulfur, the possible intermediacy of sulfuranyl radicals, tricoordinate sulfur species with a *nonet* of valence electrons (path a, Eq. 5-82), is a viable alternative to a direct synchronous displacement process (path b) (Poutsma, 1973). A similar

$$R\cdot + R'\text{—}\ddot{S}\text{—}X \xrightarrow{a} \left[R\text{—}\dot{\ddot{S}}(R')\text{—}X \right]^{\ddagger} \longrightarrow R\ddot{S}R' + X\cdot \qquad (5\text{-}82)$$

$$\xrightarrow{b} \left[R\text{---}\dot{\ddot{S}}(R')\text{---}X \right]^{\ddagger} \longrightarrow R\ddot{S}R' + X\cdot$$

possibility for nonet sulfur exists for sulfur radical anions. The ninth electron has been variously suggested to occupy an antibonding orbital or sulfur 3d orbital. Some noteworthy reactions which may involve sulfuranyl radicals are given in Eqs. 5-83 to 5-86.

$$CF_3{-}I \xrightarrow{h\nu} CF_3\cdot \xrightarrow{CH_3SSCH_3} \left[(CF_3)(CH_3)\dot{S}{-}SCH_3 \right] \xrightarrow{-CH_3\cdot} CF_3SCH_3 \ (92\%) \quad (5\text{-}83)$$

(Haszeldine *et al.*, 1972)

$$o\text{-}(PhS)C_6H_4C(=O){-}O{-}O{-}t\text{-}C_4H_9 \xrightarrow[-t\text{-}C_4H_9O\cdot]{\Delta} \left[\text{cyclic } C_6H_4C(=O)O\dot{S}Ph \right] \longrightarrow$$

$$o\text{-}(CO_2Ph)C_6H_4S\cdot \longrightarrow o\text{-}(PhO_2C)C_6H_4S{-}SC_6H_4\text{-}o\text{-}CO_2Ph \quad (52\%) \quad (5\text{-}84)$$

(Bentrude and Martin, 1962)

$$2\text{-}iodo\text{-}2'\text{-}(SCH_3)\text{biphenyl} \xrightarrow[((C_4H_9)_3SnH)]{(C_4H_9)_3Sn\cdot} 2'\text{-}(SCH_3)\text{biphenyl-2-yl radical} \longrightarrow$$

$$\text{S-methyldibenzothiophenyl radical } (\dot{S}{-}CH_3) \xrightarrow{-CH_3\cdot} \text{dibenzothiophene } (100\%) \quad (5\text{-}85)$$

(Kampmeier and Evans, 1966)

$$ArCH_2\cdot + (PhCH_2)_2S \longrightarrow (PhCH_2)_2\dot{S}CH_2Ar \longrightarrow$$
$$PhCH_2SCH_2Ar + (ArCH_2)_2 + PhCH_2CH_2Ar + (PhCH_2)_2 \quad (5\text{-}86)$$

(Schmidt *et al.*, 1970)

Radical displacement reactions of disulfides (Eq. 5-83) have been studied in great detail (Pryor and Smith, 1970) and are known to occur quite rapidly. For example, in the gas phase the rate constant for reaction of methyl radicals with dimethyl disulfide (Eq. 5-87) has the value of $6 \times 10^4\ M^{-1}\ \text{sec}^{-1}$ at 100° (Suama and Takezaki, 1962):

$$CH_3\cdot + CH_3SSCH_3 \longrightarrow CH_3SCH_3 + CH_3S\cdot \quad (5\text{-}87)$$

It has been found that radicals previously identified by ESR as RS· may actually have the structure RS$\dot{-}$SR$_2$ (the intermediate of reaction 5-83) with the unpaired electron occupying the S—S σ^* orbital (Symons, 1974). Similarly it has been found (Gara *et al.*, 1977) that photolysis of mixtures of disulfides and peroxides leads to dialkoxysulfuranyl radicals, R$\dot{S}$(OR$'$)$_2$, and not, as previously suggested (Kawamura *et al.*, 1972), alkanesulfinyl radicals, RSO·. Dialkoxysulfuranyl radicals also result from direct addition of alkyl radicals to dialkylsulfoxylates, ROSOR, from the photolysis of sulfenate esters, and from the reaction of *t*-butoxyl radicals with dialkyl sulfides (Gara *et al.*, 1977). Sulfuranyl radicals of the type $(CH_3)_2\dot{S}OSi(CH_3)_3$ result from the reaction of trimethylsiloxyl radicals with dimethyl sulfide (Gara and Roberts, 1977). ESR studies indicate that the addition of alkoxyl radicals to sulfoxylates and sulfites produces, respectively, trialkoxysulfuranyl radicals (RO)$_3$S· (*g* value 2.0068; Chapman *et al.*, 1976) and trialkoxysulfuranyloxyl radicals (RO)$_3\dot{S}$=O (*g* value 2.0053; Gilbert *et al.*, 1977b).

The rate of reaction of radicals with polysulfides and elemental sulfur is even more rapid than the rate of reaction with disulfides. Polysulfides and S_8 are useful as polymerization retarders or inhibitors while disulfides function as chain-transfer agents (Kice, 1973).

Reaction 5-84 was discovered by Martin and Koenig during the course of a study of anchimeric assistance to O—O homolysis. Thus *tert*-butyl *o*-phenylthioperbenzoate underwent homolytic decomposition at a rate 2.5×10^4 times faster than *tert*-butyl perbenzoate (Koenig, 1973). Recent work has indicated that, in contrast to the case in Eq. 5-84, the iodine abstraction step of Eq. 5-85 apparently does not involve anchimeric assistance by the methylthio group (Danen *et al.*, 1973).

General References on Free Radical Chemistry

Hay, J. M. (1974). "Reactive Free Radicals." Academic Press, New York.
Kochi, J. K. (1973). "Free Radicals," Vols. I and II. Wiley, New York.
Pryor, W. A. (1966). "Free Radicals." McGraw-Hill, New York.
Rüchardt, C. (1970). *Angew. Chem. Int. Ed. English* **9**, 830.
Stirling, C. J. M. (1965). "Radicals in Organic Chemistry." Oldbourne Press, London.

References

Aalbersberg, W. G. L., and Vollhardt, K. P. C. (1977). *J. Am. Chem. Soc.* **99**, 2792.
Abell, P. I. (1973). *In* "Free Radicals" (J. K. Kochi, ed.), Vol. II, Chapter 13. Wiley, New York.
Adams, J. Q. (1970). *J. Am. Chem. Soc.* **92**, 4535.
Alfrey, T., and Price, C. C. (1947). *J. Polym. Sci.* **2**, 101.
Amiel, Y. (1974). *J. Org. Chem.* **39**, 3867.
Ando, W. *et al.* (1972). *J. Org. Chem.* **37**, 1721.

Baarschers, W. H., and Loh, T. L. (1971). *Tetrahedron Lett.* 3483.
Bachi, M. D., Epstein, J. W., Herzberg-Minzly, Y., and Loeventhal, H. J. E. (1968). *J. Org. Chem.* **34**, 126.
Backer, H. J., Stevens, W., and Dost, N. (1948). *Rec. Trav. Chim. Pays-Bas.* **67**, 451.
Badcock, C. C., Sidebottom, H. W., Calvert, J. G., Reinhardt, G. W., and Damon, E. K. (1971). *J. Am. Chem. Soc.* **93**, 3115.
Baldock, R. W., Hudson, P., Katritzky, A. R., and Soti, F. (1974). *J. Chem. Soc. Perkin Trans.* **1**, 1422.
Baldwin, J. E., Erickson, W. F., Hackler, R. E., and Scott, R. M. (1970). *Chem. Commun.* 576; also see Baldwin, J. E., and Hackler, R. E. (1969). *J. Am. Chem. Soc.* **91**, 3646.
Baldwin, J. E., Höfle, G., and Choi, S. C. (1971). *J. Am. Chem. Soc.* **93**, 2810.
Barnard, D. (1957). *J. Chem. Soc.* 4675.
Barton, D. H. R., and McCombie, S. W. (1975). *J. Chem. Soc. Perkins Trans.* **1**, 1574.
Barton, D. H. R., Clive, D. L. J. Magnus, P. D., and Smith, G. (1971). *J. Chem. Soc. C* 2193.
Barton, D. H. R., Blair, I. A., Magnus, P. D., and Norris, R. K. (1973). *J. Chem. Soc., Perkin Trans.* **1**, 1031, 1037; Almog, J., Barton, D. H. R., Magnus, P. D., and Norris, R. K. (1974). *J. Chem. Soc., Perkin Trans.* **1**, 853.
Bechgaard, K., Parker, V. D., and Pederson, C. Th. (1973). *J. Am. Chem. Soc.* **95**, 4373; see also Pedersen, C. Th., and Lohse, C. (1975). *Acta Chem. Scand.* **B29**, 831.
Bennett, J. E., Sieper, H., and Tavs, P. (1967). *Tetrahedron* **23**, 1697.
Bentrude, W. G., and Martin J. C. (1962). *J. Am. Chem. Soc.* **84**, 1561; see also Livant, P., and Martin, J. C. *J. Am. Chem. Soc.* **98**, 7851 (1976); Nakanishi, W., *et al.*, *Tetrahedron Lett.* 81 (1977).
Bernardi, F., Epiotis, N. D., Cherry, W., Schlegel, H. G. Whangbo, M.-H., and Wolfe, S. (1976). *J. Am. Chem. Soc.* **98**, 469; see also Bernardi, F., Csizmadia, I. G., Schlegel, H. B., Tiecco, M., Whangbo, M.-H., and Wolfe, S. *Gazz. Chim. Ital.* **104**, 1101 (1974).
Biddles, I., Hudson, A., and Wiffen, J. T. (1972). *Tetrahedron* **28**, 867.
Biellmann, J. F., and Schmitt, J. L. (1973). *Tetrahedron Lett.* 4615.
Block, E. (1969). *Quart. Rep. Sulfur. Chem.* **4**, 237.
Block, E., and O'Connor, J. (1974). *J. Am. Chem. Soc.* **96**, 3929.
Block, E., and Orf, H. W. (1972). *J. Am. Chem. Soc.* **94**, 8438.
Block, E., Penn, R. E., Olsen, R. J., and Sherwin, P. F. (1976). *J. Am. Chem. Soc.* **98**, 1264.
Boekelheide, V., Reingold, I. D., and Tuttle, M. (1973). *Chem. Commun.* 406.
Boekelheide, V., Galuszko, K., and Szeto, K. S. (1974). *J. Am. Chem. Soc.* **96**, 1578.
Bonifacic, M., Schäfer, K., Mockel, H., and Asmus, K.-D. (1975). *J. Phys. Chem.* **79**, 1496.
Bordwell, F. G., and Doomes, E. (1974). *J. Org. Chem.* **39**, 2526.
Bruhin, J., Kneubuhler, W., and Jenny, W. (1973). *Chimia* **27**, 277.
Bundy, G. L., Daniels, E. G., Lincoln, F. H., and Pike, J. E. (1972). *J. Am. Chem. Soc.* **94**, 2124.
Carton, P. M., Gilbert, B. C., Laue, H. A. H., and Norman, R. O. C. (1975). *J. Chem. Soc. Perkin Trans.* **2**, 1245.
Cava, M. P., and Kuczkowski, J. A. (1970). *J. Am. Chem. Soc.* **92**, 5800.
Cooper, R. D. G. (1972). Unpublished work cited by C. F. Murphy and J. A. Webber, *in* "Cephalosporins and Penicillins" (E. F. Flynn, ed.), Chapter 4. Academic Press, New York.
Corey, E. J., and Block, E. (1969). *J. Org. Chem.* **34**, 1233.
Corey, E. J., and Hamanaka, E. (1967). *J. Am. Chem. Soc.* **89**, 2759.
Dagonneau, M., Metzner, P., and Vialle, J. (1973). *Tetrahedron Lett.* 3675.
Danen, W. C., Saunders, D. G., and Rose, K. A. (1973). *J. Am. Chem. Soc.* **95**, 1612.
da Silva Correa, C. M. M., and Waters, W. A. (1968). *J. Chem. Soc. C* 1874.

Davies, A. G., Roberts, B. P., and Sanderson, B. R. (1973). *J. Chem. Soc. Perkin Trans.* **2**, 626.

Denzer, G. C., Jr., Allen, P., Jr., Conway, P., and van der Veen, J. M. (1966). *J. Org. Chem.* **31**, 3418.

Dewar, M. J. S., and Ramsden, C. A. (1974). *J. Chem. Soc. Perkin Trans.* **1**, 1839.

Dewar, P. S., Forrester, A. R., and Thomson, R. H. (1972). *J. Chem. Soc. Perkin Trans.* **1**, 2857.

Dobbs, A. J., Gilbert, B. C., and Norman, R. O. C. (1971). *J. Chem. Soc. A* 124.

Dolling, U. H., Closs, G. L., Cohen, A. H., and Ollis, W. D. (1975). *Chem. Commun*, 545.

Gara, W. B., and Roberts, B. P. (1977). *J. Organomet. Chem.* **135**, C20.

Gara, W. B., Roberts, B. P., Gilbert, B. C., Kirk, C. M., and Norman, R. O. C. (1977). *J. Chem. Research* (*S*) 152, (*M*) 1748.

Gardner, D. M., and Fraenkel, G. K. (1956). *J. Am. Chem. Soc.* **78**, 3279.

Gilbert, B. C., Larkin, J. P., and Norman, R. O. C. (1973a). *J. Chem. Soc. Perkins Trans* **2**, 272.

Gilbert, B. C., Hodgeman, D. K. C., and Norman, R. O. C. (1973b). *J. Chem. Soc. Perkin Trans.* **2**, 1748.

Gilbert, B. C., Kirk, C. M., and Norman, R. O. C. (1977a). *J. Chem. Research* (*M*) 1974.

Gilbert, B. C., Kirk, C. M., Norman, R. O. C., and Laue, H. A. H. (1977b). *J. Chem. Soc. Perkin Trans.* **2**, 497.

Gollnick, K., and Schade, G. (1968). *Tetrahedron Lett.* 689.

Graham, D. M., Mieville, R. L., Pallen, R. H., and Sivertz, C. (1964). *Can. J. Chem.* **42**, 2250.

Griesbaum, K. (1970). *Angew. Chem. Int. Ed. English* **9**, 373.

Grütze, J., and Vögtle, F. (1977). *Chem. Ber.* **110**, 1978.

Haenel, M. W. (1974). *Tetrahedron Lett.* 3053.

Haenel, M. W., Flatow, A., Taglieber, V., and Staab, H. A. (1977). *Tetrahedron Lett.* 1733.

Harris, E. F. P., and Waters, W. A. (1952). *Nature* (*London*) **170**, 212; see also Berman, J. D., Stanley, J. H., Sherman, W. V., and Cohen, S. G. (1963). *J. Am. Chem. Soc.* **85**, 4010.

Haszeldine, R. N., Rigby, R. B., and Tipping, A, E. (1972). *J. Chem. Soc. Perkin Trans.* **1**, 159.

Heiba, E. I., and Dessau, R. M. (1967). *J. Org. Chem.* **32**, 3827.

Hepinstall, J. T., Jr., and Kampmeier, J. A. (1973). *J. Am. Chem. Soc.* **95**, 1904.

Hodgson, W. G., Buckler, S. A., and Peters, G. (1963). *J. Am. Chem. Soc.* **85**, 543.

Horner, L., and Schläfer, L. (1960). *Justus Liebigs Ann. Chem.* **635**, 31.

Horowitz, A., and Rajbenbach, L. A. (1975). *J. Am. Chem. Soc.* **97**, 10.

Howard, J. A., and Furimsky, E. (1974). *Can. J. Chem.* **52**, 555.

Inoue, H., Umeda, I., and Otsu, T. (1971). *Makromol. Chem.* **147**, 271.

Iwamura, H., Iwamura, M., Nishida, T., Yoshida, M., and Nakayama, J. (1971). *Tetrahedron Lett.* 63.

Jacobus, J. (1970). *Chem. Commun.* 709.

James, F. C., Kerr, J. A., and Simons, J. P. (1973). *J. Chem. Soc. Faraday Trans.* 2124.

Kaiser, E. T., and Mayers, D. F. (1965). *Tetrahedron Lett.* 2767.

Kampmeier, J. A., and Evans, T. R. (1966). *J. Am. Chem. Soc.* **88**, 4096.

Kawamura, T., Krusic, P. J., and Kochi, J. K. (1972). *Tetrahedron Lett.* 4075.

Kellogg, R. M. (1969). *In* "Methods in Free Radical Chemistry" (E. S. Huyser, ed.), Vol. 2, Chapter 1. Dekker, New York.

Kice, J. L. (1973). *In* "Free Radicals" (J. K. Kochi, ed.), Vol. II, Chapter 24. Wiley, New York.

Kim, K., and Shine, H. J. (1974). *Tetrahedron Lett.* 4413; also see Kim, K., Hull, V. J., and Shine, H. J. (1974). *J. Org. Chem.* **39**, 2534; Kim, K., Mani, S. R., and Shine, H. J. (1975). *ibid.* **40**, 3857.

Knittel, D., and Kastening, B. (1973). *J. Appl. Electrochem.* **3**, 291.

Koch, P., Ciuffarin, E., and Fava, A. (1970). *J. Am. Chem. Soc.* **92**, 5971.

Koelewijn, P., and Berger, H. (1972). *Recl. Trav. Chim. Pays-Bas.* **91**, 1275.

Koenig, T. (1973). *In* "Free Radicals" (J. K. Kochi, ed.), Vol. I, Chapter 3. Wiley, New York.

Kondo, K., Matsumoto, M., and Negishi, A. (1972). *Tetrahedron Lett.* 2131.

Koshar, R. J., and Mitsch, R. A. (1973). *J. Org. Chem.* **38**, 3358; also see Kelley, C. J., and Carmack, M. (1975). *Tetrahedron Lett.* 3605.

Krusic, P. J., and Kochi, J. K. (1971). *J. Am. Chem. Soc.* **93**, 846.

Kuivila, H. G. (1968). *Accounts Chem. Res.* **1**. 299.

Lagercrantz, C., and Forshult, S. (1969). *Acta Chem. Scand.* **23**, 811.

Lemal, D. M., and Chao, P. (1973). *J. Am. Chem. Soc.* **95**, 922.

Lepley, A. R. (1973). *In* "Chemically Induced Magnetic Polarization" (A. R. Lepley, and G. L. Closs, eds.), Chapter 8. Wiley, New York.

Lown, E. M., Sandhu, H. S., Gunning, H. E., and Stausz, O. P. (1968). *J. Am. Chem. Soc.* **90**, 7164.

Mackle, H. (1963). *Tetrahedron* **19**, 1159.

Majeti, S. (1971). *Tetrahedron Lett.* 2528.

Maryanoff, C. A., Maryanoff, B. E., Tang, R., and Mislow, K. (1973). *J. Am. Chem. Soc.* **95**, 5839.

Middleton, W. J. (1964). U.S. Patent 3,136,781.

Migita, T., Kosugi, M., Takayama, K., and Nakagawa, Y. (1973). *Tetrahedron* **29**, 51.

Miller, E. G., Rayner, D. R., Thomas, H. T., and Mislow, K. (1968). *J. Am. Chem. Soc.* **90**, 4861.

Miura, Y., Makita, N., and Kinoshita, M. (1975). *Tetrahedron Lett.* 127.

Miyashita, T., Iino, M., and Matsuda, M. (1977). *Bull. Chem. Soc. Jpn.* **50**, 317.

Morizur, J. P. (1964). *Bull. Soc. Chim. Fr.* 1338.

Moussebois, C., and Dale, J. (1966). *J. Chem. Soc. C* 260.

Musker, W. K., and Wolford, T. L. (1976). *J. Am. Chem. Soc.* **98**, 3055.

Nagao, Y., Ochiai, M., Kaneko, K., Maeda, A., Watanabe, K., and Fujita, E. (1977). *Tetrahedron Lett.* 1345.

Norman, R. O. C., and Pritchett, R. J. (1965). *Chem. Ind.* (*London*) 2040.

Oae, S., and Kunieda, N. (1968). *Bull. Chem. Soc. Jpn.* **41**, 696.

Ollis, W. D., Rey, M., Sutherland, I. O., and Closs, G. L. (1975). *Chem. Commun.* 543.

Ohno, A., and Ohnishi, Y. (1971). *Int. J. Sulfur Chem. A* **1** (3), 203.

Ohno, A., and Ohnishi, Y. (1972). *Tetrahedron Lett.* 339.

Ohno, A., Kito, N., and Ohnishi, Y. (1971a). *Bull. Chem. Soc. Jpn.* **44**, 463.

Ohno, A., Ohnishi, Y., and Kito, N. (1971b). *Int. J. Sulfur Chem. A* **1** (3), 151.

Oku, M., and Philips, J. C. (1973). *J. Am. Chem. Soc.* **95**, 6495.

Petrova, R. G., and Freidlina, R., Kh., Isv. (1966). *Akad. Nauk SSR Ser. Khim.* **10**, 1857.

Poutsma, M. L. (1973). *In* "Free Radicals" (J. K., Kochi, ed.), Vol. II, Chapter 14. Wiley, New York.

Pryor, W. A., and Smith, K. (1970). *J. Am. Chem. Soc.* **92**, 2731.

Rawlinson, D. J., and Sosnovsky, G. (1972). *Synthesis* 1.

Readio, P. D., and Skell, P. S. (1966). *J. Org. Chem.* **31**, 759.

Rebafka, W., and Staab, H. A. (1973). *Angew. Chem. Int. Ed. English* **12**, 776.

Rundle, W., and Scheffler, K. (1965). *Angew. Chem. Int. Ed. English* **4**, 243.

Russell, G. A. (1973). *In* "Free Radicals" (J. K. Kochi, ed.), Vol. I, Chapter 7. Wiley, New York.

Scaiano, J. C., and Ingold, K. U. (1976). *J. Am. Chem. Soc.* **98**, 4727.

Schmidt, U. (1964). *Angew. Chem.* **76**, 629.

Schmidt, U., Hochrainer, A., and Nikiforov, A. (1970). *Tetrahedron Lett.* 3677.

Schöllkopf, U., Ostermann, G., and Schossig, J. (1969). *Tetrahedron Lett.* 2619.

Schöllkopf, U., Schossig, J., and Ostermann, G. (1970). *Justus Liebigs Ann. Chem.* **737**, 158.

Seebach, D., and Beck, A. K. (1972). *Chem. Ber.* **105**, 3892; see also Roelofsen, G., Kanters, J. A., and Seebach, D. (1974). *Chem. Ber.* **107**, 253; Haas, A., and Schlosser, K. (1976). *Tetrahedron Lett.* 4631.

Seebach, D., Stegmann, H., Scheffler, K., Beck, A. K., and Geiss, K.-H. (1972). *Chem. Ber.* **105**, 3905.

Sheehan, J. C., Zoller, V., and Ben-Ishai (1974). *J. Org. Chem.* **39**, 1817.

Sherrod, S. A., da Costa, R. L., Barnes, R. A., and Boekelheide, V. (1974). *J. Am. Chem. Soc.* **96**, 1565.

Shevlin, P. B., and Greene, J. L., Jr. (1972). *J. Am. Chem. Soc.* **94**, 8447.

Shine, H. J., Rahman, M., Seeger, H., and Wu, G.-S. (1967). *J. Org. Chem.* **32**, 1901.

Siedle, A. R., and Johannesen, R. B. (1975). *J. Org. Chem.* **40**, 2002.

Stegmann, H. B., Scheffler, K., and Seebach, D. (1975). *Chem. Ber.* **108**, 64.

Strausz, O. P., Gunning, H. E., and Lown, J. W. (1972). *In* "Chemical Kinetics" (C. H. Bamford and C. F. H. Tipper, eds.), Vol. 5, Chapter 6. Elsevier, Amsterdam.

Suama, M., and Takezaki, Y. (1962). *Bull. Inst. Chem. Res. Kyoto Univ.* **40**, 229.

Sugawara, C., Iwamura, H., and Ōki, M. (1974). *Bull. Chem. Soc. Jpn.* **47**, 1496.

Surzur, J.-M., Dupuy, C., Crozet, M. P., and Airmar, N. (1969). *C.R. Acad. Sci. Paris Ser. C* **269**, 849.

Surzur, J.-M., Nouguier, R., Crozet, M. P., and Dupuy, C. (1971). *Tetrahedron Lett.* 2035.

Symons, M. C. R. (1974). *J. Chem. Soc. Perkin Trans.* **2**, 1618.

Tabushi, I., Tamaru, Y., and Yoshida, Z. (1974). *Tetrahedron* **30**, 1457; also see Dittmer, D. C., and Nelsen, T. R., *J. Org. Chem.* **41**, 3044 (1976).

Tagaki, W., Tada, T., Nomura, R., and Oae, S. (1968). *Bull. Chem. Soc. Jpn.* **41**, 1696.

Thyrion, F. C., and Debecker, G. (1973). *Int. J. Chem. Kinet.* **5**, 583.

Timberlake, J. W., Garner, A. W., and Hodges, M. L. (1973). *Tetrahedron Lett.* 309; Bandlish, B. K., Garner, A. W., Hodges, M. L., and Timberlake, J. W. (1975). *J. Am. Chem. Soc.* **97**, 5856.

Topping, R. M., and Kharasch, N. (1962). *J. Org. Chem.* **27**, 4353.

Trost, B. M., Schinski, W. L., Chen, F., and Mantz, I. B. (1971). *J. Am. Chem. Soc.* **93**, 676.

Umemoto, T., Otsubo, T., and Misumi, S. (1974). *Tetrahedron Lett.* 1573.

Uneyama, K., and Torii, S. (1972). *J. Org. Chem.* **37**, 367; also see Torii, S., Matsuyama, Y. Kawasaki, K., and Uneyama, K. (1973). *Bull. Chem. Soc. Jpn.* **46**, 2912.

Uneyama, K., Namba, H., and Oae, S. (1968). *Bull. Chem. Soc. Jpn.* **41**, 1928.

Uneyama, K., Sadakage, T., and Oae, S. (1969). *Tetrahedron Lett.* 5193.

Uneyama, K., Torii, S., and Oae, S. (1971). *Bull. Chem. Soc. Jpn.* **44**, 815.

Vögtle, F. (1969). *Angew. Chem. Int. Ed. English* **8**, 274.

Vögtle, F., and Rossa, L. (1977). *Tetrahedron Lett.* 3577.

Ward, H. R. (1973). *In* "Free Radicals" (J. K. Kochi, ed.), Vol. I, Chapter 6, Wiley, New York.

Weseman, J. K., Williamson, R., Greene, J. L., Jr., and Shevlin, P. B. (1973). *Chem. Commun.* 901.

Wilson, R. M., and Wunderly, S. W. (1974). *J. Am. Chem. Soc.* **96**, 7350.

Woodward, R. B., and Hoffmann, R. (1970). "The Conservation of Orbital Symmetry" Verlag Chemie, Weinheim and Academic Press, New York.

Young, L. J. (1961). *J. Polymer Sci.* **54**, 411.

Zweig, A., and Hoffmann, A. K. (1965). *J. Org. Chem.* **30**, 3997.

Chapter 6 | Organosulfur Carbenes and Carbenoids

6.1 Introduction

Carbenes are short-lived bivalent species containing a carbon atom with only six valence electrons. Four of these electrons are associated with the two covalent bonds leaving two electrons to distribute themselves between two nonbonding orbitals. If these two orbitals are close in energy, in the ground state they will each be singly occupied by electrons with parallel spins (Hund's rule; this configuration minimizes electrostatic interaction). This species, possessing triplet multiplicity, shows some of the characteristics of a diradical. An alternative singlet configuration places both electrons, with spins paired, in one of the two nonbonding orbitals leaving a p orbital vacant. Paradoxically, singlet carbenes possess the characteristics of both carbonium ions (electron deficient; vacant p orbital) and carbanions (nonbonding electron pair). The electrophilic or nucleophilic character of singlet carbenes is dependent on the electron-withdrawing or electron-supplying ability of adjacent groups. The schizoid behavior of singlet carbenes is well tolerated by divalent sulfur which can stabilize both carbonium ions and carbanions. Thus, bis(phenylthio)methylene, $(PhS)_2C$:, is sufficiently stable to warrant the alternative name carbon monoxide bis(phenylthio)acetal (Seebach and Beck, 1969).

The choice of singlet or triplet ground state for a carbene has been shown by molecular orbital methods to be dependent on the energy difference between the two nonbonding levels of the carbene (Gleiter and Hoffmann, 1968). Through interaction of one of the carbene nonbonding orbitals with adjacent vacant or filled orbitals, there may result an energy separation between the carbene nonbonding levels which is greater than the repulsive electrostatic energy associated with electron pairing ($\sim$30 kcal/mole). In this event, as Hoffmann predicts for the incorporation of a nonbonding carbene orbital into a cyclic $4n + 2$ π-electron system as **6-1**, a ground state singlet carbene should be favored. With total suppression of its "vacant" p orbital, 1,3-dithiolium carbene **6-1** behaves as a nucleophilic via its carbenic lone pair (Hartzler, 1973). A low-lying singlet state also seems likely (based on the stereospecificity of addition to diastereomeric olefins) for the strongly electrophilic α-sulfonylcarbenes **6-2** in which the carbenic lone pair should be delocalized as in the case of α-sulfonylcarbanions.

$RSO_2CH:$

6-1 **6-2**

Reactions considered diagnostic for carbenes, i.e., addition across multiple bonds, insertion into σ bonds, and formation of dimeric olefins, may actually involve either "unimolecular" or "bimolecular" processes (see Eqs. 6-1 and 6-2, respectively), akin to the S_N1 and S_N2 processes of carbonium ion chemistry (Kirmse and Kapps, 1971). Where the groups A and B are weakly

"Unimolecular" $$R_2C(A)(B) \xrightarrow{-AB} R_2C: \xrightarrow{X—Y} X—CR_2—Y \quad (6\text{-}1)$$

"Bimolecular" $$R_2C(A)(B) \xrightarrow{X—Y} \text{[A,B···CR}_2\text{···X,Y transition state] or } A—B—CR_2—X—Y \rightarrow X—CR_2—Y + A—B \quad (6\text{-}2)$$

bonded to carbon in the transition state, the differences between the two processes is minimized. Closs and Moss (1964) have coined the term "carbenoid" to describe intermediates which exhibit reactions qualitatively similar to those of carbenes without necessarily being free divalent carbon

$$PhSCH_2Cl \xrightarrow{KO\text{-}t\text{-}C_4H_9} PhSCH(K)(Cl) \xrightarrow{\text{cyclohexene}} \text{7-(phenylthio)norcarane (}\sim SPh\text{)} \quad (6\text{-}3)$$

species. In most cases where metal-containing reagents are required, as in Eq. 6-3, the evidence points to the formation of a carbenoid rather than free carbene. Particularly interesting systems which, on the basis of their reactions, may or may not be considered carbenoids, are the transition metal–

$$(OC)_5CrC(SR)(CH_3)$$

6-3

carbene complexes, such as **6-3** (Fischer and Kreis, 1973). It should be recognized that organometallic carbenoids are capable of aggregating to varying degrees, depending on solvent, etc., and should show some differences from "free" carbenes in reactivity (particularly with respect to stereoselectivity). Another point worth noting is that all of the reactions considered diagnostic for carbenes are shown, under the proper circumstances, by ylides. In many cases it is possible, fortunately, to distinguish between processes involving ylides and carbenes (carbenoids) on mechanistic grounds.

Organosulfur carbenes and carbenoids of varying structure have been generated and several undergo unique and novel rearrangements. We shall consider the reactions of sulfur-containing carbenes following a summary of methods available for their generation.

6.2 Generation

Various approaches to carbenes and carbenoids are diagrammed in Eq. 6-4. Consonant with the versatile stabilizing ability of sulfur, all of these

$$RR'\bar{C}—X \xrightarrow{-X^-} RR'C: \qquad RR'\dot{C}—X \xrightarrow{-X\cdot} RR'C: \qquad RR'\overset{+}{C}—X \xrightarrow{-X^+} RR'C: \qquad RR'C{=}X \xrightarrow{-X:} RR'C: \qquad RR'C(X)(Y) \xrightarrow{-X—Y} RR'C: \tag{6-4}$$

approaches as well as a few other novel routes have been used to generate organosulfur carbenes and carbenoids. Specific examples are given in Scheme 6-1.

cheme 6-1. Generation of Sulfur-Containing Carbenes and Carbenoids

eprotonation of carbonium ions

$(CH_3S)_2CH^+ \xrightarrow{R_3N} (CH_3S)_2C:$	Olofson *et al.* (1968)
4,5-diphenyl-1,3-dithiolium (+)—H $\xrightarrow{R_3N}$ 4,5-diphenyl-1,3-dithiol-2-ylidene :	Pazdro and Polaczkowa (1970)

adical homolysis

$(PhS)_3C\cdot \longrightarrow PhS\cdot + (PhS)_2C:$	Seebach *et al.* (1972b)

(continued)

Scheme 6-1. *(Continued)*

α-Elimination

Reaction	Reference
$CH_3SCH(Cl)CO_2CH_3 \xrightarrow{KOt\text{-}C_4H_9} CH_3S\ddot{C}(Cl)CO_2CH_3 \xrightarrow{-Cl^-} CH_3S\ddot{C}CO_2CH_3$	Moore (1961)
$(PhS)_3CLi$ + cyclohexene oxide ⟶ $(PhS)_2C:$ + 2-(phenylthio)cyclohexanol (SPh, OH)	Seebach (1972)
$RSCHCl_2 \xrightarrow{KOt\text{-}C_4H_9} RS\ddot{C}Cl_2^- \longrightarrow RS\ddot{C}Cl + Cl^-$	Schöllkopf *et al.* (1966)

Fragmentation

Reaction	Reference
$(RS)_2C{=}NN(Na)Ts \xrightarrow{\Delta} (RS)_2C: + N_2 + NaOTs$	Lemal *et al.* (1965); Schöllkopf and Wiskott (1966)
spiro-dithiolane norbornadiene ⟶ benzene + 1,3-dithiolan-2-ylidene (S, S)	
bis(thiadiazolinylidene) (R, N, N, S, R′) ⇌ thiadiazolinylidene (R, N, N, S, R′)	
$RSO_nCHN_2 \xrightarrow{h\nu \text{ or } \Delta} RSO_nCH: + N_2$ $(n = 1 \text{ or } 2)$	Van Leusen *et al.* (1967); Venier *et al.* (1975)
$(CH_3)_2\overset{+}{S}{-}\overset{-}{C}HSPh \xrightarrow{h\nu} (CH_3)_2S + PhSCH:$	Hayasi and Nozaki (1971)

Addition to CS_2

Reaction	Reference
$RC{\equiv}CR + CS_2 \longrightarrow$ 4,5-dialkyl-1,3-dithiol-2-ylidene (R, R, S, S)	Hartzler (1973); Coffen (1970)
benzyne + $CS_2 \longrightarrow$ benzo-1,3-dithiol-2-ylidene (S, S)	Fields and Meyerson (1970); Nakayama (1975)
tetramethylcycloheptyne + $CS_2 \longrightarrow$ fused 1,3-dithiol-2-ylidene (S, S)	Krebs and Kimling (1971)
norbornene + $CS_2 \xrightarrow{\Delta}$ fused 1,3-dithiolan-2-ylidene (S, S)	Plieninger and Heuck (1972)

6.3 Reactions of α-Thiomethylenes: Cycloaddition to Multiple Bonds

Perhaps the most characteristic reaction of carbenes and carbenoids is their addition to multiple bonds. The formation of cyclopropanes from *unactivated* olefins and various sulfur-containing reagents is generally cited as *prima facie* evidence for the involvement of sulfur-containing carbenes or carbenoids in these processes, although in some instances non-carbene/carbenoid routes may also be available (such as involving the intermediacy of pyrazolines when the precursors are diazo compounds). Doering and Henderson were apparently the first to realize cyclopropane formation from thiocarbenes (Eq. 6-5; Henderson, 1959). Recently, Moss has argued that $CH_3S\ddot{C}Cl$ must be a *free carbene* rather than a carbenoid when generated by *t*-butoxide-induced α-elimination (Eq. 6-5) since the reactivity pattern of $CH_3S\ddot{C}Cl$ toward a series of olefins is identical whether or not the elimination

$$CH_3SCHCl_2 \xrightarrow{KO\textit{t}\text{-}C_4H_9} CH_3S\ddot{C}Cl_2^- \xrightarrow{-Cl^-} CH_3S\ddot{C}Cl$$

$$CH_3S\ddot{C}Cl \xrightarrow{(CH_3)_2C{=}C(CH_3)_2} \text{1-chloro-1-(methylthio)-2,2,3,3-tetramethylcyclopropane } (44\%)$$

$$CH_3S\ddot{C}Cl \xrightarrow{(CH_3)_2C{=}CHCH_3} \text{1-chloro-1-(methylthio)-2,2,3-trimethylcyclopropane } (59\%) \qquad (6\text{-}5)$$

is conducted in the presence of the potassium-complexing macrocyclic polyether, 18-crown-6† (Moss *et al.*, 1975). Table 6-1 indicates some representative examples of sulfur-substituted cyclopropanes which may be prepared from olefins and appropriate organosulfur precursors. While carbenes/carbenoids are indicated in Table 6-1 as the intermediates in each case, other types of intermediates cannot be excluded in certain of these cases. In these carbene additions the olefin functions as a nucleophile. Simple olefins such as cyclohexene and 2,3-dimethyl-2-butene are not sufficiently nucleophilic to trap carbenes with a high degree of internal stabilization such as $(CH_3S)_2C$: (Lemal and Banitt, 1964) and $CH_3S\ddot{C}CO_2CH_3$ (Moore, 1961; particular stability for this carbene may reflect a significant contribution by the resonance structure $CH_3S^+{=}C{=}C(O^-)OCH_3$). In the presence of weakly nucleophilic olefins these carbenes are more efficiently trapped by their carbanionic precursors, ultimately affording carbene "dimers" (see Eq. 6-19).

† See structure **2-1** in Chapter 2.

Table 6-1. Three-Membered Ring Formation with Thiocarbenes/Carbenoids

Entry	Carbene/ carbenoid	Source	Olefin	Product(s)	Yield	Reference
1	$CH_3S\ddot{C}H$	CH_3SCH_2Cl[a]	$(CH_3)_2C{=}CHN(CH_3)_2$	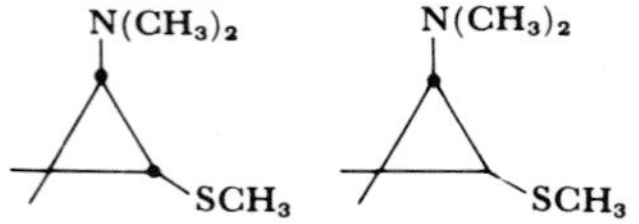 $N(CH_3)_2$, SCH_3; $N(CH_3)_2$, SCH_3	5:1; 15%	Rynbrandt and Dutton (1972)
2	$CH_3S\ddot{C}SCH_3$	$(CH_3S)_2C{=}NNHTs$[a]	$CH_2{=}C(OC_2H_5)_2$	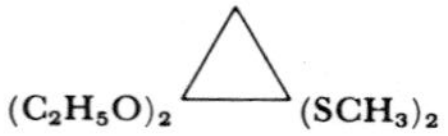 $(C_2H_5O)_2$, $(SCH_3)_2$	30%	Schöllkopf and Wiskott (1966)
3	$CH_3S\ddot{C}SCH_3$	$(CH_3S)_2C{=}NNHTs$[a]	*cis*-$CH_3CH{=}CHOC_3H_7$	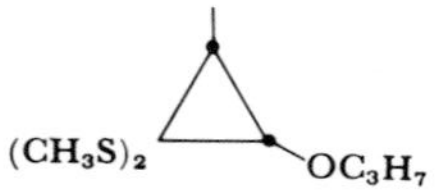 $(CH_3S)_2$, OC_3H_7	11%	Schöllkopf and Wiskott (1966)
4	$PhS\ddot{C}H$	$PhSCH_2Cl$[a]	$(CH_3)_2C{=}CH_2$	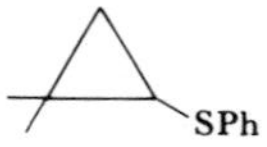 SPh	80%	Schöllkopf *et al.* (1964)
5	$PhS\ddot{C}H$	$PhSCH_2Cl$[a]	*trans*-$CH_3CH{=}CHCH_3$	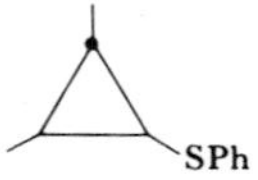 SPh	68%	Schöllkopf *et al.* (1964)
6	$PhS\ddot{C}H$	$PhSCH_2Cl$[a]	*cis*-$CH_3CH{=}CHCH_3$	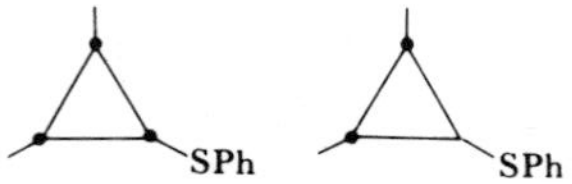 SPh; SPh	7:1; 60%	Schöllkopf *et al.* (1964)
7	$PhS\ddot{C}Cl$	$PhSCHCl_2$[a]	$CH_2{=}CHCH{=}CH_2$	 PhS, $CH{=}CH_2$	—	Corey and Walinsky (1972)
8	$PhS\ddot{C}H$	$PhSCH_2Cl$[a]	 (cyclohexadiene)	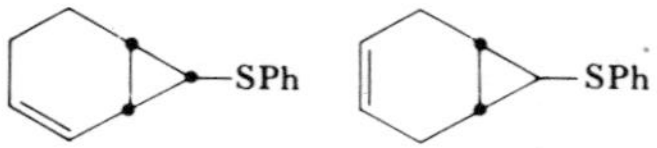 SPh; SPh	2.4:1; 60%	Schöllkopf *et al.* (1964)

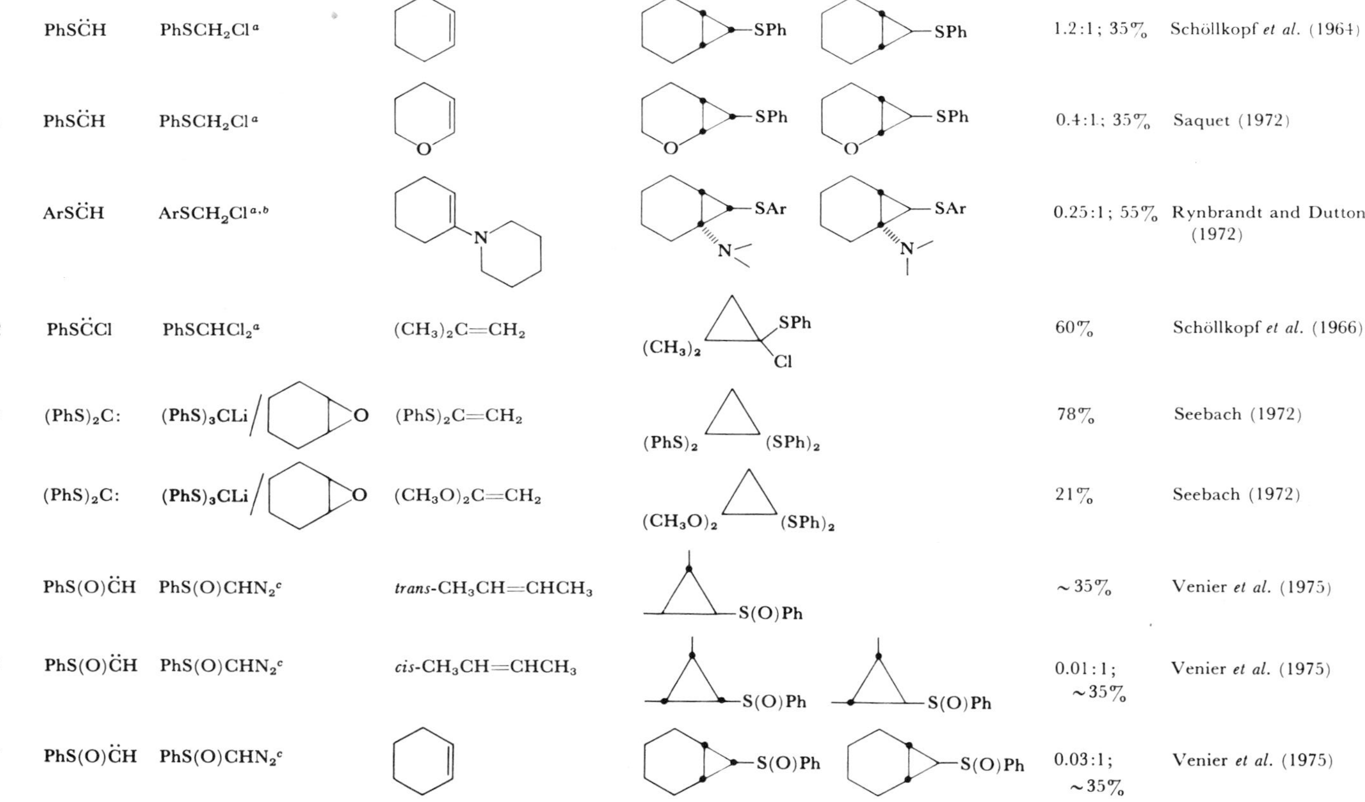

9	PhS$\ddot{C}$H	$PhSCH_2Cl$[a]		1.2:1; 35%	Schöllkopf *et al.* (1964)
10	PhS$\ddot{C}$H	$PhSCH_2Cl$[a]		0.4:1; 35%	Saquet (1972)
11	ArS$\ddot{C}$H	$ArSCH_2Cl$[a,b]		0.25:1; 55%	Rynbrandt and Dutton (1972)
12	PhS$\ddot{C}$Cl	$PhSCHCl_2$[a]	$(CH_3)_2C{=}CH_2$	60%	Schöllkopf *et al.* (1966)
13	$(PhS)_2C$:	$(PhS)_3CLi$/	$(PhS)_2C{=}CH_2$	78%	Seebach (1972)
14	$(PhS)_2C$:	$(PhS)_3CLi$/	$(CH_3O)_2C{=}CH_2$	21%	Seebach (1972)
15	PhS(O)$\ddot{C}$H	$PhS(O)CHN_2$[c]	*trans*-$CH_3CH{=}CHCH_3$	~35%	Venier *et al.* (1975)
16	PhS(O)$\ddot{C}$H	$PhS(O)CHN_2$[c]	*cis*-$CH_3CH{=}CHCH_3$	0.01:1; ~35%	Venier *et al.* (1975)
17	PhS(O)$\ddot{C}$H	$PhS(O)CHN_2$[c]		0.03:1; ~35%	Venier *et al.* (1975)

(*continued*)

Table 6-1. (*Continued*)

Entry	Carbene/ carbenoid	Source	Olefin	Product(s)	Yield	Reference
18	$PhSO_2\ddot{C}H$	$PhSO_2CHN_2$[d]	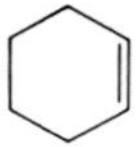	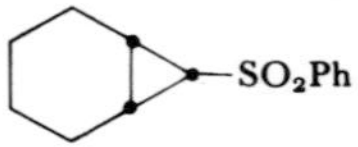 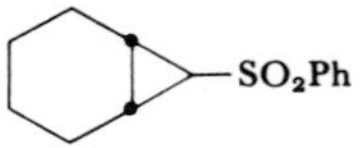 SO_2Ph; SO_2Ph	0.47:1; 50%	Abramovitch and Roy (1965)
19	$ArSO_2\ddot{C}H$	$ArSO_2CHN_2$[e,f]	$(CH_3)_2C{=}C(CH_3)_2$	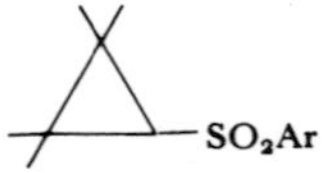 SO_2Ar	75%	Van Leusen *et al.* (1967)
20	$ArSO_2\ddot{C}H$	$ArSO_2CHN_2$[e,f]	*trans*-$CH_3CH{=}CHCH_3$	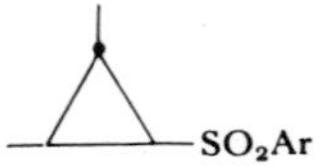 SO_2Ar	79%	Van Leusen *et al.* (1967)
21	$ArSO_2\ddot{C}H$	$ArSO_2CHN_2$[e,f]	*cis*-$CH_3CH{=}CHCH_3$	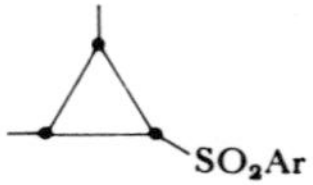 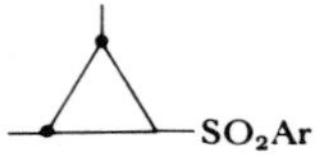 SO_2Ar; SO_2Ar	0.33:1; 83%	Van Leusen *et al.* (1967)
22	$ArSO_2\ddot{C}H$	$ArSO_2CHN_2$[e,f]	$CH_3C{\equiv}CCH_3$	 SO_2Ar	36%	Van Leusen and Strating (1970)
23	$ArSO_2\ddot{C}H$	$ArSO_2CHN_2$[e,f]	$C_2H_5OCH{=}CH_2$	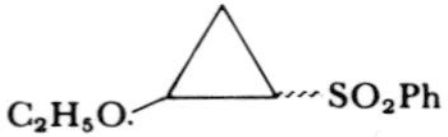 C_2H_5O, SO_2Ph	—	Van Leusen and Srating (1970)
24	$PhSO_2\ddot{C}Cl$	$PhHgCCl_2SO_2Ph$		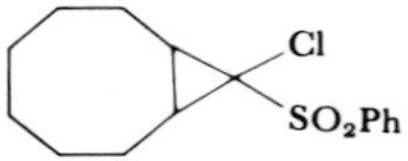 Cl, SO_2Ph	49%	Seyferth and Woodruff (1974)
25	$PhSO_2\ddot{C}Cl$	$PhHgCCl_2SO_2Ph$	$CH_2{=}CHCH_2Si(CH_3)_3$	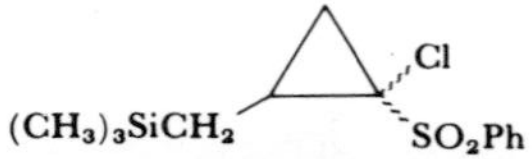 $(CH_3)_3SiCH_2$, Cl, SO_2Ph	47%	Seyferth and Woodruff (1974)

[a] Base = $KOt\text{-}C_4H_9$.
[b] Ar = *p*-chlorophenyl.
[c] Pyrolysis.
[d] Pyrolysis or UV irradiation.
[e] Ar = *p*-methoxyphenyl.
[f] UV irradiation.

However, with ketene acetals and thioacetals and other more nucleophilic olefins, cyclopropane formation can be achieved.

Sulfur-substituted cyclopropanes have synthetic utility. For example, Seebach has developed a novel ketone synthesis based on the hydrolysis of 1,1-bis(methylthio)cyclopropanes (Eq. 6-6; Seebach *et al.*, 1973a). The

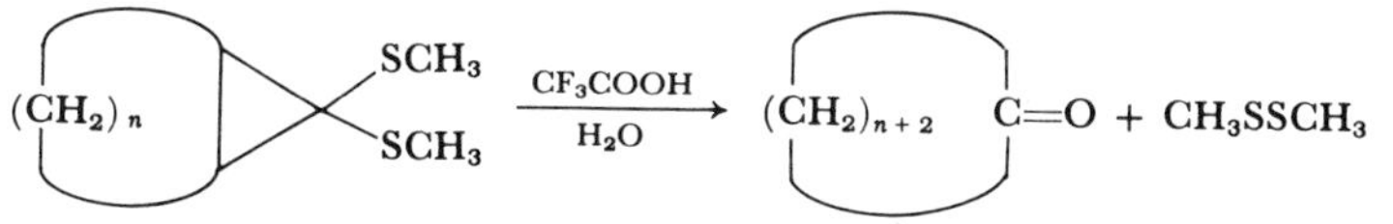

$$(CH_2)_n\text{-cyclopropane-}(SCH_3)_2 \xrightarrow[H_2O]{CF_3COOH} (CH_2)_{n+2}\,C{=}O + CH_3SSCH_3 \qquad (6\text{-}6)$$

$n = 3$–10, yields 50–90%

1,1-bis(methylthio)cyclopropanes were prepared from the gem-dibromocyclopropanes (dibromomethylene adducts) via the 1-bromo-1-methylthiocyclopropanes. Chlorothiomethylmethylene and chlorothiophenylmethylene adducts should serve equally well as starting points for this synthesis. A synthesis of cyclopentenes and Δ^3-cyclopentenones by Corey and Walinsky is based on the pyrolytic rearrangement of sulfur-substituted vinyl cyclopropanes (Eq. 6-7; Corey and Walinsky, 1972).† Other reactions involving

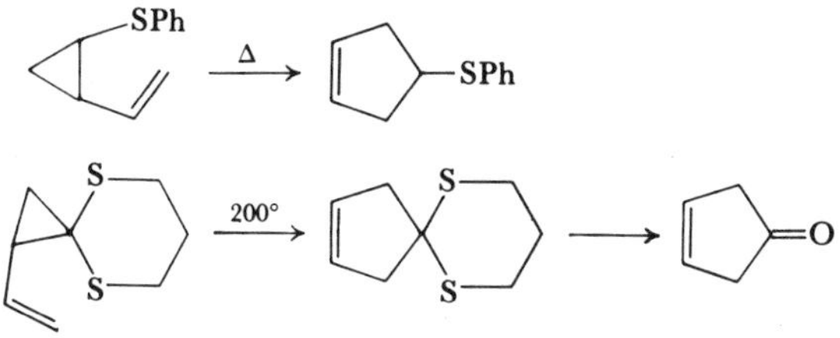

SPh, Δ, SPh

S, S, 200°, S, S, O

(6-7)

sulfur-substituted cyclopropanes (such as Trost's spiroannelation procedure (Eq. 3-23)) are considered elsewhere in this book.

6.4 Other Reactions of α-Thiomethylenes: General Survey

Besides adding to olefins, α-thiomethylenes also undergo the unimolecular processes of fragmentation and rearrangement and the bimolecular processes of "dimerization" and reaction with nucleophiles and electrophiles. Scheme 6-2 illustrates these processes.

Fragmentation Reactions: the Corey–Winter Olefin Synthesis

In 1963, Corey and Winter reported a new stereospecific olefin synthesis based on the fragmentation of heterocyclic carbenes generated by desulfurization of cyclic thiocarbonates and trithiocarbonates with tervalent

† See Chapter 4, p. 150, for an unusual preparation of the spiro 1,3-dithiane.

Scheme 6-2. Reactions of α-Thiomethylenes

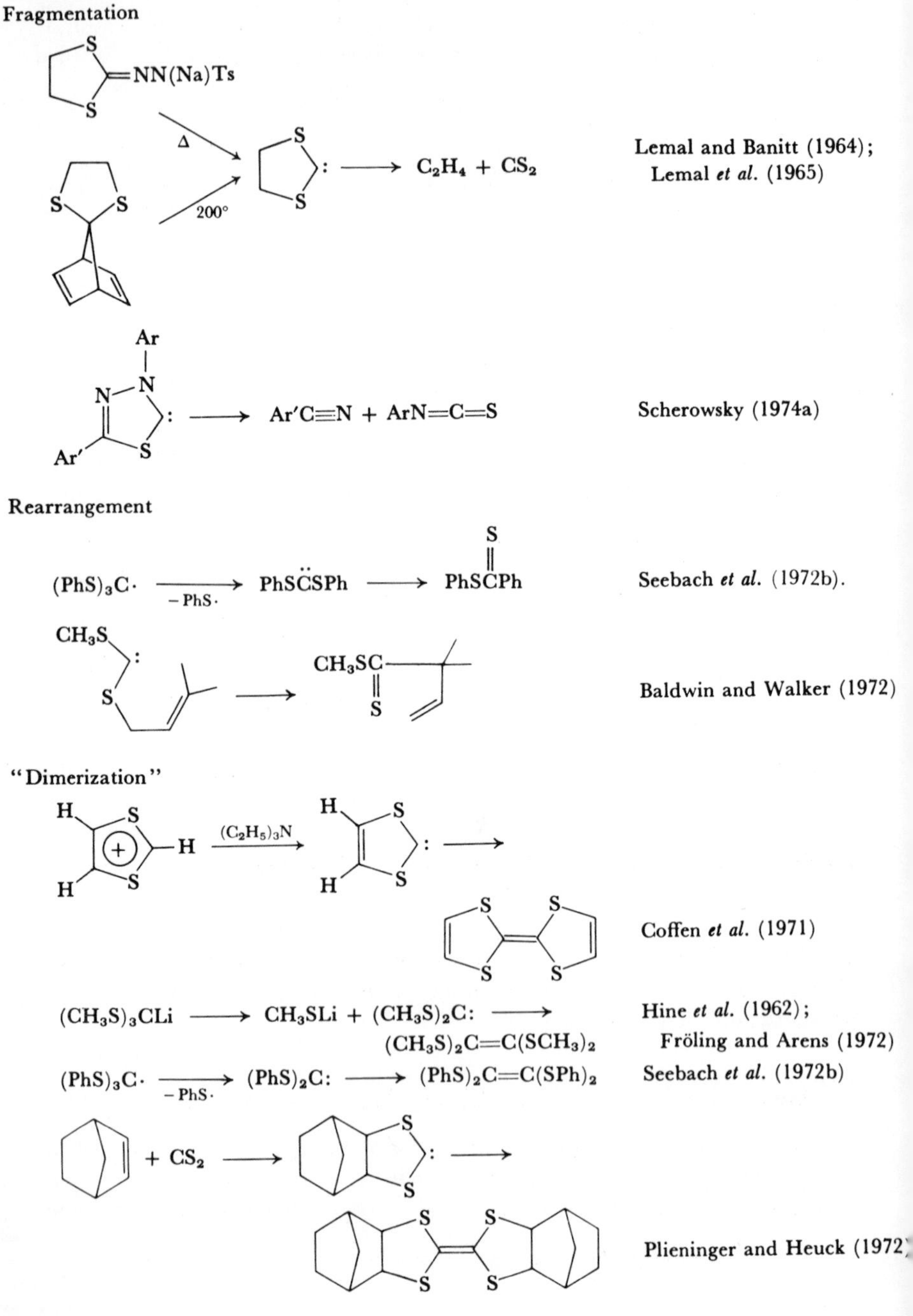

Fragmentation

1,3-dithiolan-2-ylidene-NN(Na)Ts →(Δ) 1,3-dithiolan-2-ylidene: → $C_2H_4 + CS_2$ — Lemal and Banitt (1964); Lemal *et al.* (1965)

spiro(1,3-dithiolane)-norbornadiene →(200°) 1,3-dithiolan-2-ylidene:

Ar/Ar′ thiadiazolinylidene: → $Ar'C{\equiv}N + ArN{=}C{=}S$ — Scherowsky (1974a)

Rearrangement

$(PhS)_3C\cdot \xrightarrow{-PhS\cdot} PhS\ddot{C}SPh \longrightarrow PhSC(=S)Ph$ — Seebach *et al.* (1972b).

$CH_3S(C:)S{-}CH_2CH{=}C(CH_3)_2 \longrightarrow CH_3SC(=S){-}C(CH_3)_2CH{=}CH_2$ — Baldwin and Walker (1972)

"Dimerization"

1,3-dithiolium (H, H, +, H) $\xrightarrow{(C_2H_5)_3N}$ 1,3-dithiol-2-ylidene: → tetrathiafulvalene — Coffen *et al.* (1971)

$(CH_3S)_3CLi \longrightarrow CH_3SLi + (CH_3S)_2C: \longrightarrow (CH_3S)_2C{=}C(SCH_3)_2$ — Hine *et al.* (1962); Fröling and Arens (1972)

$(PhS)_3C\cdot \xrightarrow{-PhS\cdot} (PhS)_2C: \longrightarrow (PhS)_2C{=}C(SPh)_2$ — Seebach *et al.* (1972b)

norbornene + CS_2 → norbornane-fused 1,3-dithiolan-2-ylidene: → bis(norbornane-fused dithiolanylidene) — Plieninger and Heuck (1972)

Reaction with nucleophiles

$(C_2H_5S)_2C: + R_3P \longrightarrow (C_2H_5S)_2\bar{C}\text{—}\overset{+}{P}R_3$ Lemal and Banitt (1964)

$(PhS)_2C: + PhSLi \longrightarrow (PhS)_3CLi$ Seebach and Beck (1969)

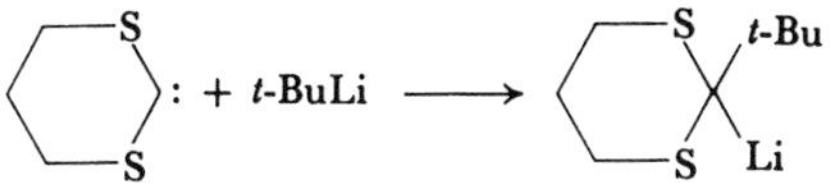

1,3-dithian-2-ylidene : + t-BuLi ⟶ 2-t-Bu-2-Li-1,3-dithiane Carlson and Helquist (1969)

$PhS\ddot{C}H + PhCH_2N(CH_3)_2 \longrightarrow PhCH_2\overset{+}{N}(CH_3)_2\bar{C}HSPh$ Julia *et al.* (1972)

$PhS\ddot{C}H + PhCH_2SPh \longrightarrow PhCH_2\overset{+}{S}(Ph)\bar{C}HSPh$ Julia *et al.* (1972)

Reaction with electrophiles

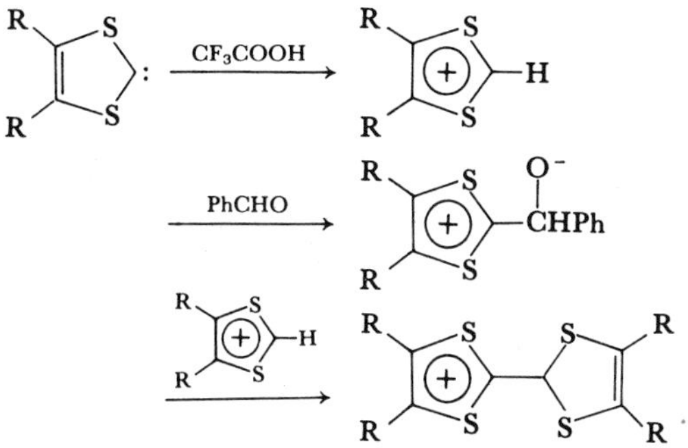

4,5-R₂-1,3-dithiol-2-ylidene $\xrightarrow{CF_3COOH}$ 4,5-R₂-1,3-dithiolium (2-H) Hartzler (1973)

$\xrightarrow{PhCHO}$ 4,5-R₂-1,3-dithiolium-2-CH(O⁻)Ph

$\xrightarrow{\text{4,5-R}_2\text{-1,3-dithiolium (2-H)}}$ 2-(4,5-R₂-1,3-dithiol-2-yl)-4,5-R₂-1,3-dithiolium

---

phosphorus compounds (Eq. 6-8; Corey and Winter, 1963; Corey *et al.*, 1965; Corey and Shulman, 1968). As we shall see, the originally proposed

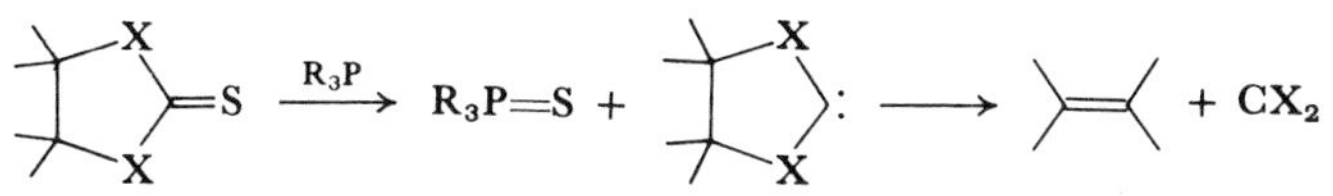

$$\text{cyclic thiocarbonate (C=S)} \xrightarrow{R_3P} R_3P\text{=}S + \text{cyclic carbene (:)} \longrightarrow \text{olefin} + CX_2 \qquad (6\text{-}8)$$

(X = S or O)

mechanism turned out to be an oversimplification. Table 6-2 indicates some applications of this olefin synthesis. Several methods for preparing the requisite thiocarbonates and trithiocarbonates are shown in Eqs. 6-9 to 6-11.

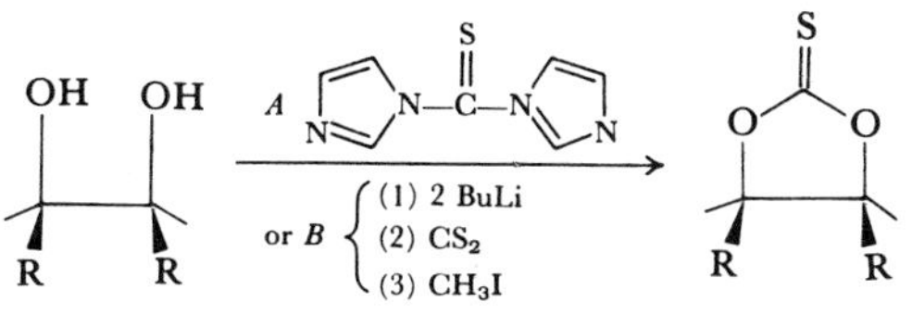

$$\text{RCH(OH)–CH(OH)R} \xrightarrow[\text{or } B \begin{cases}(1)\ 2\ \text{BuLi}\\(2)\ CS_2\\(3)\ CH_3I\end{cases}]{A\ \text{(imidazolyl)–C(=S)–(imidazolyl)}} \text{cyclic thiocarbonate (O–C(=S)–O), R, R} \qquad (6\text{-}9)$$

$$\text{cis-epoxide (H, R; H, R)} \xrightarrow{KSC(S)OCH_3} \text{trans-1,3-dithiolane-2-thione (H, R; R, H)} \quad (6\text{-}10)$$

$$\text{cyclooctene} \xrightarrow{(SCN)_2} \text{1,2-bis(thiocyanato)cyclooctane} \xrightarrow[(2)\ H_2S]{(1)\ HBr;\ Na_2CO_3} \text{cyclooctane-fused 1,3-dithiolane-2-thione} \quad (6\text{-}11)$$

The mechanism of the Corey–Winter reaction (Eq. 6-12) is thought to involve thiophilic addition of the phosphorus to the thiocarbonyl group, α-elimination of the pentavalent

$$\rangle P{=}S$$

compound, trapping of the carbene by tervalent phosphorus to give an ylide, and finally fragmentation of the ylide to the olefin and the phosphine and carbon disulfide (or to the known phosphine–carbon disulfide adducts (Davies and Jones, 1929)). The observation that the addition of benzaldehyde surpresses olefin formation, giving instead the ketene thioacetal has been used

Table 6-2. The Corey–Winter Olefin Synthesis

Entry	Substrate	Product	(Yield (%)	Reference
1	cis-4,5-diphenyl-1,3-dithiolane-2-thione	cis-PhCH=CHPh	94	Corey *et al.* (1965)
2	trans-4,5-diphenyl-1,3-dithiolane-2-thione	trans-PhCH=CHPh	100	Corey *et al.* (1965)
3	cyclooctane-fused 1,3-dithiolane-2-thione[a]	trans-cyclooctene[a]	99	Corey and Shulman (1968)
4	1,3-oxathiolane-2-thione	ethylene	42	Jones and Andreades (1969)
5	cycloheptane-fused 1,3-dioxolane-2-thione	[trans-cycloheptene][b]	—	Corey *et al.* (1965)

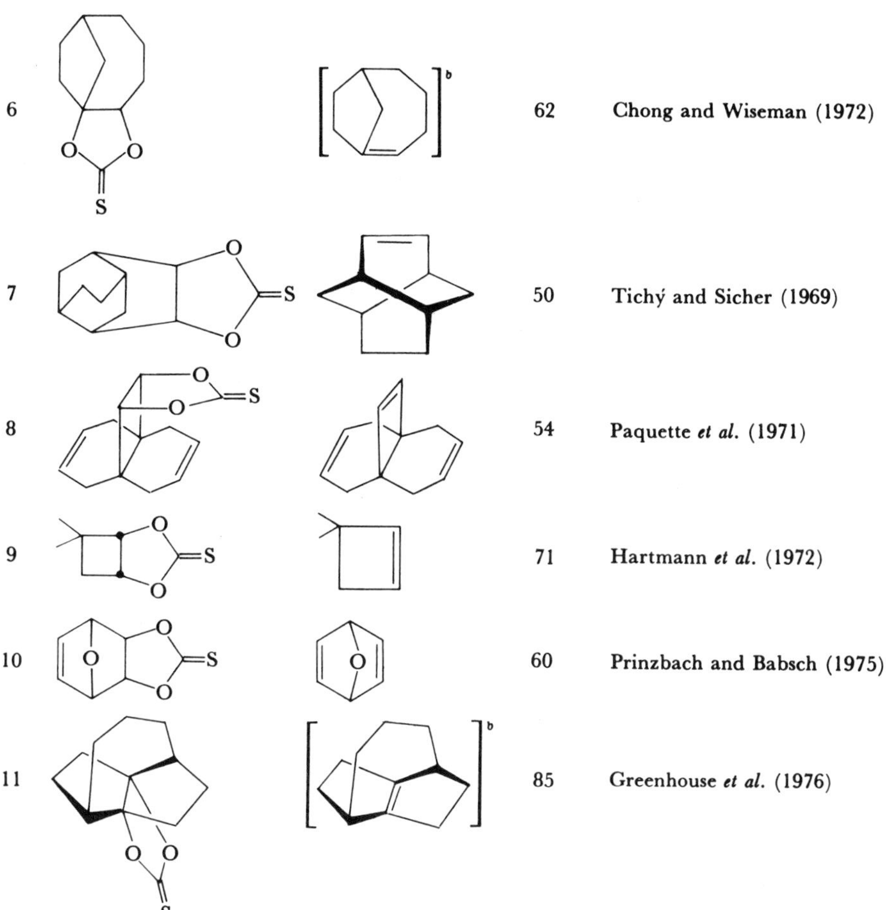

6	[ ][b]	62	Chong and Wiseman (1972)
7		50	Tichý and Sicher (1969)
8		54	Paquette *et al.* (1971)
9		71	Hartmann *et al.* (1972)
10		60	Prinzbach and Babsch (1975)
11	[ ][b]	85	Greenhouse *et al.* (1976)

[a] Optically active.
[b] As 2,5-diphenyl-3,4-isobenzofuran adduct.

as an argument in favor of the intermediacy of either an ylide or a 1,3-dipole (Eq. 6-12; Corey and Märkl, 1967; Scherowsky and Weiland, 1974). While detection of small amounts of an ortho ester during the application of the Corey–Winter reaction to a carbohydrate 1,3-thionocarbonate has been cited as direct evidence for the intermediacy of a carbene (Horton and Tindall, 1970), the result can also be rationalized using the 1,3-dipolar intermediate (Eq. 6-13; Scherowsky and Weiland 1974c). The isolation of the carbene "dimer" during the attempted preparation of *trans*-cyclohexene may also be explained by reactions involving either a 1,3-dipole or a carbene (Eq. 6-14; Yoshida *et al.*, 1975; Scherowsky and Weiland, 1974).

R_3P

$\overset{-}{C}$—$\overset{+}{S}PR_3$ $\xrightarrow{-R_3PS}$: $\xrightarrow{R_3P}$ $\overset{-}{C}$—$\overset{+}{P}R_3$

1,3 dipole

Carbene

Ylide

?

PhCHO

+ $R_3\overset{+}{P}$—C(S)(S)$^-$

(or R_3P + CS_2)

PhCHO

H

Ph

(6-12)

CH_2=CH

OH

$(CH_3O)_3P$

OH

+

H

(6-13)

=S + —$\overset{+}{S}PR_3$ $\xrightarrow{-R_3PS}$

(and isomer)

R_3P

(6-14)

+ R_3PS

(and isomer)

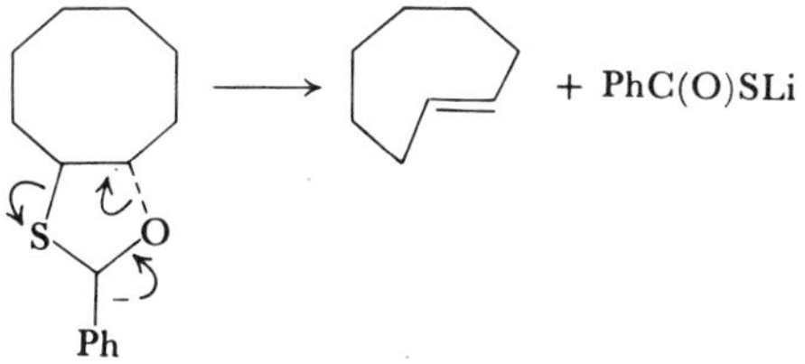

$$\longrightarrow \quad + \ PhC(O)SLi \qquad (6\text{-}15)$$

Precedence for the ylide fragmentation step is found in the preceding "heteroanion fragmentation" (Eq. 6-15; Jones *et al.*, 1974). Ketene thioacetals, prepared for example from trimethylene trithiocarbonate (as shown in Eq. 6-16), are in themselves useful synthetic intermediates since they can be converted to carboxylic acids by hydrolysis, to homologated aldehydes by reduction followed by hydrolysis, or to ketones on reaction with alkyllithiums followed by alkylation (Eq. 6-16; Carey and Court, 1972; Seebach *et al.*, 1972, 1973b).

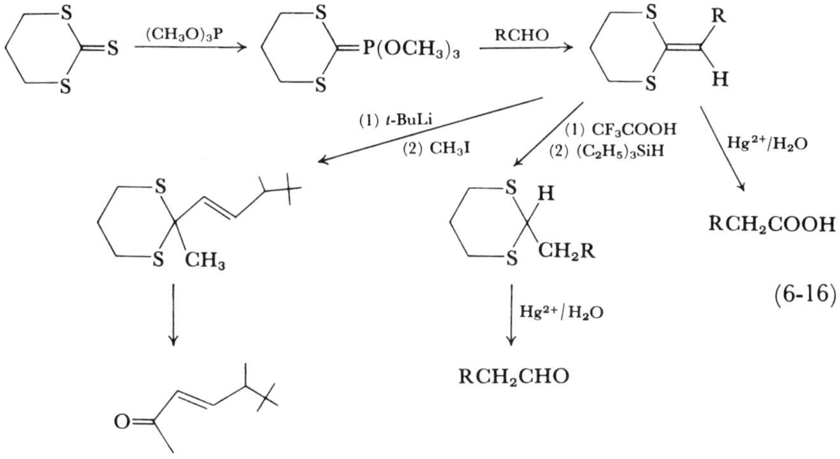

$$\xrightarrow{(CH_3O)_3P} \quad =P(OCH_3)_3 \quad \xrightarrow{RCHO}$$

(1) t-BuLi (2) CH_3I ; (1) CF_3COOH (2) $(C_2H_5)_3SiH$; Hg^{2+}/H_2O

CH_3 ; H, CH_2R ; RCH_2COOH (6-16)

Hg^{2+}/H_2O

RCH_2CHO

O=

($R = CH_3CH{=}CH$ in above)

Rearrangement Reactions

The [2,3] sigmatropic rearrangement of a sulfur-stabilized carbenoid forms the basis for a novel synthesis of carboxylic acid derivatives (Eq. 6-17;

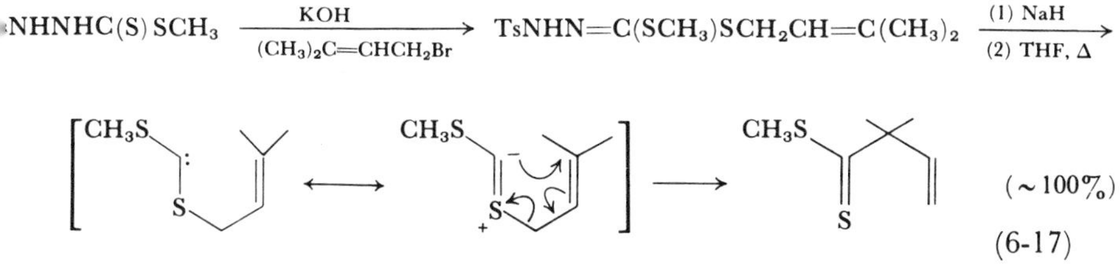

$$NHNHC(S)SCH_3 \xrightarrow[(CH_3)_2C{=}CHCH_2Br]{KOH} TsNHN{=}C(SCH_3)SCH_2CH{=}C(CH_3)_2 \xrightarrow[(2)\ THF,\ \Delta]{(1)\ NaH}$$

CH_3S ; S ; CH_3S ; S ; CH_3S ; S (~100%)

(6-17)

Baldwin and Walker, 1972). A most elegant application of this rearrangement to the total synthesis of the eremophilane sesquiterpene bakkenolide-A has been published (Eq. 6-18; Evans *et al.*, 1977).†

$H_2C{=}C(CH_3)Li$ / $(C_2H_5)_2O$; PBr_3; $TsNHN{=}C(SCH_3)SNa$; $TsNHN{=}C(SCH_3)S$; NaH / THF; SCH_3; [2,3] shift; CH_3S; (1) $HgO/HgCl_2$ (2) selenious acid

bakkenolide-A

(6-18)

"Dimerization": Tetrathiofulvenes

One of the most ubiquitous products of reactions thought to involve α-thiomethylenes are the carbene "dimers." In a majority of cases these "dimers" are most likely formed by reactions of the carbene with their precursors (or occasionally by reactions not involving carbenes at all) rather

$(RS)_3C^-$ $\xrightarrow{-RS^-}$ $(RS)_2C:$; $(RS)_3C\cdot$ $\xrightarrow{-RS\cdot}$ $(RS)_2C:$; $(RS)_2C{=}N_2$ $\xrightarrow{-N_2}$ $(RS)_2C:$; $(RS)_3CH^+$ $\xrightarrow{-H^+}$ $(RS)_2C:$

$(RS)_2C:$ $\xrightarrow{(RS)_3C^-}$ $(RS)_2\bar{C}{-}C(SR)_3$ $\xrightarrow{-RS^-}$ $(RS)_2C{=}C(SR)_2$

$(RS)_2C:$ $\dashrightarrow$ $(RS)_2C{=}C(SR)_2$ (6-19)

$(RS)_2C:$ $\xrightarrow{(RS)_2CH+}$ $(RS)_2\overset{+}{C}{-}CH(SR)_2$ $\xrightarrow{-H^+}$ $(RS)_2C{=}C(SR)_2$

† It should be noted that a reasonable mechanism for this rearrangement can also be written using the carbanionic precursor of the carbene (Evans *et al.*, 1977).

than by bimolecular carbene coupling. Some of the possible routes to a tetrathioethylene from carbene/carbenoid precursors are illustrated in Eq. 6-19.

Direct carbene dimerization has been invoked to explain the results of crossover experiments involving 2,2′-bi(3,5-diaryl-1,3,4-thiadiazolines) (Eq. 6-20; Scherowsky, 1974a).

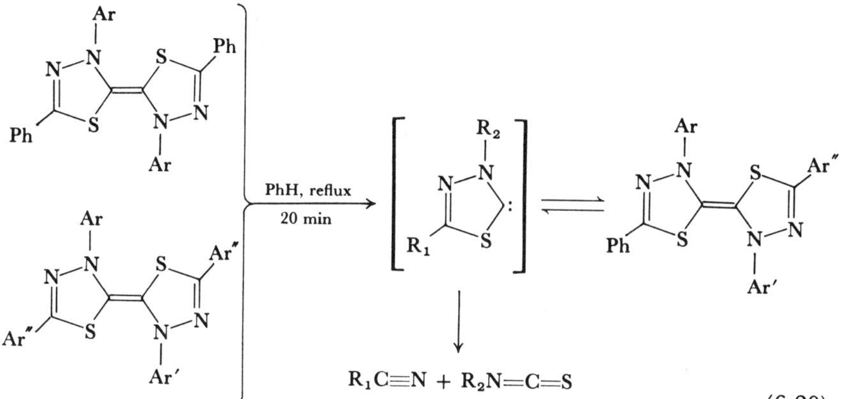

$$R_1C{\equiv}N + R_2N{=}C{=}S \tag{6-20}$$

Tetrathiofulvalenes, most conveniently prepared in high yield by the reaction shown in Eq. 6-21 (cf. Hartzler, 1973, Narita and Pittman, 1976; Yoshida *et al.*, 1975, and references therein), have been the subject of considerable recent research interest following the discovery that donor-acceptor complexes of tetrathiofulvalene–tetracyanoquinodimethane (TTF–TCNQ) show remarkably high electrical conductivity (Ferraris *et al.*, 1973; Coleman *et al.*, 1973). For example the observed conductivity at 66°K of 1.47×10^4 $\text{ohm}^{-1}\,\text{cm}^{-1}$ for TTF–TCNQ may be compared to the conductivity of metallic copper at 298°K of 6×10^5 $\text{ohm}^{-1}\,\text{cm}^{-1}$ (Ferraris *et al.*, 1973).†

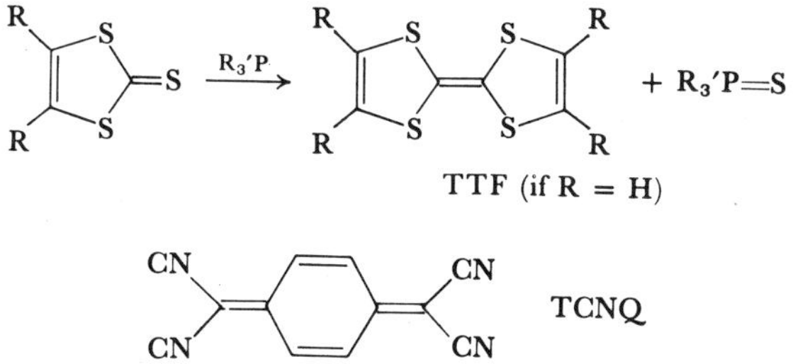

$$\xrightarrow{R_3'P} \quad + R_3'P{=}S \tag{6-21}$$

TTF (if R = H)

TCNQ

The 1:1 complex is thought to involve the TTF cation radical (highly polarizable because of the presence of sulfur) and the TCNQ anion radical.

† Another electrically conducting sulfur compound is polymeric sulfur nitride, $(SN)_x$, polythiazyl (see, for example, Cohen *et al.*, 1976, and references therein), and its brominated derivative poly(thiazyl bromide), $(SNBr_{0.7})_x$ (see Akhtar *et al.*, 1977, and Street *et al.*, 1977).

Reaction with Nucleophiles

We have already indicated that the addition of an α-thiomethylene to an olefin can be considered as the reaction of a nucleophile with the carbene. A variety of other nucleophiles such as carbanions, phosphines, or thiolate anions are also effective carbene-trapping agents, as summarized in Eq. 6-22.

$$(PhS)_2C: \underset{}{\overset{PhSLi}{\rightleftharpoons}} (PhS)_3CLi$$

$$\xrightarrow{(PhS)_3CLi} (PhS)_2\bar{C}{-}C(SPh)_3 \longrightarrow (PhS)_2C{=}C(SPh)_2$$

$$\xrightarrow{PhLi} (PhS)_2C(Ph)Li \xrightarrow{H^+} (PhS)_2CHPh$$

$$\downarrow (PhS)_2C:$$

$$(PhS)_2C{=}C(Ph)SPh$$

$$\xrightarrow{Ph_3P} (PhS)_2C{=}PPh_3 \qquad (6\text{-}22)$$

(Seebach, 1972a; Bos *et al.*, 1967; Beak and Worley, 1972)

Reaction with Electrophiles

While simple bis(alkylthio)methylenes may react with electrophiles, as indicated by reaction 6-23 (Olofson *et al.*, 1968), most of the examples that may be interpreted as electrophilic trapping of carbenes are found with heterocyclic carbenes for which ylidic resonance structures are thought to be particularly important.

$$(CH_3S)_2\overset{+}{C}H \xrightarrow{(i\text{-}C_3H_7)_2NC_2H_5} (CH_3S)_2C: \xrightarrow{(CH_3S)_2\overset{+}{C}H}$$

$$(CH_3S)_2\overset{+}{C}CH(SCH_3)_2 \xrightarrow{-H^+} (CH_3S)_2C{=}C(SCH_3)_2 \qquad (6\text{-}23)$$

Reaction of a carbene with an electrophile becomes particularly favorable when the resultant cation is highly stabilized as in the case of the dithiolium cation in reactions 6-24 (Hartzler, 1973; Pazdro and Polaczkowa, 1972).

The thiazolium "carbene" has a very important role in biochemistry. In 1957, Breslow reported that hydrogen H_a in thiamine (vitamin B_1) was remarkably acidic (Breslow, 1957). Deprotonation affords a species analogous to the previously discussed dithiolium carbene for which a variety of resonance structures can be written (Eq. 6-25; Elvidge *et al.*, 1974). Among other biochemical functions, thiamine acts in conjunction with the enzyme carboxylase in catalyzing the decarboxylation of pyruvate by a sequence of the sort shown (Eq. 6-26; Lowe and Ingraham, 1975). The preparation of "dimers" of thiazolium and thiadiazolium "carbenes" has been reported

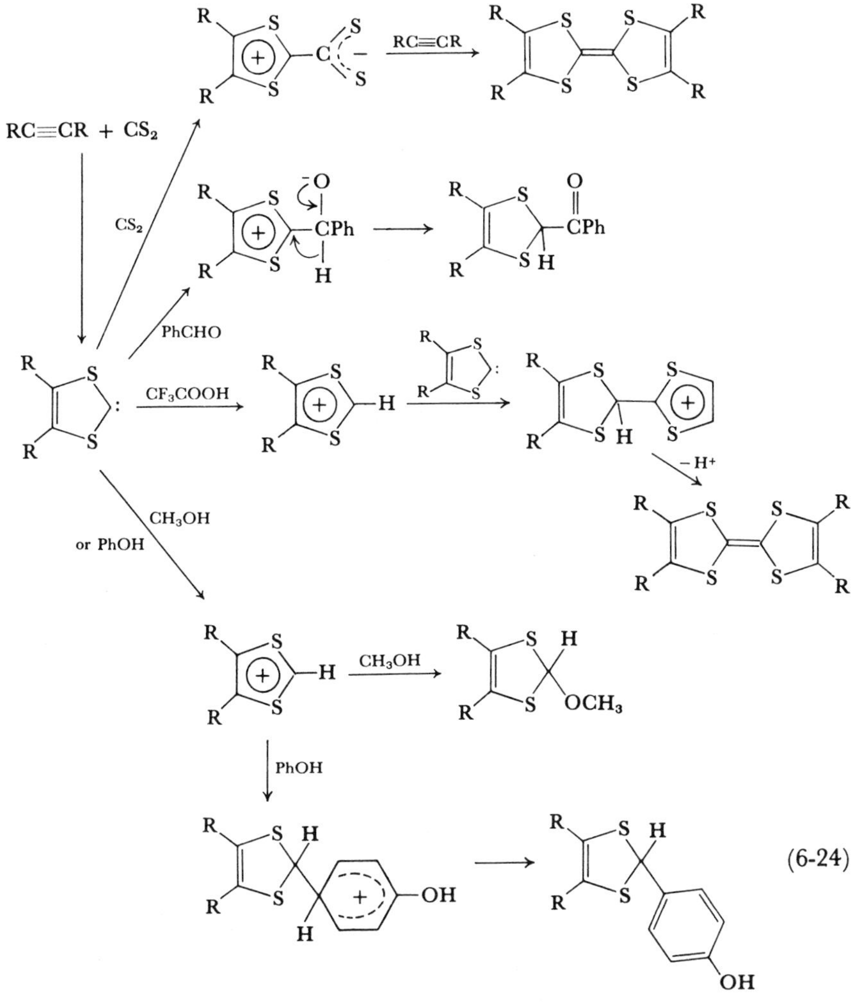

$RC{\equiv}CR + CS_2$

CS_2

$RC{\equiv}CR$

PhCHO

CPh

CF_3COOH

$-H^+$

CH_3OH or PhOH

CH_3OH

OCH_3

PhOH

OH

(6-24)

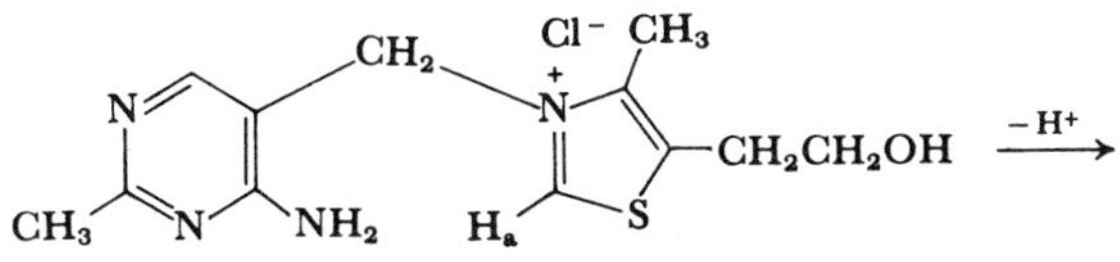

Cl^- CH_3

CH_2

CH_2CH_2OH $\xrightarrow{-H^+}$

CH_3 NH_2 H_a

Thiamine chloride

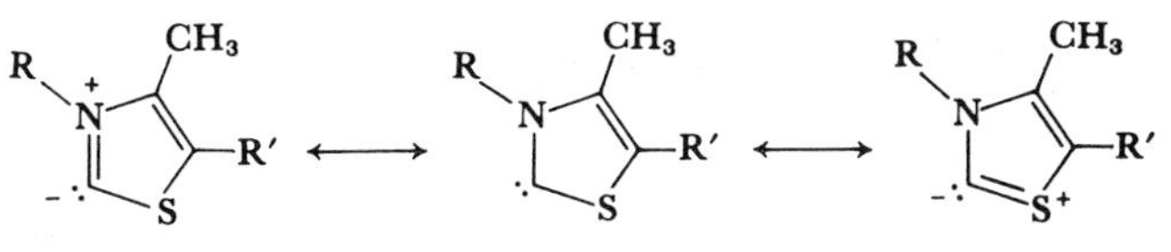

CH_3 R R′ ⟷ CH_3 R R′ ⟷ CH_3 R R′ (6-25)

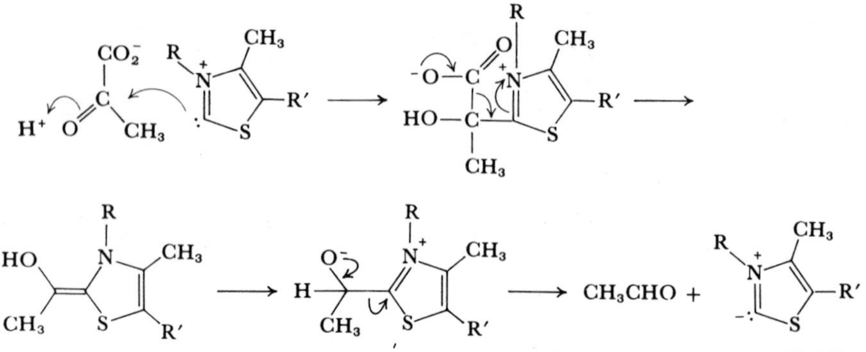

(6-26)

(Eq. 6-27; Wanzlich, 1962; Eq. 6-28; Scherowsky, 1974a,b). The formation of these products says nothing about the carbene character of the deprotonated heterocycles because an ionic mechanism can be presented as an alternative here (Eq. 6-29).

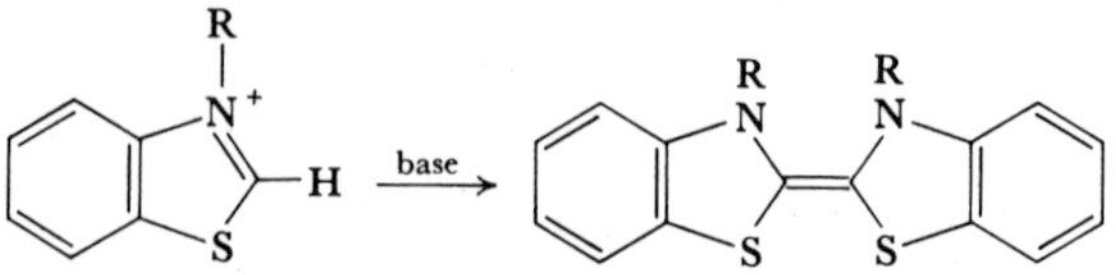

(6-27)

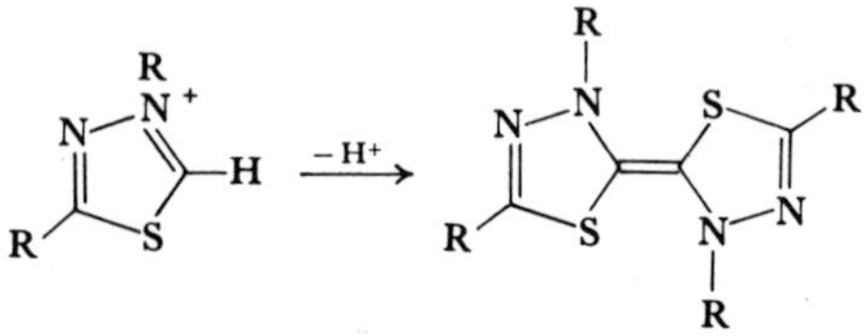

(6-28)

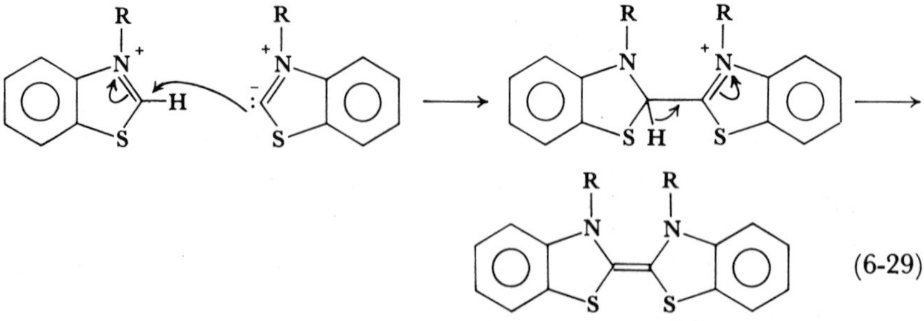

(6-29)

6.5 Reactions of β-, γ-, and δ-Thiomethylenes

Sulfides are excellent singlet carbene traps and novel intramolecular reactions might be anticipated for carbenes containing a sterically accessible sulfide substituent. As evidence for the superior intermolecular carbene-trapping capability of sulfides, Ando reports that bis(carbomethoxy)-methylene reacts with dimethyl sulfide four times faster than cyclohexene (Ando *et al.*, 1969). In the reaction of allyl sulfides with dimethyl diazomalonate under conditions of direct irradiation or copper catalysis, capture of the carbene by sulfur is favored over addition to the double bond (Ando *et al.*, 1972). Only in the case of benzophenone-sensitized decomposition of the diazo compound in the presence of allylic sulfides did double bond addition predominate over carbene capture by sulfur, presumably because of the inability of sulfur to capture triplet carbenes (Eq. 6-30; Ando *et al.*, 1972). In the reaction of allyl sulfide with diazomethane under conditions of copper catalysis, capture of the carbene/carbenoid by sulfur is favored over addition to the double bond as demonstrated by Kirmse (Eq. 6-31; Kirmse and Kapps, 1968).

$C(CO_2CH_3)_2$

$RSCH_2CH—CH_2$

(11%) (37%)

$RSCH_2CH{=}CH_2$ $\xrightarrow[h\nu]{(CH_3O_2C)_2CN_2}$ $R = C_4H_9$ $R = C_2H_5$ $\xleftarrow[h\nu,\ Ph_2CO]{(CH_3O_2C)_2CN_2}$ $RSCH_2CH{=}CH_2$

(57%) (18%) (6-30)

$(CH_3O_2C)_2CN_2$, Δ, Cu(II)

SR

$:\underline{C}(CO_2CH_3)_2$

$R = C_4H_9$

(93%) → $RSC(CO_2CH_3)_2CH_2CH{=}CH_2$

Table 6-3 indicates the type of products formed from β-, γ-, and δ-thiomethylenes, generated photochemically or thermally from tosylhydrazone salts or diazo compounds. Most of the more unusual products can be explained by invoking rearrangement of the sulfur ylides formed by intramolecular nucleophilic trapping of the carbenes by sulfur. β-Thiomethylenes

S, CD_2, CD_2 $\xrightarrow[Cu_2Cl_2]{CH_2N_2}$ CD_2, $\overset{+}{S}$, CD_2, CH_2^- ⟶ S, CD_2, CD_2

(29%)

S, CD_2, CD_2

(6%)

(6-31)

Table 6-3. Reactions of β-, γ-, and δ-Thiomethylenes

Entry	Conditions	Position of S	Carbene	Product(s) (Yield %)	Reference
1	Δ	β	$Ph\ddot{C}CH_2SC_2H_5$[b]	$PhC(SC_2H_5){=}CH_2$ $PhCH{=}CHSC_2H_5$	Robson and Shechter (1967)
2	Δ	β	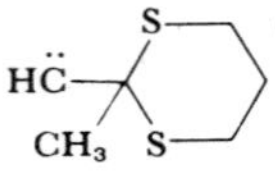 $H\ddot{C}$, S, S, CH_3[b]	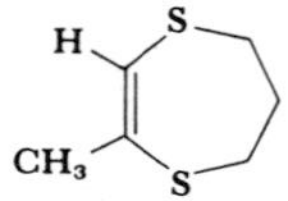 H, S, S, CH_3	Robson and Shechter (1967)
3	$h\nu/-70°$	β	$Ph\ddot{C}CH_2SCH_2CH{=}CH_2$[b]	$CH_2\ \ C(Ph)SCH_2CH\ \ CH_2$ (8) $PhCH{=}CHSCH_2CH{=}CH_2$ (11)	Ojima and Kondo (1973)
4	$h\nu/CH_3OH$	β	$H\ddot{C}C(O)SCH_3$[a]	$CH_3SCH_2CO_2CH_3$ (86)	Hixson and Hixson (1972)
5	$h\nu/CH_3OH$	β	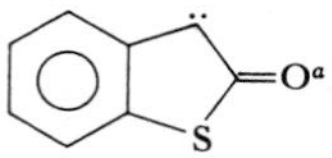 S, =O[a]	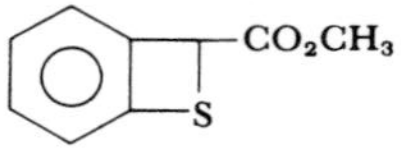 CO_2CH_3, S	Voigt and Meier (1976)
6	Δ	β	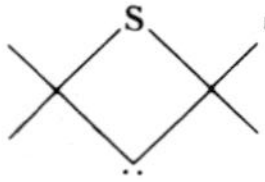 S[b]	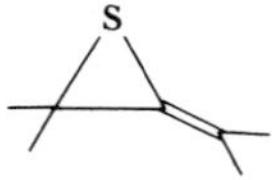 S (40–65)	Hortmann and Bhattacharjya (1976)

7	$h\nu$	γ	$Ph\ddot{C}CH_2C(CH_3)_2SCH_2CH{=}CHCH_3$[b]	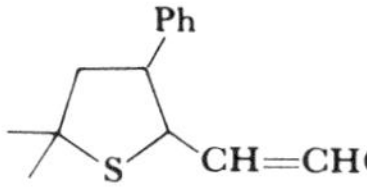 $PhCH{=}CHC(CH_3)_2SCH_2CH{=}CHCH_3$ (42) 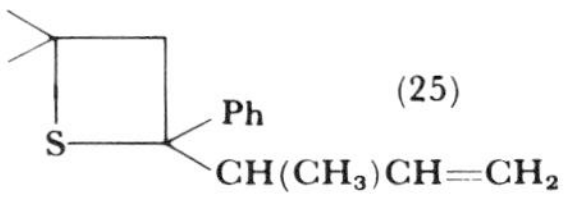 S, Ph, $CH(CH_3)CH{=}CH_2$ (25) Ph, S, $CH{=}CHCH_3$ (33)	Kondo and Ojima (1975)
8	Δ	γ	$Ph\ddot{C}CH_2C(CH_3)_2SPh$[b]	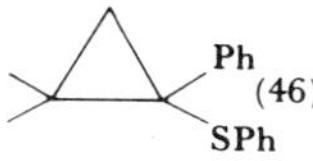 Ph, SPh (46) $PhCH{=}CHC(CH_3)_2SPh$ (55)	Kondo and Ojima (1975)
9	Δ/Cu(II)	δ	$H\ddot{C}C(O)CH_2CH_2SCH_2Ph$[a]	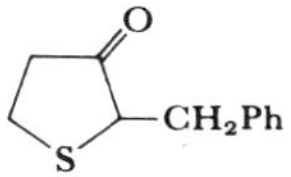 O, S, CH_2Ph (51)	Kondo and Ojima (1972)
10	Δ/Cu(II)	δ	$H\ddot{C}C(O)CH_2CH_2SCH_2CH{=}CH_2$[a]	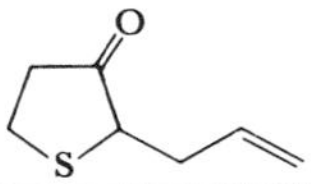 O, S (78)	Kondo and Ojima (1972)

[a] Precursor is diazo compound.
[b] Precursor is tosylhydrazone salt.

$PhC̈CH_2SC_2H_5$ ⟶ Ph ... ^+S ... CH_2CH_3 ⟶ $Ph{-}C(SC_2H_5){=}CH_2$ (6-32)

$HC̈(=O)SCH_3$ ⟶ H ... O ... ^+S ... CH_3 ⟷ H ... O^- ... ^+S ... CH_3 ⟶ $CH_3SCH{=}C{=}O$ (6-33)

N_2 ... O ... S —$h\nu$, $-N_2$⟶ O ... S ⟷ O ... S^+ ⟶

C=O ... S —CH_3OH⟶ H ... CO_2CH_3 ... S (6-34)

S ... N—N(Na)Ts —Δ⟶ S ⟷ $\overset{+}{S}$ ⟶

S (6-35)

S ... Ph ⟶ S^+ ... Ph —[2,3] sigmatropic shift⟶ S ... Ph

(6-36)

S^+ ... Ph —Stevens⟶ S ... Ph ⟶ $CH_3CH{=}CH$ ... S ... Ph

(E and Z isomers at double bond)

give episulfonium ylides which rearrange to olefins or ketenes as seen in Eqs. 6-32 to 6-35. The ylides derived from γ- and δ-thiomethylenes can undergo Stevens rearrangements or, with suitable substituents, [2,3] sigmatropic shifts, e.g., Eq. 6-36.

6.6 Sulfinyl- and Sulfonylmethylenes

α-Sulfinylmethylenes have only very recently been generated and relatively little is known about them (Venier *et al.*, 1975). Thus, reaction of phenyl diazomethyl sulfoxide with cyclohexene at room temperature gave the corresponding norcaranes with remarkable stereoselectivity (the anti–syn ratio was 34:1). The intermediacy of pyrazolines in this reaction could be ruled out through study of the reaction of phenyl diazomethyl sulfoxide with a 1:1 mixture of cyclohexene and diethyl fumarate. Although diethyl fumarate is an excellent dipolarophile, no adducts of diethyl fumarate were found. In this experiment, a slight reduction in the yield of the norcarane was accompanied by the formation of bis(phenylsulfinyl)ethene (Eq. 6-37; Venier *et al.*, 1975). Intramolecular cycloaddition of α-sulfinyl carbenes has

PhS(O)CHN$_2$ ⟶ PhS(O)CH: ⟶ (cyclohexene) norcarane–S(O)Ph + norcarane–S(O)Ph (34:1)

PhS(O)CHN$_2$ ⟶ (cyclohexene + $C_2H_5O_2C$–CH=CH–$CO_2C_2H_5$) norcarane–S(O)Ph + PhS(O)CH=CHS(O)Ph only

(6-37)

S(O)Cl azetidinone (R, H H, N, COOR) ⟶ (CH_2N_2) S(O)CHN$_2$ azetidinone (R, H H, N, COOR) ⟶ tricyclic sulfoxide (R, H H, O, S, H, CH_3, N, COOR) and isomers

(6-38)

been employed by Lammert and Kukolja in the preparation of antibacterial 2,3-methylenecepham derivatives (Eq. 6-38; Lammert and Kukolja, 1975).

Until recently the only satisfactory sources of α-sulfonylmethylenes were α-diazosulfones. The α-elimination route from α-halosulfones generally fails because of the substantial stability of α-sulfonyl carbanions. Thus, the carbanion $(CF_3SO_2)_2C(Br)^-Na^+$ is stable up to 250°! (Koshar and Mitsch,

1973.) The carbanion from α-halosulfones can be readily trapped with electrophiles (Stetter and Steinbeck, 1972); with cyclohexene, there is no evidence for adduct formation (Eq. 6-39; Paquette, 1964). Seyferth has found

$$ArSO_2CH_2Cl \xrightarrow{RMgX} ArSO_2\ddot{C}^{-}HCl \qquad (6\text{-}39)$$

$ArSO_2\ddot{C}^{-}HCl \xrightarrow{CO_2} ArSO_2CHClCO_2^-$

$ArSO_2\ddot{C}^{-}HCl \xrightarrow{D_2O} ArSO_2CHDCl$

$ArSO_2\ddot{C}^{-}HCl$ + cyclohexene ⇸ 7-(SO_2Ar)bicyclo[4.1.0]heptane

that phenyl(phenylsulfonyldichloromethyl)mercury can function as a $PhSO_2\ddot{C}Cl$ transfer agent although unusually high temperatures and long reaction times are required (Eq. 6-40; Seyferth and Woodruff, 1974).

$$PhHgCCl_2SO_2Ph \xrightarrow[\text{cyclooctene}]{140°,\ 8\ \text{days}} \text{(bicyclic: Cl, SOPh)} \quad (49\%) \qquad (6\text{-}40)$$

2-biphenylyl-SO_2CHN_2 $\xrightarrow{120°}$ [2-biphenylyl-SO_2CH:] → dibenzothiopyran (S, O_2)

[2-biphenylyl-SO_2CH:] → norcaradiene adduct (S, O_2, H) → cycloheptatriene-fused sulfone (S, O_2, H)

[2-biphenylyl-SO_2CH:] $\xrightarrow{ArSO_2\bar{C}H-\overset{+}{N}\equiv N}$ $[ArSO_2\bar{C}H-CH(^+N_2)SO_2Ar]$ $\xrightarrow{-N_2}$ $ArSO_2CH{=}CHSO_2Ar$ $\xrightarrow{ArSO_2CH:\ (\text{or } ArSO_2CHN_2)}$ tris(SO_2Ar)cyclopropane (SO_2Ar, $ArSO_2$, SO_2Ar)

(6-41)

(Abramovitch *et al.*, 1972)

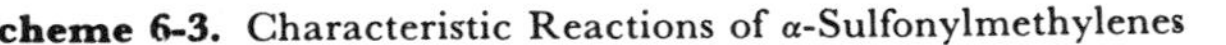

Scheme 6-3. Characteristic Reactions of α-Sulfonylmethylenes

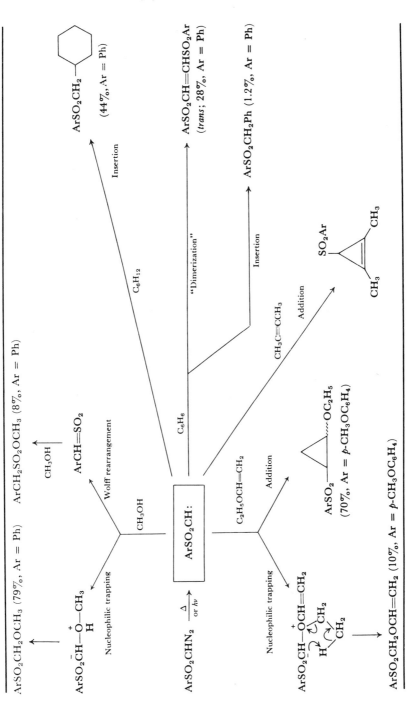

The presence of the strongly electron-withdrawing sulfonyl group and the absence of lone pairs on the sulfur make α-sulfonylmethylenes considerably more electrophilic than α-thiomethylenes. The characteristic reactions of α-sulfonylmethylenes include addition to olefins and even to acetylenes (less nucleophilic than olefins), trapping with nucleophiles, "dimerization," insertion into C—H bonds, and Wolff rearrangement, as summarized in Scheme 6-3 (van Leusen and Strating, 1970). Some additional reactions of interest are shown in Eqs. 6-41 to 6-44.

$$(PhSO_2)_2CN_2 \xrightarrow[(CH_3)_2S]{h\nu} (PhSO_2)_2\overset{-}{C}\text{—}\overset{+}{S}(CH_3)_2$$
$$\xrightarrow[(CH_3)_2SO]{h\nu} (PhSO_2)_2\overset{-}{C}\text{—}\overset{+}{S}(O)(CH_3)_2 \qquad (6\text{-}42)$$

(Dieckmann, 1965)

$$ArSO_2C(N_2)C(O)Ph \xrightarrow[C_2H_5OH]{h\nu} ArSO_2\ddot{C}C(O)Ph$$
$$ArSO_2(Ph)CHCO_2C_2H_5\ (33\%) \xleftarrow{C_2H_5OH} ArSO_2(Ph)C{=}C{=}O \qquad \not\rightarrow \quad ArCC(O)Ph\ (\text{with } {=}SO_2) \qquad (6\text{-}43)$$

(Illiger *et al.*, 1972)

$$ArSO_2C(N_2)Ph \xrightarrow{350^\circ} ArSO_2\ddot{C}Ph \longrightarrow (Ar)(Ph)C{=}SO_2 \xrightarrow{-SO} ArCOPh\ (13\%)$$
$$(Ar)(Ph)C{=}SO_2 \xrightarrow{-SO_2} Ar\ddot{C}Ph \longrightarrow \text{3-methylfluorene-type product}\ (3\%) \qquad (6\text{-}44)$$

(Ar = *m*-tolyl)

(Sarver *et al.*, 1975)

Pirkle and Kalish have reported novel allene-forming reactions which may involve cyclic β-sulfonyl- and β-sulfinylmethylenes as intermediates (Eq. 6-45; Kalish, 1969; Kalish and Pirkle, 1967).

$$\text{2,2,4,4-tetramethyl-3-diazothietane } S\text{-}O_n\ (n = 1 \text{ or } 2) \xrightarrow{\Delta} \left[\text{2,2,4,4-tetramethylthietan-3-ylidene } S\text{-}O_n\right] \longrightarrow (CH_3)_2C{=}C{=}C(CH_3)_2 + SO_n \qquad (6\text{-}45)$$

$$\begin{pmatrix} n = 1, 35\% \\ n = 2, 56\% \end{pmatrix}$$

General References on Carbene Chemistry

Bethell, D. (1973). *In* "Organic Reactive Intermediates" (S. P. McManus, ed.), Chapter 2. Academic Press, New York.

Jones, M., Jr., and Moss, R. A. (ed.) (1973). "Carbenes," Vol. I. Wiley (Interscience), New York.

Kirmse, W. (1971). "Carbene Chemistry," 2nd ed., Academic Press, New York.

References

Abramovitch, R. A., and Roy, J. (1965). *Chem. Commun.* 542.

Abramovitch, R. A., Alexanian, V., and Smith, E. M. (1972). *Chem. Commun.* 893; Abramovitch, R. A., and Alexanian, V. *Heterocycles* **2**, 595 (1974).

Akhtar, M., Kleppinger, J., MacDiarmid, A. G., Millikan, J., Moran, M. J., Chiang, C. K., Cohen, M. J., Heeger, A. J., and Peebles, D. L. (1977). *Chem. Commun.* 473.

Ando, W., Yagihara, T., Tozune, S., and Migita, T. (1969). *J. Am. Chem. Soc.* **91**, 2786.

Ando, W., Kondo, S., Nakayama, K., Ichibori, K., Kohoda, H., Yamato, H., Imai, I., Nakaido, S., and Migita, T. (1972). *J. Am. Chem. Soc.* **94**, 3870.

Baldwin, J. E., and Walker, J. A. (1972). *Chem. Commun.* 354.

Beak, P., and Worley, J. W. (1972). *J. Am. Chem. Soc.* **94**, 597.

Bos, H. J. T., Brandsma, L., and Arens, J. F. (1967). *Monatsh. Chem.* **98**, 1043.

Breslow, R. (1957). *J. Am. Chem. Soc.* **79**, 1762; **80**, 3719 (1958).

Carey, F. A., and Court, A. S. (1972). *J. Org. Chem.* **37**, 1926.

Carlson, R. M., and Helquist, P. M. (1969). *Tetrahedron Lett.* 173.

Chong, J. A., and Wiseman, J. R. (1972). *J. Am. Chem. Soc.* **94**, 8627.

Closs, G. L., and Moss, R. A. (1964). *J. Am. Chem. Soc.* **86**, 4042.

Coffen, D. L. (1970). *Tetrahedron Lett.* 2633.

Coffen, D. L., Chambers, J. Q., Williams, D. R., Garrett, P. E., and Canfield, N. D. (1971). *J. Am. Chem. Soc.* **93**, 2258.

Cohen, M. J., Garito, A. F., Heeger, A. J., MacDiarmid, A. G., Mikulski, C. M., Saran, M. S., and Kleppinger, J. (1976). *J. Am. Chem. Soc.* **98**, 3844.

Coleman, L. B., Cohen, M. J., Sandman, P. J., Yamagishi, F. G., Garito, A. F., and Heeger, A. J. (1973). *Solid State Commun.* **12**, 1125.

Corey, E. J., and Märkl, G. (1967). *Tetrahedron Lett.* 3201.

Corey, E. J., and Shulman, J. I. (1968). *Tetrahedron Lett.* 3655.

Corey, E. J., and Walinsky, S. W. (1972). Unpublished results.

Corey, E. J., and Winter, R. A. E. (1963). *J. Am. Chem. Soc.* **85**, 2677.

Corey, E. J., Carey, F. A., and Winter, R. A. E. (1965). *J. Am. Chem. Soc.* **87**, 934.

Davies, W. C., and Jones, W. J. (1929). *J. Chem. Soc.* (*London*) 33.

Diekmann, J. (1965). *J. Org. Chem.* **30**, 2272.

Evans, D. A., Sims, C. L., and Andrews, G. C. (1977). *J. Am. Chem. Soc.* **99**, 5453.

Elvidge, J. A., Jones, J. R., O'Brien, C. O., Evans, E. A., and Sheppard, H. C. (1974). *In* "Advances in Heterocyclic Chemistry" (A. R. Katritzky, and A J. Boulton, eds.), Vol. 16. Academic Press, New York.

Ferraris, J., Cowan, D. O., Walatka, V., Jr., and Perlstein, J. H. (1973). *J. Am. Chem. Soc.* **95**, 948.

Fields, E. K., and Meyerson, S. (1970). *Tetrahedron Lett.* 629.

Fischer, E. O., and Kreis, G. (1973). *Chem. Ber.* **106**, 2310.

Fröling, A., and Arens, J. F. (1962). *Rec. Trav. Chim. Pays-Bas.* **87**, 1009.

Gleiter, R., and Hoffmann, R. (1968). *J. Am. Chem. Soc.* **90**, 5457.

Greenhouse, R., Ravindranathan, T., and Borden, W. T. (1976). *J. Am. Chem. Soc.* **98**, 6738.
Hartmann, W., Schrader, L., and Wendisch, D. (1973). *Chem. Ber.* **106**, 1076.
Hartzler, H. D. (1973). *J. Am. Chem. Soc.* **95**, 4379.
Hayasi, Y., and Nozaki, H. (1971). *Bull. Chem. Soc. Jpn.* **45**, 198.
Henderson, W. A. (1959). Ph.D. Thesis, Yale Univ.
Hine, J., Bayer, R. P., and Hammer, G. G. (1962). *J. Am. Chem. Soc.* **84**, 1751.
Hixson, S. S., and Hixson, S. H. (1972). *J. Org. Chem.* **37**, 1279.
Hortmann, A. G., and Bhattacharjya, A. (1976). *J. Am. Chem. Soc.* **98**, 7081.
Horton, D., and Tindall, C. G. Jr. (1970). *J. Org. Chem.* **35**, 3558.
Illger, W., Liedhegener, A., and Regitz, M. (1972). *Justus Liebigs Ann. Chem.* **760**, 1.
Jones, F. N., and Andreades, S. (1969). *J. Org. Chem.* **34**, 3011.
Jones, M., Temple, P., Thomas, E. J., and Whitham, G. H. (1974). *J. Chem. Soc. Perkin Trans.* **1**, 433.
Julia, S., Huynh, C., and Michelot, D. (1972). *Tetrahedron Lett.* 3587.
Kalish, R. (1969). Ph.D. Thesis, Univ. of Illinois, Urbana, Illinois.
Kalish, R., and Pirkle, W. H. (1967). *J. Am. Chem. Soc.* **89**, 2782.
Kirmse, W., and Kapps, M. (1968). *Chem. Ber.* **101**, 994.
Kondo, K., and Ojima, I. (1972). *Chem. Commun.* 860.
Kondo, K., and Ojima, I. (1975). *Bull. Chem. Soc. Jpn.* **48**, 1490.
Koshar, R. J., and Mitsch, R. A. (1973). *J. Org. Chem.* **38**, 3358.
Krebs, A., and Kimling, H. (1971). *Angew. Chem. Int. Ed. English* **10**, 509.
Lammert, S. R., and Kukolja, S. (1975). *J. Am. Chem. Soc.* **97**, 5583.
Lemal, D. M., and Banitt, E. H. (1964). *Tetrahedron Lett.* 245.
Lemal, D. M., Lovald, R. A., and Harrington, R. W. (1965). *Tetrahedron Lett.* 2779.
Lowe, J. N., and Ingraham, L. L. (1974). "An Introduction to Biochemical Reaction Mechanism," Chapter 5. Academic Press, New York, and references therein.
Moore, G. G. (1961). Ph.D. Thesis, Yale Univ.
Moss, R. A., Joyce, M. A., and Pikiewicz, F. G. (1975). *Tetrahedron Lett.* 2425.
Nakayama, J. (1975). *J. Chem. Soc. Perkin Trans.* **1**, 525.
Narita, M., and Pittman, C. U., Jr. (1976). *Synthesis* 489.
Ojima, I., and Kondo, K. (1973). *Bull. Chem. Soc. Jpn.* **46**, 1539.
Olofson, R. A., Walinsky, S. W., Marino, J. P., and Jernow, J. L. (1968). *J. Am. Chem. Soc.* **90**, 6554.
Paquette, L. A. (1964). *J. Am. Chem. Soc.* **86**, 4085.
Paquette, L. A., Philips, J. C., and Wingard, R. E., Jr. (1971). *J. Am. Chem. Soc.* **93**, 4516.
Pazdro, K. M., and Polaczkowa, W. (1970). *Roczniki Chem.* **44**, 1823.
Pazdro, K. M., and Polaczkowa, W. (1972). *Roczniki Chem.* **46**, 839.
Plieninger, H., and Heuck, C. C. (1972). *Tetrahedron* **28**, 73.
Prinzbach, H., and Babsch, H. (1975). *Angew. Chem. Int. Ed. English* **14**, 753.
Robson, J. H., and Shechter, H. (1967). *J. Am. Chem. Soc.* **89**, 7112.
Rynbrandt, R. H., and Dutton, F. E. (1972). *Tetrahedron Lett.* 1933.
Saquet, M. (1972). *C.R. Acad. Sci. Paris C* **275**, 283.
Sarver, B. E., Jones, M., Jr., and Van Leusen, A. M. (1975). *J. Am. Chem. Soc.* **97**, 4771.
Scherowsky, G. (1974a). *Tetrahedron Lett.* 1813.
Scherowsky, G. (1974b). *Chem. Ber.* **107**, 1092.
Scherowsky, G., and Weiland, J. (1974). *Chem. Ber.* **107**, 3155.
Schöllkopf, V., and Wiskott, E. (1966). *Justus Liebigs Ann. Chem.* **694**, 44.
Schöllkopf, V., Lehmann, G. J., Paust, J., and Härtl, H.-D. (1964). *Chem. Ber.* **97**, 1527.
Schöllkopf, V., Woerner, F. P., and Wiskott, E. (1966). *Chem. Ber.* **99**, 806.

Seebach, D. (1972). *Chem. Ber.* **107**, 487.

Seebach, D., and Beck, A. K. (1969). *J. Am. Chem. Soc.* **91**, 1540.

Seebach, D., Gröbel, B. Th., Beck, A. K., Braun, M., and Geiss, K.-H. (1972a). *Angew Chem. Int. Ed. English* **11**, 443.

Seebach, D., Stegmann, H., Scheffler, K., Beck, A. K., and Geiss, K.-H. (1972b). *Chem. Ber.* **105**, 3905.

Seebach, D., Braun, M., and DuPreez, N. (1973a). *Tetrahedron Lett.* 3509; Braun, M., and Seebach, D. *Chcm. Ber.* **109**, 669 (1976).

Seebach, D., Kolb, M., and Gröbel, B.-T. (1973b). *Angew. Chem. Int. Ed. Engl.* **12**, 69.

Seyferth, D., and Woodruff, R. A. (1974). *J. Organomet. Chem.* **71**, 335.

Stetter, H., and Steinbeck, K. (1972). *Justus Liebigs Ann. Chem.* **766**, 89.

Street, G. B., Bingham, R. L., Crowley, J. I., and Kuyper, J. (1977). *Chem. Commun.* 464.

Tichý, M., and Sicher, J. (1969). *Tetrahedron Lett.* 4609.

van Leusen, A. M., and Strating, J. (1970). *Quant. Rep. Sulfur Chem.* **5**, 67.

van Leusen, A. M., Mulder, R. J., and Strating, J. (1967). *Rec. Trav. Chim. Pays-Bas.* **86,** 225.

Venier, C. G., Barager, H. J. III, and Ward, M. A. (1975). *J. Am. Chem. Soc.* **97**, 3228; see also Franck-Neumann, M., and Lohmann, J.-J. *Angew. Chem. Int. Ed. English* **16,** 323 (1977).

Voigt, E., and Meier, H. (1976). *Angew. Chem.* **88**, 94.

Wanzlich, H. W. (1962). *Angew. Chem. Int. Ed. English* **1**, 75.

Yoshida, Z., Kawase, T., and Yoneda, S. (1975). *Tetrahedron Lett.* 331.

Chapter 7 | Pericyclic Reactions of Organosulfur Compounds

7.1 Introduction

Pericyclic reactions may be defined as reactions in which electronic reorganization occurs in some sort of *cyclic* array of the participating atomic centers. These reactions are concerted, involving neither radical nor ionic intermediates, and may be subclassified as sigmatropic, electrocyclic, cycloaddition, cheletropic, and cycloelimination reactions. Pericyclic reactions may be initiated by thermal energy (Δ) or by photochemical excitation ($h\nu$).

1. *Sigmatropic* changes are those in which a σ-bond, flanked by one or more π-electron systems migrates in an uncatalyzed process to a new position in the molecule ($1\sigma + m\pi \rightleftharpoons 1\sigma + m\pi$). An important example of this sequence is the Claisen rearrangement, a [3,3] sigmatropic process (Eq. 7-1).

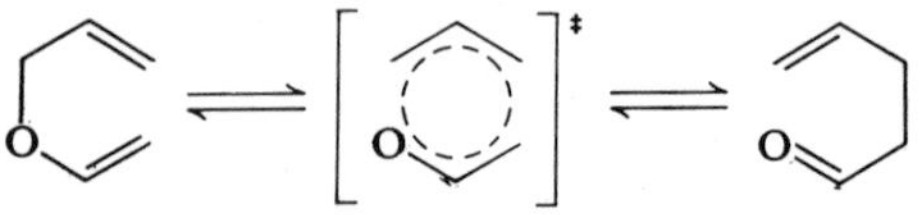

(7-1)

2. *Electrocyclic* processes involve the formation of a single bond between the termini of a conjugated polyene (or heteroatom counterpart) and the reverse reaction (1 π-bond $\rightleftharpoons$ 1 σ-bond). An example is the ring opening reaction of cyclobutenes (Eq. 7-2) (Winter, 1965).

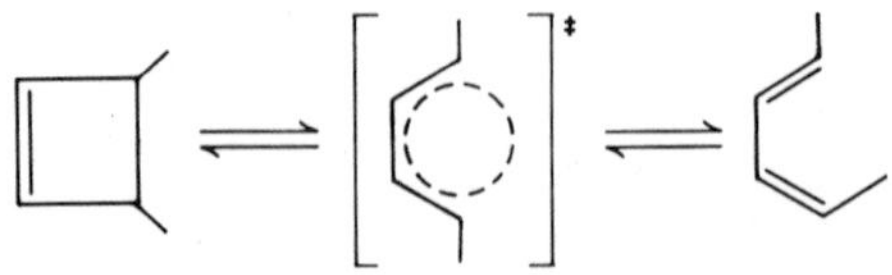

(7-2)

3. *Cycloaddition* reactions are ring-forming reactions involving most often formation of two or three new σ-bonds at the expense of two or three π-bonds ($m\sigma$-bonds $\rightleftharpoons$ $m\pi$-bonds; $m > 1$). The reverse processes, which result in the formation of two or three new π-bonds at the expense of two or three σ-bonds, have been termed cycloretrogression or reverse-cycloaddition

reactions. Illustrative of these processes are the Diels–Alder and retro-Diels–Alder reactions (Eq. 7-3).

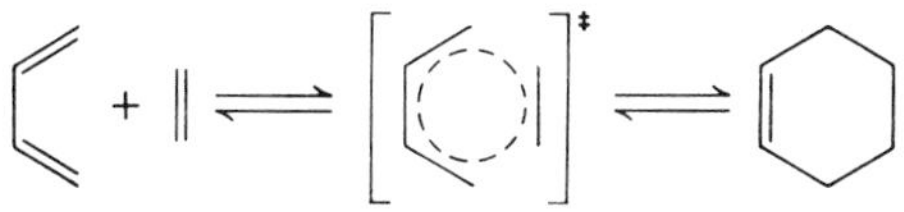

(7-3)

4. *Cheletropic* reactions are defined as processes in which two σ-bonds that terminate at a single atom, and which are separated by $m\pi$-bonds ($m \geq 0$), are made (or broken) in a concerted fashion, for example as seen in Eq. 7-4. In this chapter we shall consider together both cycloaddition and cheletropic reactions.

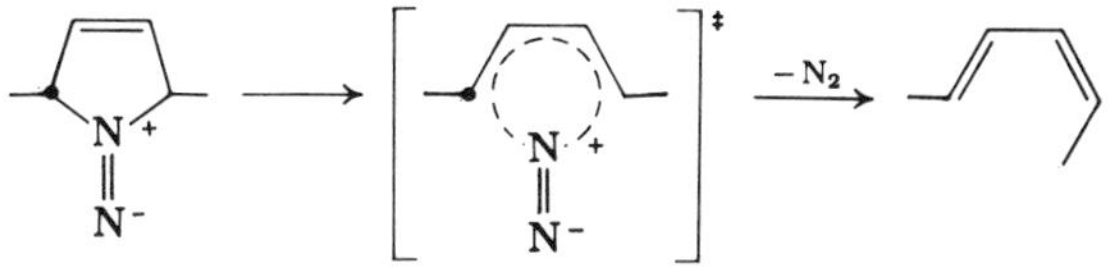

(7-4)

5. *Cycloelimination* reactions involve the transformation of $n\sigma$-bonds and $m\pi$-bonds into $(n - 1)$ σ-bonds and $(m + 1)$ π-bonds as illustrated by the retro-ene reaction (Eq. 7-5). The reverse process (right to left in Eq. 7-5) is an important process known as the ene reaction.

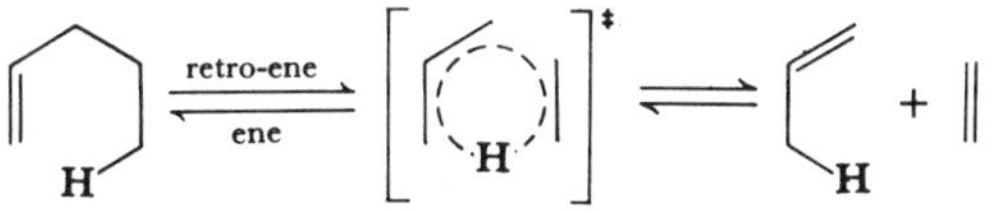

(7-5)

A theoretical basis for the interpretation of these several categories of pericyclic reactions was provided in 1965 by Woodward and Hoffmann when they enunciated the principle of conservation of orbital symmetry (Woodward and Hoffmann, 1965, 1970; Hoffmann and Woodward, 1965). A number of excellent volumes and reviews are available which deal at various levels of sophistication with the Woodward–Hoffman rules and related theoretical approaches to pericyclic reactions (Woodward and Hoffmann, 1970; Gilchrist and Storr, 1972; Gill and Willis, 1974; Lehr and Marchand, 1972; Simmons and Bunnett, 1974; Marchand and Lehr, 1977). We shall here survey the rich variety of pericyclic reactions which involve organosulfur compounds with particular emphasis, as in previous chapters, on those reactions which are of synthetic utility. Also included in this survey will be a limited number of related processes which involve polar or radical intermediates.

7.2 Sigmatropic Rearrangements

[1,*j*] Sigmatropic Processes

Relatively few examples of [1,*j*] sigmatropic reactions of organosulfur compounds have been reported. We have previously considered the anionic [1,2] sigmatropic processes commonly known as the Stevens and Wittig rearrangements (Eqs. 7-6 and 7-7, respectively; see Sections 3.5, 5.10, and 5.11) and have indicated the likelihood that radical pair mechanisms may

$$\mathrm{Ph\overset{-}{C}H{-}\overset{+}{S}(CH_2Ph)CH_3} \xrightarrow{[1,2]} \mathrm{PhCH(CH_2Ph){-}SCH_3} \tag{7-6}$$

$$\mathrm{Ph\overset{-}{C}HSCH_2Ph} \xrightarrow{[1,2]} \mathrm{PhCH(CH_2Ph){-}S^-} \tag{7-7}$$

operate for these reactions together with, or to the exclusion of, concerted mechanisms.

As with anionic [1,2] sigmatropic shifts, neutral [1,3] sigmatropic rearrangements are orbital symmetry forbidden to occur suprafacially with respect to the migrating center; the transition state geometry for the allowed antarafacial process at sulfur is very unfavorable (see Woodward and Hoffmann, 1965). Thus, there is evidence that 1,3-thioallylic rearrangement of allyl aryl sulfides involves a dipolar intermediate (Eq. 7-8; Kwart and George, 1977) while the analogous 1,3-migration shown in Eq. 7-9, on the basis of the activation parameters (compare the values given in Eq. 7-9 with the data in Table 7-1), is thought to involve diradicals (Anastassiou *et al.*, 1975).

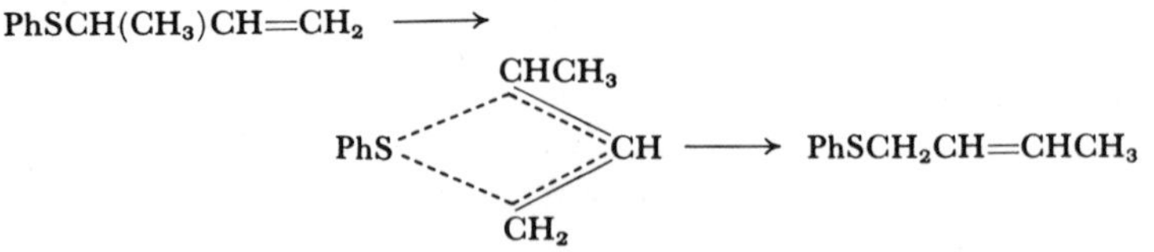

$$\mathrm{PhSCH(CH_3)CH{=}CH_2} \longrightarrow [\mathrm{PhS}\cdots\mathrm{CHCH_3}\cdots\mathrm{CH}\cdots\mathrm{CH_2}\cdots] \longrightarrow \mathrm{PhSCH_2CH{=}CHCH_3} \tag{7-8}$$

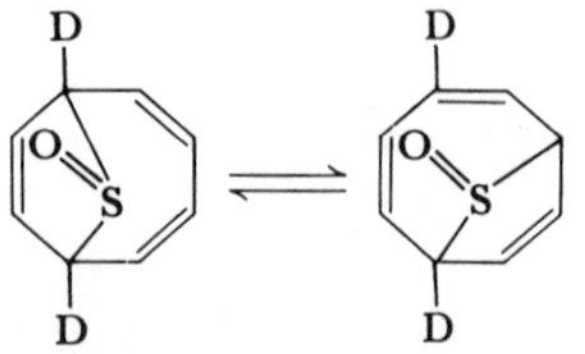

$$\Delta H^{\ddagger} = 27.8 \text{ kcal/mole}$$
$$\Delta S^{\ddagger} = -4 \text{ e.u.} \tag{7-9}$$
$$\Delta G^{\ddagger} = 29.5 \text{ kcal/mole at } 110.7°$$

Very recently, Lemal has developed a new description for [1,3] sigmatropic processes involving allylic sulfoxides such as the reactions shown in Eqs. 7-10

Table 7-1. Activation Parameters for Racemization of Alkyl Aryl Sulfoxides by Various Mechanisms[a]

Entry	Compound[b]	Process	$\Delta H^{\ddagger}$ (kcal/mole)	$\Delta S^{\ddagger}$ (e.u.)	$\Delta G^{\ddagger}$ (kcal/mole) at 100°C
1	$ArS(O)CH_2CH{=}CH_2$	Racemization via Eq. 7-35	23.1	−4.9	24.9
2	$Ar'S(O)CH_2CH{=}CH_2$	Racemization via Eq. 7-35	21.2	−8.0	24.2
3	$Ar'S{-}O{-}CH_2CH{=}CH_2$	Rearrangement via Eq. 7-35 to $ArS(O)CH_2CH{=}CH_2$	18.8	−4.8	20.6
4	$ArS(O)CH_2Ph$	Racemization via $PhCH_2\cdot/ArSO\cdot$ (Eq. 5-64)	43.0	+24.6	33.8
5	$ArS(O)CH_3$	Racemization via pyramidal inversion	37.4	−8.0	40.4
6	ArS(O)-1-adamantyl	Racemization via pyramidal inversion	42.0	+3.8	40.6

[a] Tang and Mislow (1970), Bickart *et al.* (1968), and Rayner *et al.* (1968).
[b] Ar = *p*-tolyl; Ar′ = *p*-trifluoromethylphenyl.

to 7-12 (Ross *et al.*, 1976). Lemal suggests (as indicated in Eq. 7-12) that the sulfur lone pair forms the new bond to carbon, and the electrons of the cleaving C—S bond become the new lone pair. A reaction such as Eq. 7-12 (a degenerate process) is termed a "pseudopericyclic reaction . . ., a concerted transformation whose primary changes in bonding occur within a

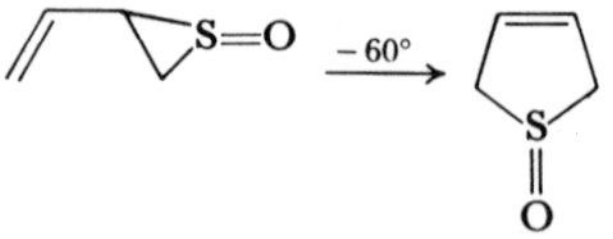

(7-10)

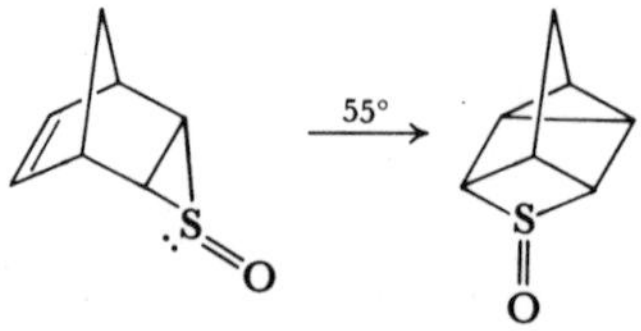

(7-11)

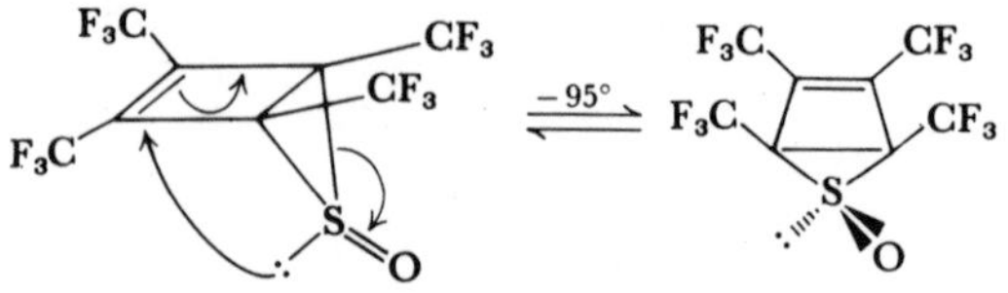

(7-12)

cyclic array of atoms, at one (or more) of which nonbonding and bonding atomic orbitals interchange roles" (Ross *et al.*, 1976). Because there is a disconnection in the cyclic array of overlapping orbitals (the atomic orbitals switching functions are mutually orthogonal), "pseudopericyclic reactions cannot be symmetry forbidden" (Ross *et al.*, 1976).†

Anionic [1,4] sigmatropic shifts are orbital symmetry allowed as suprafacial processes at the migrating center. Mislow has recently reported a [1,4] sigmatropic shift involving sulfur (Eq. 7-13; Ogura *et al.*, 1976).

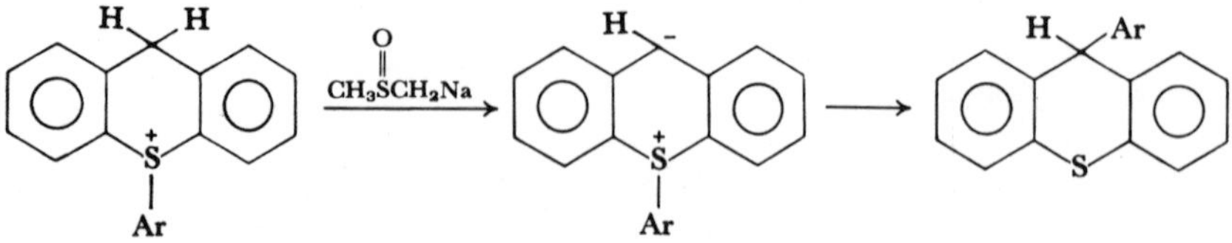

(7-13)

[3,3] Sigmatropic Processes

Among the processes classified as hetero-Cope reactions (see Eq. 7-14) are a number of variants involving one or more sulfur atoms in the rearrang-

† For an alternative mechanism for reaction (7-12), see Kwart and George (1977).

ing skeleton (Winterfeldt, 1970; Rhoads and Raulins, 1975). The first

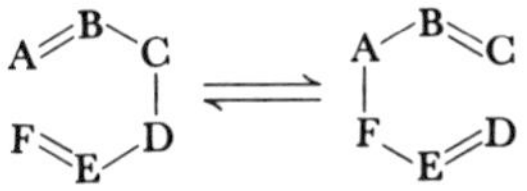

(7-14)

example of a [3,3] sigmatropic process involving sulfur (termed a *thio-Claisen reaction*) was reported in 1962 by Kwart, who has since investigated this reaction in great detail (Eq. 7-15; Kwart and Schwartz, 1974). Kwart

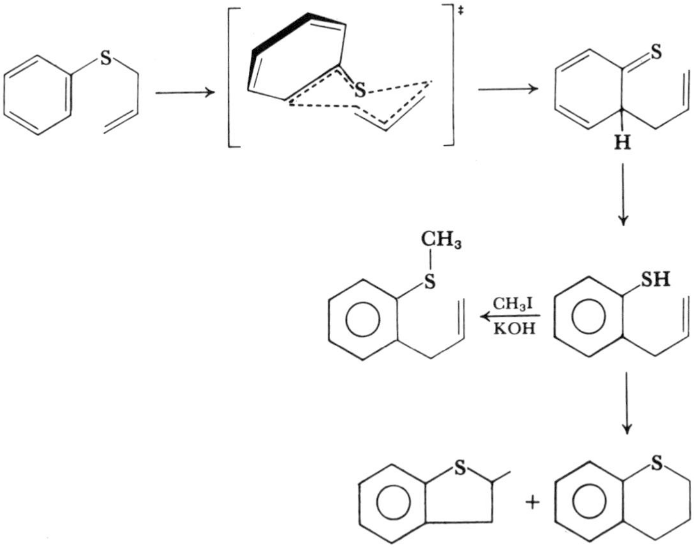

CH_3I / KOH

(7-15)

suggests that the driving force of the thio-Claisen rearrangement derives from electron displacement in the direction of the heteroatom producing trigonal hybridization of the allylic carbon. An unusual feature of this reaction is its susceptibility to catalysis by *nucleophiles* (in contrast to the oxy-Claisen reaction where *electrophilic* catalysis is observed) (Kwart and Schwartz, 1974).

While the thio-Claisen reaction of allyl phenyl sulfides is primarily of interest for mechanistic reasons, the thio-Claisen rearrangement of allyl vinyl sulfides is of considerable synthetic utility, particularly since it can be conducted at substantially lower temperatures, since the intermediate thiocarbonyl group can be hydrolyzed to a carbonyl function *in situ*, and since electrophilic substitution on the carbanion derived from the original allyl vinyl sulfide is possible prior to rearrangement. Selected synthetic applications include preparation of "propylure," the sex attractant of the pink bollworm moth (Eq. 7-16; Oshima *et al.*, 1973a), synthesis of *cis*-jasmone (Eq. 7-17; Oshima *et al.*, 1973b), a spiroannelation method (Eq. 7-18; Corey and Shulman, 1970), and rearrangement under unusually mild conditions (Eq. 7-19; Kondo and Ojima, 1972). Also of interest is a sequence which is equivalent to mono- and dialkylation of a thioketone (Eq. 7-20; Morin and Paguer, 1976). The *trans*-stereospecificity seen in the propylure synthesis (Eq. 7-16) can be attributed to the preference of substituents for the equatorial position in the chair conformation favored as the transition state

(1) *sec*-C_4H_9Li, THF
(2) $I(CH_2)_4OTHP$

DME
H_2O, $CaCO_3$,
heat

THPO CHO

several steps

AcO

"propylure"

(7-16)

C_2H_5O

(1) *sec*-C_4H_9Li, THF
(2) Br

OC_2H_5

DME
reflux
$CaCO_3$

CHO NaOH

O

(1) CH_3Li
(2) CrO_3

cis-jasmone

(7-17)

+ $P(OC_2H_5)_2$

HgO
190°

CHO

several steps

(7-18)

25°

Ph

O₂

(7-19)

Br / $NaCH_2S(O)CH_3$ / C_6H_6 ; 150° [3,3] ; (1) Br, base / (2) 170°

(70%) (90%) (7-20)

for [3,3] sigmatropic processes (see Eq. 7-16). Other variants of the thio-Claisen reaction include rearrangement of 1-alkenyl allenyl sulfides (Eq. 7-21; Brandsma and Verkruijsse, 1974), of S-allyldithiocarbamates employed in the synthesis of *E*-nuciferal (Eq. 7-22; Nakai *et al.*, 1975) and enones (Eq. 7-23; Nakai *et al.*, 1974), of S-alkylketenethioacetals used in the synthesis of α,β-unsaturated esters (Eq. 7-24; Takahashi *et al.*, 1973), of dialkenyl disulfides to give 1,4-dithiones (Eq. 7-25; Campbell and Evgenios, 1973), of thionocarbonates (Eq. 7-26; Faulkner and Petersen, 1973), of N-allylthiocarbostyril (Eq. 7-27; Makisumi and Sasatani, 1969), which latter reaction represents a *reverse* thio-Claisen rearrangement, and of allyl vinyl sulfone (Eq. 7-28; King and Harding, 1976) giving a "sulfo-Cope" rearrangement.

$$CH_2{=}C(CH_3){-}S{-}C(C_4H_9){=}C{=}CH_2 \xrightarrow[CaCO_3,\ 130^\circ]{DMSO{-}H_2O} C_4H_9C{\equiv}CCH_2CH_2COCH_3\ (54\%) \quad (7\text{-}21)$$

$N(CH_3)_2$, S ; $(i\text{-}C_3H_7)_2NLi$; Li, S, $N(CH_3)_2$; Ar, Br ; Ar, S, S, $N(CH_3)_2$; Ar, S, $CSN(CH_3)_2$; (1) RLi / (2) CH_3SSCH_3 ; Ar, $SCSN(CH_3)_2$, SCH_3 ; Hg(II) ; Ar, CHO (7-22)

α-nuciferal (Ar = *p*-tolyl)

$SCSN(CH_3)_2$ (1) RLi (2) CH_3SSCH_3 →

SCH_3, S, S, $N(CH_3)_2$ → SCH_3, S, S, $N(CH_3)_2$ (1) RLi (2) $C_5H_{11}I$ →

CH_3S, $SCSN(CH_3)_2$, C_5H_{11} — Hg(II), CH_3CN-H_2O → C_5H_{11}, CHO (68%) (7-23)

S, S — *sec*-C_4H_9Li, − 78° → S, Li, SLi — (1) $C_8H_{17}X$ (2) CH_3I →

CH_3S, S, C_8H_{17} — − 25° → C_8H_{17}, SCH_3, S — $CuCl_2$, CuO, C_2H_5OH, 25° →

C_8H_{17}, $CO_2C_2H_5$ (70%) (7-24)

S—S — C_6H_6, Δ →

H, S, S (antara–antara) → S S, H H (88%) (7-25)

O, S, OPh → O, S, OPh (67%) (7-26)

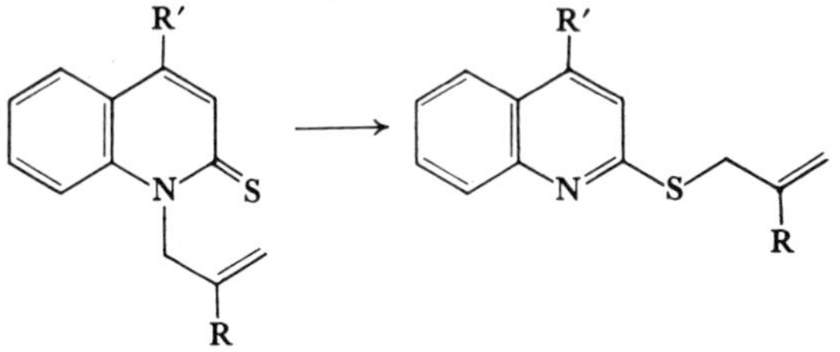

(7-27)

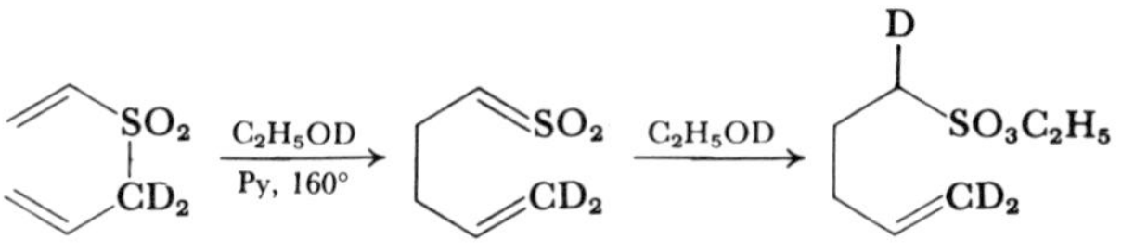

(7-28)

[2,3] Sigmatropic Processes

A general representation of a [2,3] sigmatropic rearrangement is given by Eq. 7-29. In Chapters 2, 3, 5, and 6 we have already discussed examples of

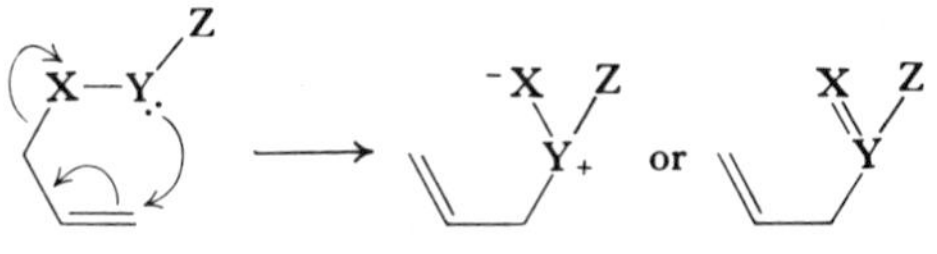

(7-29)

this ubiquitous reaction and shall here summarize some common features of this process. The Sommelet–Hauser reaction, e.g., Eq. 7-30, has been known for almost 50 yr (see Eq. 3-71, for example) and represents one variant of the [2,3] sigmatropic process. Here the initial allylic double bond (see Eq. 7-29)

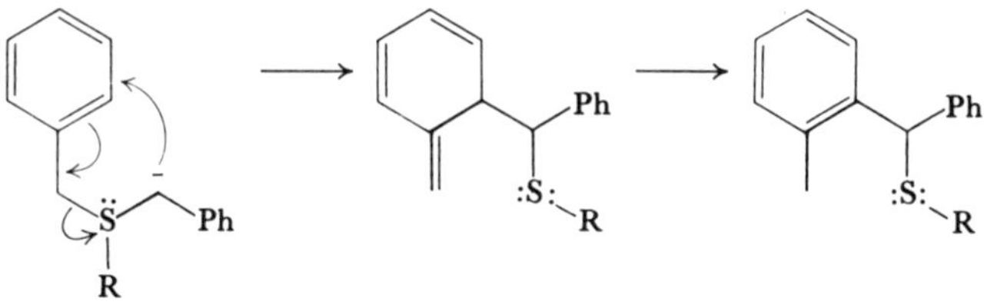

(7-30)

(R = lone pair or alkyl group)

is part of a benzene ring; the reaction proceeds readily despite the fact that it requires sacrifice of benzene resonance at the critical stage. Sommelet reactions of azasulfonium and aryloxysulfonium ylides (Eq. 7-31) are key steps in useful procedures for ortho alkylation of aromatic amines and phenols developed by Gassman and Amick (1974) as previously discussed in Chapter 3. This latter chapter also provides a number of examples of [2,3] sigmatropic

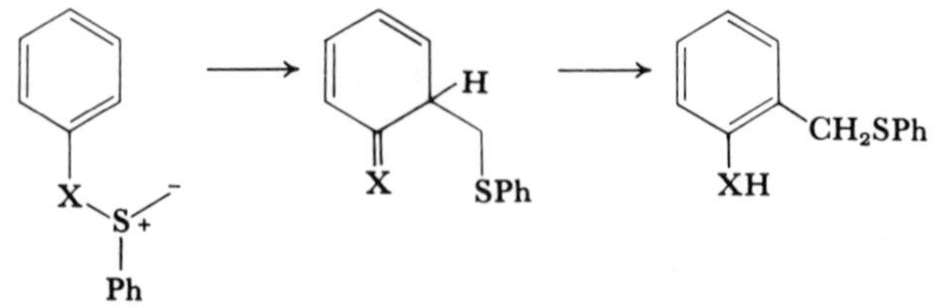

(7-31)

(X = O or NH)

processes involving allylic sulfonium ylides and variants involving propargylic and allenic ylides (see Eqs. 3-73 to 3-78). Additional examples of [2,3] sigmatropic rearrangement of allylic sulfonium ylides are given in Chapter 6 along with examples involving allylthiocarbenes (see Eqs. 6-17 and 6-18).

An indication of the scope of [2,3] sigmatropic processes as applied to organosulfur compounds is provided by the collection of reactions shown in Eqs. 7-32 to 7-40. Included are rearrangements of anions of allyl sulfides and thioacetals (Eqs. 7-32 and 7-33, respectively; Rautenstrauch, 1971; Leger *et al.*, 1975), of β-ketosulfonium ylides (Eq. 7-34; Ratts and Yao, 1968), of allyl sulfoxides and thiosulfinates (Eqs. 7-35 and 7-36, respectively; Tang and Mislow, 1970; Baldwin *et al.*, 1971), of propargyl sulfoxylates and sulfites (Eqs. 7-37 and 7-38, respectively; Braverman and Segev, 1974; Beetz *et al.*, 1975), and allyl disulfides and trisulfides (Eqs. 7-39 and 7-40, respectively; Höfle and Baldwin, 1971; Barnard *et al.*, 1969). In Eqs. 7-35, 7-37, and 7-38, the equilibria favor the higher valence state of sulfur possessing strong $S{=}O \leftrightarrow S^{+}{-}O^{-}$ double/polar bonds, e.g., sulfoxide is favored over sulfenate in Eq. 7-35, sulfone over sulfinate over sulfoxylate in Eq. 7-37, and sulfonate over sulfite in Eq. 7-38. Equation 7-36 is an exception to this trend, perhaps attributable to electron-withdrawing effects of the second sulfur (Baldwin *et al.*, 1971). In contrast to sulfoxides, thiosulfoxides (Eqs. 7-39 and 7-40) are generally unstable with respect to isomeric structures with

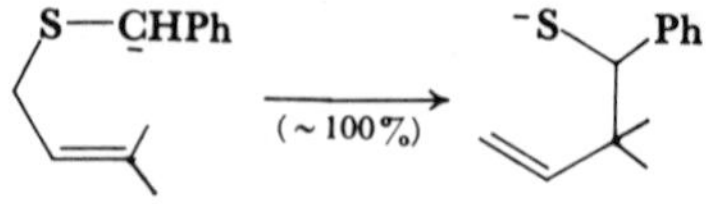

(~100%) (7-32)

RMgBr / THF; CH_3I (90%) (7-33)

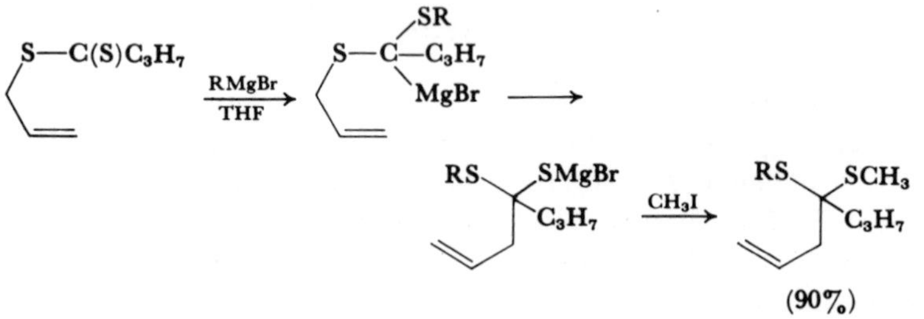

divalent sulfur (exceptions such as $F_2S{=}S$ and thiono sulfites are known; Kuczkowsky and Wilson, 1963 and Thompson *et al.*, 1965, respectively). Baldwin and Barnard have demonstrated the intermediacy of thiosulfoxides in [2,3] sigmatropic processes involving allyl disulfides and trisulfides (Eqs. 7-39 and 7-40) and point out the great tendency of thiosulfoxides to expel the thionosulfur either spontaneously or with interception by a phosphine (Baldwin *et al.*, 1971; Barnard *et al.*, 1969).

Reaction 7-35 is of particular interest since it represents a mechanism for the racemization of chiral allylic sulfoxides through reversible isomerization to achiral sulfenate esters, and since it represents one of the most thoroughly studied [2,3] sigmatropic processes. Table 7-1 summarizes the activation parameters for reaction 7-35 (Tang and Mislow, 1970). These values (entries 1–3) may be compared to the analogous parameters for racemization

S O Ar → S O Ar (7-34)

ArS=O ⇌ ArS—O (7-35)

S S O ⇌ S S O (7-36)

OH $\xrightarrow[-70°]{SCl_2}$ S O O → S=O O →

O S+ -O H $\xrightarrow[reaction]{ene}$ O_2S (7-37)

S O O O → O S+ -O O → O_2S O (7-38)

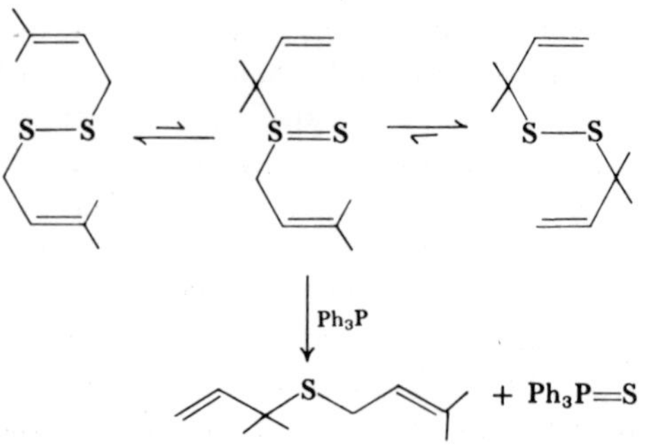

$$\text{S—S} \rightleftharpoons \text{S=S} \rightleftharpoons \text{S—S} \quad \xrightarrow{Ph_3P} \quad \text{allyl sulfide} + Ph_3P{=}S \tag{7-39}$$

$$\text{S—S—S} \rightleftharpoons \text{S=S—S} \tag{7-40}$$

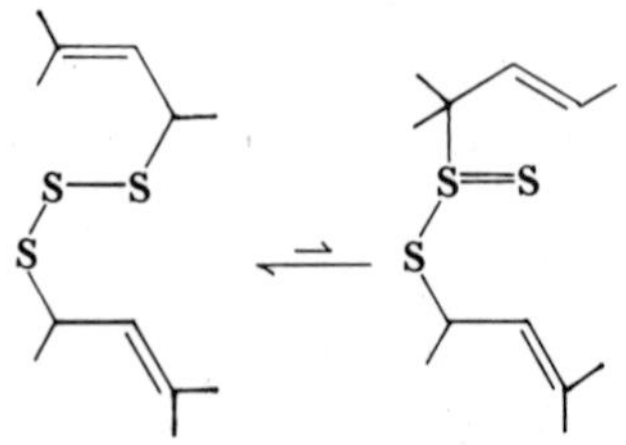

of benzyl *p*-tolyl sulfoxide (entry 4), a process known to proceed via C—S bond homolysis (see reaction 5-64), and for racemization of methyl and 1-adamantyl *p*-tolyl sulfoxides (entries 5 and 6, respectively), processes which involve pyramidal inversion (Bickart *et al.*, 1968; Rayner *et al.*, 1968). The negative values of $\Delta S^\ddagger$ for entries 1–3 reflect an ordered transition state for reaction 7-35 in contrast to the disordered transition state for the homolytic process in entry 4 which is reflected in the large positive $\Delta S^\ddagger$. In a study of the disulfide isomerization indicated in Eq. 7-39, Baldwin has found activation parameters which are very similar to those seen in reaction 7-35, e.g., $\Delta H^\ddagger = 18.7$ kcal/mole, $\Delta S^\ddagger = -9.7$ e.u. and $\Delta G^\ddagger$ (100°C) $= 22.3$ kcal/mole (Höfle and Baldwin, 1971).

Evans and Grieco have developed a useful synthetic procedure which utilizes the allylic sulfoxide–sulfenate interconversion (Evans and Andrews, 1974; Grieco, 1972; Grieco and Finkelhor, 1973). Thus an allylic sulfoxide may be converted to its carbanion, alkylated, and rearranged in the presence of a sulfenate trapping agent giving an allylic alcohol (Eq. 7-41; Evans and Andrews, 1974). Among the sulfenate trapping (thiophilic) reagents employed are $LiBH_3CN$, piperidine, diethylamine, thiophenoxide, and trimethylphosphite (Evans and Andrews, 1974). It should be noted that the overall reaction results in the 1,3-transposition of the sulfoxide and hydroxyl functions. The reversible nature and high stereospecifity of this [2,3] sigmatropic rearrangement is nicely illustrated by Grieco's synthesis of *E*-nuciferol (Eq. 7-42; Grieco and Finkelhor, 1973), among other examples (Evans and Andrews, 1974; Lansbury and Britt, 1976; Trost and Stanton, 1975;

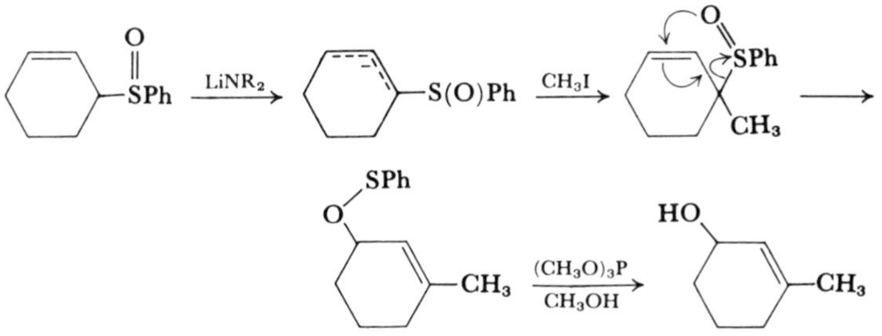

(7-41)

Hoffmann and Maak, 1976). Grieco's synthesis begins with methyallyl alcohol which is converted to an unstable sulfenate ester through reaction with *p*-toluenesulfenyl chloride as described in Eq. 7-42. The overall synthetic sequence may be envisioned as involving the 2-methyl-1-propen-3-ol anion synthon, **7-1**. A particularly ingenious exploitation of the preference of the sulfenate-sulfoxide transformation for *trans* olefin geometry is found in a recent total synthesis of prostaglandin E_1 (Eq. 7-43; Miller *et al.*, 1974).

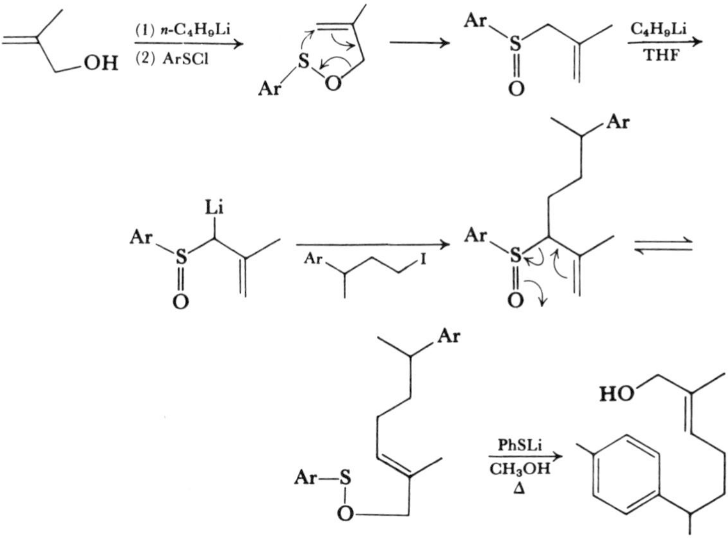

(7-42)

(Ar = *p*-tolyl)

E-nuciferol (58%)

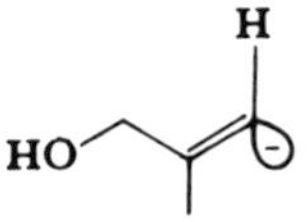

7-1

Baldwin has recently employed the [2,3] sigmatropic rearrangement of allyl sulfinate esters in a manner similar to that shown in Eq. 7-42 in the preparation of α-hydroxy ketones (Eq. 7-44; Baldwin *et al.*, 1976).

THPO, $(CH_2)_6CO_2CH_3$, + LiCu, n-C_5H_{11}, OR, 2, C_5H_{11}-n, OH, ArSCl, $(C_2H_5)_3N$, (81%), S=O, Ar, $(CH_3O)_3P$, CH_3OH, (88%), CH_3COOH, HO

(±)-PGE$_1$-methyl ester

(7-43)

OCH_3, Li, OH, ArSOCl, S=O, Ar, SO_2Ar, RCO_3H, Al(Hg)

(7-44)

7.3 Electrocyclic Processes

The selection rules for electrocyclic reactions indicate that thermally, $4n$ systems will open or close in a conrotatory manner and $4n + 2$ systems in a disrotatory fashion. The rules are reversed for photochemical processes so that $4n$ systems open or close in a disrotatory manner and $4n + 2$ systems in a conrotatory manner (Woodward and Hoffmann, 1965). An example of an electrocyclic ring closure reaction of an organosulfur compound in which the stereochemistry is known is the *conrotatory* closure of the 4π electronic thiocarbonyl ylides (Eq. 7-45; Buter *et al.*, 1972). Other representative examples of electrocyclic processes involving organosulfur compounds are shown in Eqs. 7-46 through 7-52.

(7-45)

$(R = t\text{-}C_4H_9)$

(7-46)

(Küsters and de Mayo, 1974)

(7-47)

(Dittmer *et al.*, 1972)

$Ph_2C{=}S + CH_3C{\equiv}CNR_2 \longrightarrow$

$\longrightarrow Ph_2C{=}C(CH_3)C({=}S)NR_2$ (7-48)

$(60\%, R = C_2H_5)$

(Brouwer and Kos, 1976)

(7-49)

(Burger *et al.*, 1975)

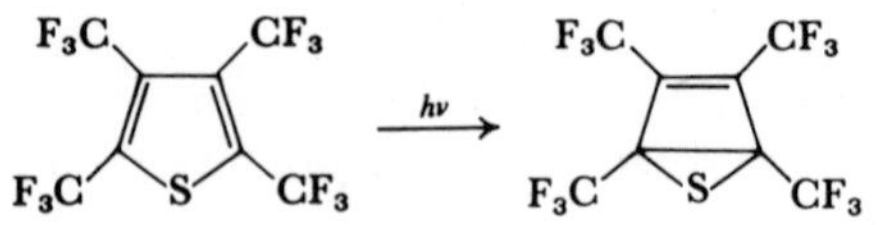

(7-50)

(Wiebe *et al.*, 1972)

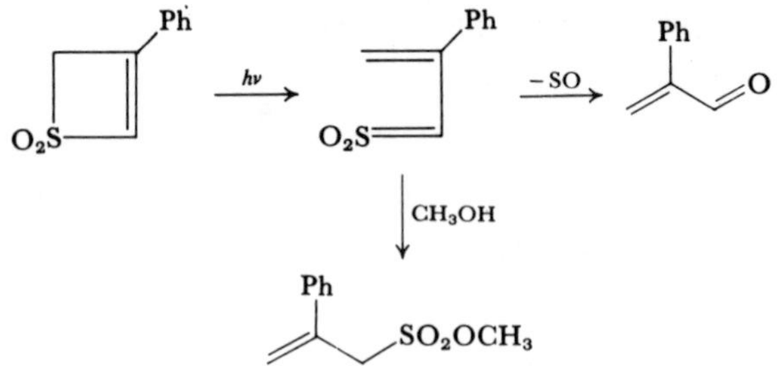

(7-51)

(Langendries and de Schryver, 1972)

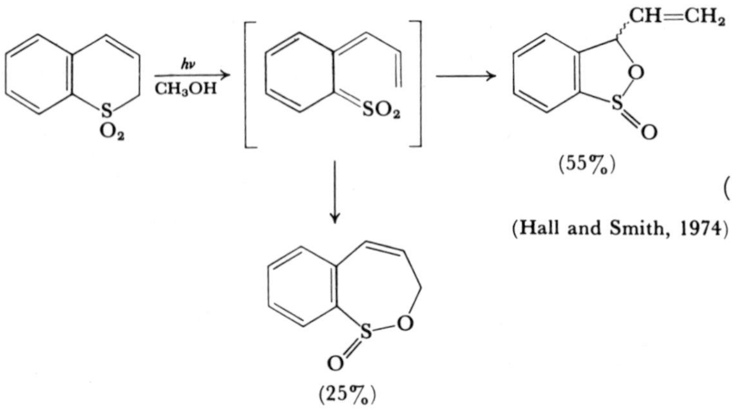

(55%)

(7-52)

(Hall and Smith, 1974)

(25%)

7.4 Cycloaddition Reactions

Reactions Involving Sulfur Dioxide

Ring-forming and -breaking reactions involving sulfur dioxide constitute a fascinating chapter in organosulfur chemistry dating back to the discovery of the butadiene–sulfur dioxide adduct sulfolene (Eq. 7-53) by de Bruin in 1914. Reaction 7-53 is a typical cheletropic transformation, described in

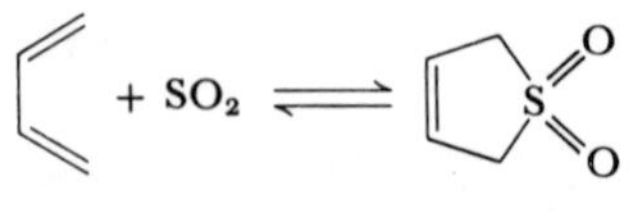

$$+ SO_2 \rightleftharpoons \qquad (7\text{-}53)$$

shorthand notation as a $_{\pi}4_s + _{\omega}2_s$ process. An alternative mode of combination of butadiene and sulfur dioxide leads to a sultine by a $_{\pi}4_s + _{\pi}2_s$ cycloaddition reaction (Eq. 4-54). In this section we shall consider together both cheletropic and cycloaddition ring-forming reactions and their reversals.

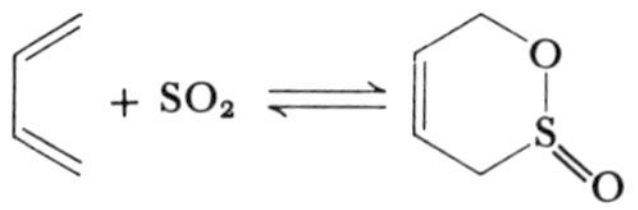

$$+ SO_2 \rightleftharpoons \qquad (7\text{-}54)$$

The "sulfolene reaction" (Eq. 7-53) is fully reversible so that sulfolenes are useful intermediates for the modification, purification, and storage of dienes (Turk and Cobb, 1967). In a novel variant of this reaction, polymer-bound 1,3-diene systems have been examined as a means of removal of sulfur dioxide from waste and flue gases for control of air pollution. These polymer systems were able to bind sulfur dioxide as the polymer-bound sulfolene. Heating the latter resulted in liberation of sulfur dioxide with concomitant recovery of the original 1,3-diene system (Nieuwstad *et al.*, 1976). The stereochemistry of sulfur dioxide elimination from sulfolenes has been shown by Mock and Lemal to be cleanly (i.e., >99.9%) disrotatory under thermal conditions (Eq. 7-55; Mock, 1966; McGregor and Lemal, 1966). The corresponding photochemical process (Eq. 7-55) is not completely stereospecific but does favor the conrotatory path as predicted by Woodward–Hoffmann orbital symmetry considerations (Saltiel and Metts, 1967). Cheletropic transformations have also been studied for episulfones (Eq. 7-56) and 2,7-dihydrothiepin 1,1-dioxides (Eq. 7-57). In considering all of these reactions

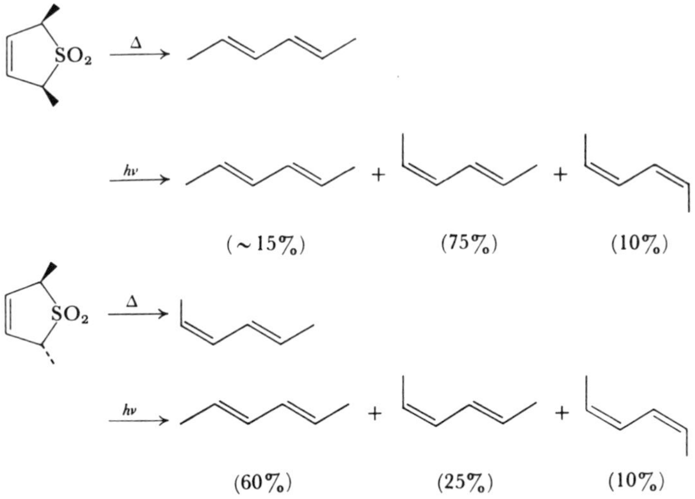

SO_2 $\xrightarrow{\Delta}$

$\xrightarrow{h\nu}$ + +

(~15%) (75%) (10%)

(7-55)

SO_2 $\xrightarrow{\Delta}$

$\xrightarrow{h\nu}$ + +

(60%) (25%) (10%)

$$\triangleright SO_2 \longrightarrow \| + SO_2 \quad (7\text{-}56)$$

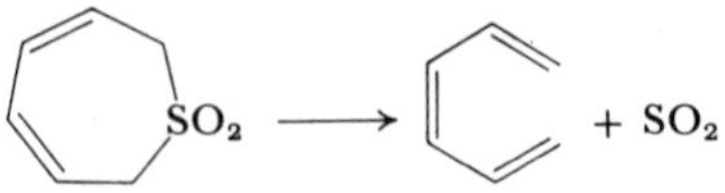

$$SO_2 \longrightarrow + SO_2 \quad (7\text{-}57)$$

using a frontier orbital approach (see Gilchrist and Storr, 1972), we may consider the interaction of the highest occupied molecular orbital (lone pair orbital; HOMO) of sulfur dioxide (see **7-2**) with the lowest unoccupied orbital (LUMO) of the olefin as well as the LUMO (vacant *p* orbital) of

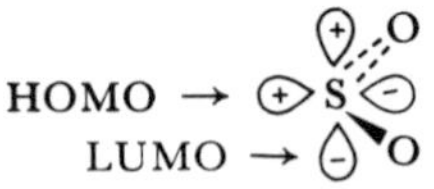

HOMO →
LUMO →

7-2

sulfur dioxide with the olefin HOMO. Thus, an orbital representation of reaction 7-53 is shown in **7-3** and the suprafacial, suprafacial character of this so-called *linear* cheletropic process can be seen.

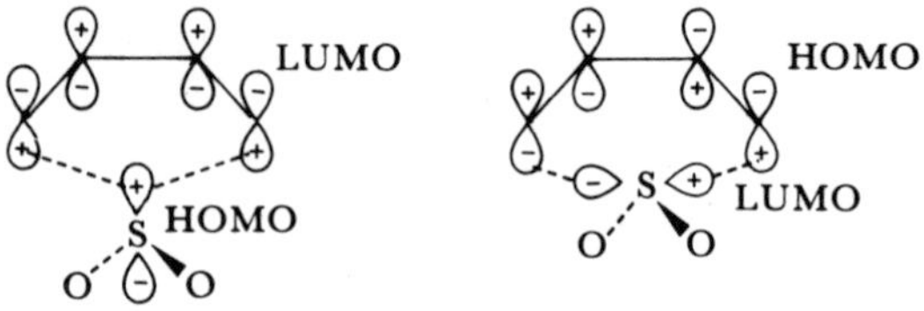

LUMO HOMO
HOMO LUMO

7-3

Episulfone decomposition, which we have previously considered in Chapter 2 in connection with the Ramberg–Bäcklund reaction, is stereospecific and cleanly suprafacial with respect to olefin. However a concerted suprafacial, suprafacial ${}_{\pi}2_s + {}_{\omega}2_s$ interaction is a disallowed process as seen in **7-4**. For the reaction to be concerted the sulfur dioxide must depart in an antarafacial manner as illustrated in **7-5** so that this ${}_{\pi}2_s + {}_{\omega}2_a$ process becomes analogous to an allowed ${}_{\pi}2_s + {}_{\pi}2_a$ olefin dimerization or its reversal. Woodward and Hoffmann indicate that expulsion of sulfur dioxide from episulfones involves a "nonlinear" reaction path whereby the sulfone is deformed prior to scission (Eq. 7-58a; Woodward and Hoffmann, 1970). It may be noted that for episulfones a continuum of reaction paths is likely to be available since the molecular distortion leading to the nonlinear reaction path (Eq. 7-58a) readily accommodates itself to sequential bond cleavage in that the motion of the sulfonyl group in the nonlinear mode

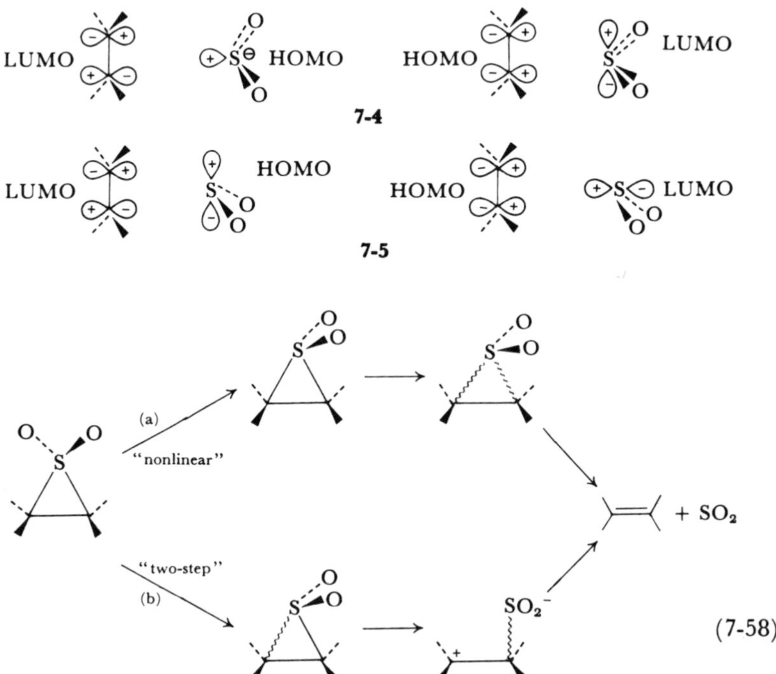

inherently differentiates between the C—S bonds favoring nonsynchrony (Eq. 7-58b) (Mock, 1975). Indeed, on the basis of solvent and substituent effects, Bordwell argues for nonconcerted decomposition of substituted episulfones (Bordwell *et al.*, 1968).

The concerted fragmentation of a 2,7-dihydrothiepin 1,1-dioxide (Eq. 7-57) could formally be either a linear or a nonlinear cheletropic reaction (Gilchrist and Storr, 1972). Symmetry considerations require that the thermal ${}_{\pi}6 + {}_{\omega}2$ process be suprafacial, antarafacial. If the sulfur dioxide is the suprafacial component, it is a ${}_{\pi}6_{a} + {}_{\omega}2_{s}$ linear cheletropic reaction and the triene must rotate in a conrotatory sense (see **7-6a**); if the sulfur dioxide is the antarafacial component, it is a ${}_{\pi}6_{s} + {}_{\omega}2_{a}$ nonlinear cheletropic process and the triene must rotate in a disrotatory sense as in **7-6b**. The experimental results

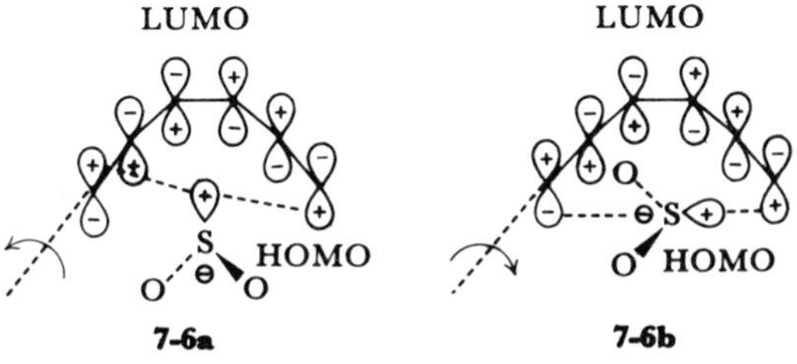

provided by Mock indicate that the reaction proceeds greater than 97% in an antarafacial manner with respect to the triene, i.e., conrotatory rotation via a $_{\pi}6_a + _{\omega}2_s$ process (Eq. 7-59; Mock, 1975; Mock and McCausland, 1976). An indication of the unfavorability of the nonlinear $_{\pi}6_s + _{\omega}2_a$ cheletropic process **7-6b** is the relatively slow rate of sulfur dioxide elimination

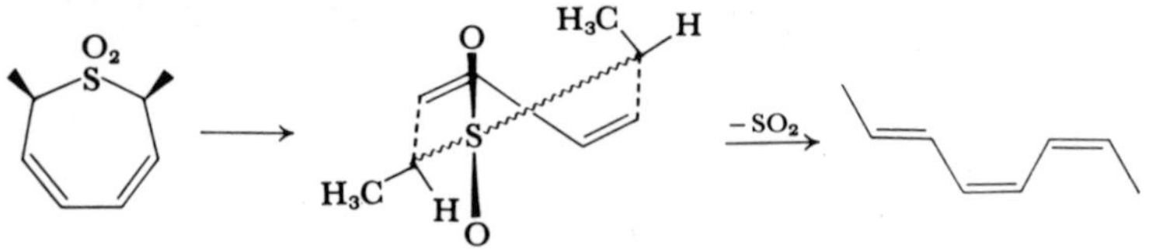

$$\xrightarrow{\quad} \qquad \xrightarrow{-SO_2} \qquad (7\text{-}59)$$

from 9-thiabicyclo[4.2.1]nona-2,4-diene 9,9-dioxide (entry 5, Table 7-2). The geometry of this compound permits only suprafacial elimination with respect to the triene with the consequence that this compound decomposes 60,000 times more slowly than the isomeric structure in entry 4, Table 7-2, for which a $_{\pi}4_s + _{\pi}2_s$ fragmentation is possible. Returning to the decomposition of thiirane 1,1-dioxide (entry 1, Table 7-2), it seems remarkable that the rate acceleration, which might be expected on the basis of ring strain, is so small. Mock suggests that there may be a counterbalancing stabilization of this compound due to the unfavorability of the nonlinear process which neutralizes any potential rate acceleration due to ring strain (Mock, 1975).

Table 7-2. Comparative Fragmentation Rates for Some Cyclic Sulfones

Entry	Sulfone	k_{rel} fragmentation (125°)
1	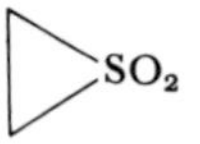 SO_2	15.0
2	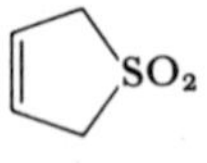 SO_2	1.0
3	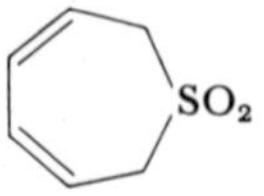 SO_2	1.2
4	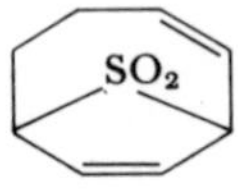 SO_2	5.9
5	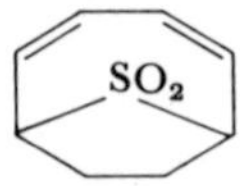 SO_2	0.0001

Recently, evidence has been presented that in the reaction of sulfur dioxide with certain dienes, thermodynamic control leads to sulfolene formation while kinetic control can lead to δ-sultine formation (Eq. 7-60; Heldeweg and Hogeveen, 1976). Durst had previously demonstrated that $\Delta^{4,5}$-δ-sultines such as **7-7** undergo concerted loss of sulfur dioxide (Jung *et al.*, 1974).

SO_2 + ⇌ (< 20°) → (20°) O_2S (7-60)

Ph ... Δ → Ph + SO_2

7-7

It is known that β-sultines, generated according to the method of Durst (Jung *et al.*, 1973), also undergo facile loss of sulfur dioxide providing a useful synthesis of olefinic compounds (Eq. 7-61; Nokami *et al.*, 1975a).†

=O + —S(=O)—CH(MgX)$CO_2C_2H_5$ ⟶ OH —$CHCO_2C_2H_5$, S=O $\xrightarrow{SO_2Cl_2}$

[$CO_2C_2H_5$, OH Cl—$\overset{+}{S}$=O] ⟶ [$CO_2C_2H_5$, O—$\overset{+}{S}$=O, Cl^-] $\xrightarrow{-t\text{-}C_4H_9Cl}$

$CO_2C_2H_5$, O—S=O $\xrightarrow{-SO_2}$ $CO_2C_2H_5$ (85%) (7-61)

† For a computational study of the retrocycloaddition of β-sultines, see Carlsen and Snyder (1977).

Synthetic processes employing the "sulfolene reaction" and its reversal are numerous. Sulfolene itself is routinely used as a convenient source of 1,3-butadiene in preparative reactions (Sample and Hatch, 1968). Other selected applications of the "sulfolene reaction" include the separation of *cis* and *trans* isomers of the red bollworm moth sex pheromone by selective reaction with sulfur dioxide (Eq. 7-62; Nesbitt *et al.*, 1973), preparations of α-diketones (Eq. 7-63; Chaykovsky *et al.*, 1972), deuteriobutadienes (Eq. 7-64; Cope *et al.*, 1961), and bromoisoprene (Eq. 7-65; Krug and Yen, 1956; this reaction takes advantage of the unreactivity of the α-position of sulfones in free radical halogenations as discussed in Section 5.3), cycloheptatrienes (Eq. 7-66; van Tilborg *et al.*, 1975), 2-thioalkyl (or thioaryl)-1,3-butadienes (Eq. 7-67, Gundermann and Holtmann, 1966), and variously substituted cyclooctatetraenes (Eqs. 7-68 and 7-69; Paquette, 1975). In Eq. 7-69, the novel Lewis acid promoted addition of sulfur dioxide to cyclooctatetraene, discovered by Huisgen (Gasteiger and Huisgen, 1973), should be noted. A useful variant of the sulfolene reaction, in which the double bond of a sulfolene is replaced by a three-membered ring, provides a route to 1,4-dienes, divinyl ethers, and divinyl carbamates (Eq. 7-70; Mock, 1970; Meyers and Takaya, 1971). 1,4-Dienes are also formed quantitatively and

OAc

OAc

$\xrightarrow[-20^\circ]{SO_2}$

OAc (7-62)

SO_2 + unreacted *cis*-diene

Δ

trans-diene (>99% purity)

$$ArCH_2SO_2CH_2Ar \xrightarrow[NaOC_2H_5]{(CO_2C_2H_5)_2}$$

O O Ar SO_2 Ar $\xrightarrow[C_2H_5OH,\ THF]{Zn,\ HOAc}$

OH OH Ar SO_2 Ar $\xrightarrow[(98\%)]{\Delta \quad HOAc,\ NaOAc}$ HO OH Ar Ar $\rightleftharpoons$ $ArCH_2\overset{O}{\overset{\|}{C}}\overset{O}{\overset{\|}{C}}CH_2Ar$ (7-63)

SO_2 $\xrightarrow[NaOD]{D_2O}$ D_2 SO_2 D_2 $\xrightarrow[-SO_2]{\Delta}$ $CD_2{=}CH{-}CH{=}CD_2$ (7-64)

$+ SO_2 \longrightarrow$ SO_2 $\xrightarrow{NBS}$

Br SO_2 $\xrightarrow{\Delta}$ Br $+ SO_2$ (7-65)

SO_2 + $\longrightarrow$ SO_2 $\xrightarrow[-SO_2]{\Delta}$ $\longrightarrow$

(7-66)

$\xrightarrow{RSCl}$ Cl SR $\xrightarrow[90°]{R_3N}$ SR $\xrightarrow[-SO_2]{110°}$ RS $+ SO_2$ (7-67)

S O_2 S O_2 S O_2

SO_2 Cl $\xrightarrow[DMSO]{KO\textit{t}\text{-}C_4H_9}$ SO_2 $\xrightarrow{\Delta}$ (7-68)

$\xrightarrow{O{=}\overset{+}{S}{=}\overset{-}{O}{-}SbF_5}$ + O S O SbF_5 H $\xrightarrow{(95\%)}$

SO_2 $\xrightarrow[(2)\ 2CH_3I]{(1)\ 2C_4H_9Li}$ CH_3 SO_2 CH_3 $\xrightarrow{h\nu}$ CH_3 CH_3 (7-69)

CH_3 X SO_2 CH_3 $\xrightarrow[SO_2]{\Delta}$ X

X = CH_2, O

(7-70)

Ph CH_3 O S O H **7-8** $\xrightarrow[reflux]{CHCl_3}$ Ph $+ SO_2$

(100%)

stereospecifically on decomposition ($_{\pi}2_s + _{\pi}2_s + _{\pi}2_s$ cycloreversion) of 3,4-oxathiabicyclo[4.1.0]heptane 4-oxides **7-8** in refluxing chloroform (Jung, 1976).

Reactions Involving Other Sulfur Oxides

Cycloaddition reactions have been reported for SO_3, S_2O, and SO. Sulfur trioxide undergoes 2 + 2 cycloadditions, probabily via an ionic intermediate (Eq. 7-71, Nagayama *et al.*, 1973). Dodson has found that disulfur monoxide undergoes an inefficient 4 + 2 cycloaddition with dienes, e.g., as shown in Eq. 7-72 (Dodson *et al.*, 1972). Reactions involving sulfur

$SO_3 \cdot$dioxane (7-71)

Ph, Ph + S=S⁺–O⁻ ⟶ (7-72)

monoxide are complicated by the fact that this species has a triplet ground state with the lowest singlets lying 18.2 and 30.05 kcal/mole above the ground state (Saito, 1968). Decomposition of thiirane 1-oxide most likely involves homolysis followed by intersystem crossing of the singlet diradical to a triplet diradical which may then decompose with conservation of spin to triplet sulfur monoxide and singlet ethylene (Eq. 7-73). In contrast to episulfones,

S=O ⟶ SO·↑ ·↓ $\xrightarrow{ISC}$ SO·↓ ·↓ ⟶ $C_2H_4 + SO(^3\Sigma^-)$ (7-73)

episulfoxides generally decompose thermally with loss of stereochemistry; the diradical intermediate can be trapped with thioketones, which are known to be excellent radical scavengers (Eq. 7-74; Kondo, 1972). Neither the addition of thioketone nor change in the solvent polarity has any significant effect on the rate of decomposition of episulfoxides as expected for a process involving diradical formation in the rate-determining step (Kondo *et al.*, 1972). In contrast to the results shown in Eq. 7-74, recent studies of the pyrolysis of *cis*- and *trans*-2,3-dideuteriothiirane oxide indicate that deuterated ethylenes are formed with 95% retention of stereochemistry, leading the authors to speculate that there may be a significant contribution of a concerted process to thiirane oxide decomposition (Aalbersberg and Vollhardt,

1977). The sulfur monoxide produced by the decomposition of episulfoxides can be trapped with dienes or trienes (Eq. 7-75; Dodson and Sauers, 1967; Dodson and Nelson, 1969), but unlike the parallel reaction with sulfur

Ph Ph SO_2 Δ $-SO_2$ Ph Ph only

Ph Ph S O Ph SO· $-SO$ Ph Ph Ph Ph (55:45 at 50°C) (7-74)

$Ar_2C{=}S$

Ar_2 S Ph S O Ph mixture of isomers

Ph Ph S O (25%) Δ, $PhCH_3$ S=O Δ Ph Ph S O (35%) (7-75)

dioxide, the cycloadducts are not formed in a stereospecific manner (Eq. 7-76; Chao and Lemal, 1972, Lemal and Chao, 1972). Decomposition of the cycloadducts (e.g., 2,5-dihydrothiophene 1-oxides) shows little stereoselectivity and extensive isomerization of the starting material is observed (Chao and Lemal, 1972). Sulfur monoxide will undergo direct addition to cyclooctatraene, (Eq. 7-77; Anastassiou and Chao, 1971), in contrast to sulfur dioxide (see Eq. 7-69).

S=O + Δ $-C_2H_4$ H SO· + H SO· (7-76)

S^+ O^- (87:13) S^+ O^-

$$\text{(thiirane S-oxide)} + \text{(cyclooctatetraene)} \xrightarrow{140°} \text{(bicyclic sulfoxide, SO bridge)} \quad (30\%) \qquad (7\text{-}77)$$

Reactions Involving C—S Double Bonds

Thiocarbonyl compounds participate in a wide variety of ring-forming reactions. The superiority of the thiocarbonyl group over the carbonyl group in these processes can be attributed to the diminished stability and greater polarizability of the former group. It should be underscored that many of these "cycloaddition" reactions and their reversals involve discrete ionic or diradical intermediates which in some cases can be intercepted. Representative examples of ring-forming reactions of thiocarbonyl compounds and formally related structures (sulfenes, thiocarbonyl ylides, and α-thiocarbonium ions) are shown in Eqs. 7-78 to 7-89. These various reactions and related variants are of synthetic utility, mechanistic interest, or in the case of

$$Cl_2C{=}S \underset{\Delta}{\overset{h\nu}{\rightleftharpoons}} \text{(1,3-dithietane, } Cl_2 \text{ at C-2 and C-4)} \qquad (7\text{-}78)$$

(Schönberg and Stephenson, 1933)

$$\text{(adamantanethione)}{=}S \xrightarrow[\text{or } H^+]{h\nu} \text{(dispiro-1,3-dithietane of two adamantanes)} \qquad (7\text{-}79)$$

(Lawrence *et al.*, 1976)

$$CH_3SO_2Cl \xrightarrow[\text{THF, } -20°]{(CH_3)_3N} [CH_2{=}SO_2] \longrightarrow \text{(1,3-dithietane 1,1,3,3-tetroxide)} \qquad (7\text{-}80)$$

(Opitz and Mohl, 1969)

$$\text{(1,3-dithietane 1-oxide)} \longrightarrow CH_2{=}S{=}O + CH_2{=}S \qquad (7\text{-}81)$$

(Block *et al.*, 1976)

$(CF_3)_2C{=}S + CH_2{=}CHOCH_3 \xrightarrow{\Delta}$ [thietane: CH_3O, S, $(CF_3)_2$] (7-82)

(Middleton, 1965)

[cytosine: NH_2, N, O, N, R] + [thiouracil: S, H, N, O, N, R] $\xrightarrow[335\ nm]{h\nu}$ [NH_2, H, O, N, NR, N, O, N, R, S] $\longrightarrow$

[NH_2, O, N, NR, N, O, N, R, SH] $\xrightarrow{-H_2S}$ [NH_2, O, N, NR, N, O, N, R] (7-83)

(Bergstrom and Leonard, 1972)

[adamantanethione =S] $\xrightarrow{CH_2N_2/C_6H_6\text{ or pet ether}}$ [S, N=N]

$\xrightarrow{CH_2N_2/ROH\text{ or }CH_3CN}$ [S, N, N] (7-84)

(Krapcho *et al.*, 1974)

[H, $\overset{+}{S}$, −, H] $\xrightarrow{RC{\equiv}CR}$ [S, R, R] (7-85)

(Buter *et al.*, 1972)

$(CF_3)_2C{=}S$ + [butadiene] $\xrightarrow{-78°}$ [S, $(CF_3)_2$] (7-86)

(Middleton, 1965)

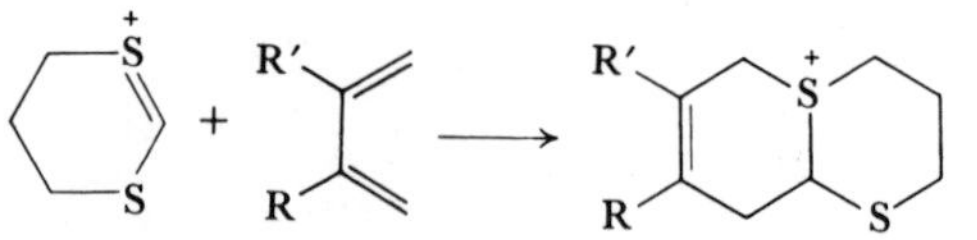

(7-87)

(Corey and Walinsky, 1972)

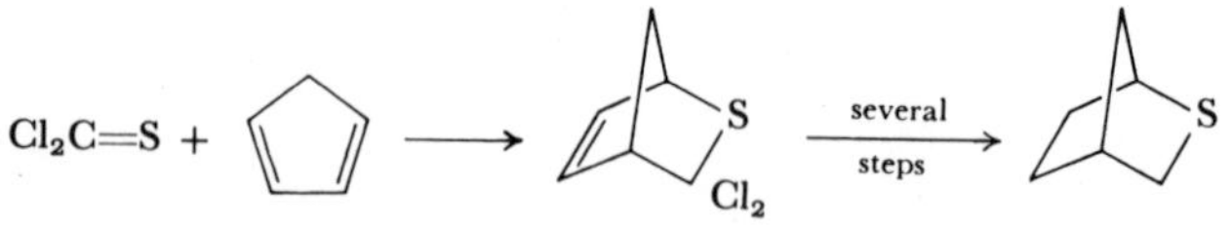

$Cl_2C{=}S$ + ... several steps (7-88)

(Johnson *et al.*, 1969)

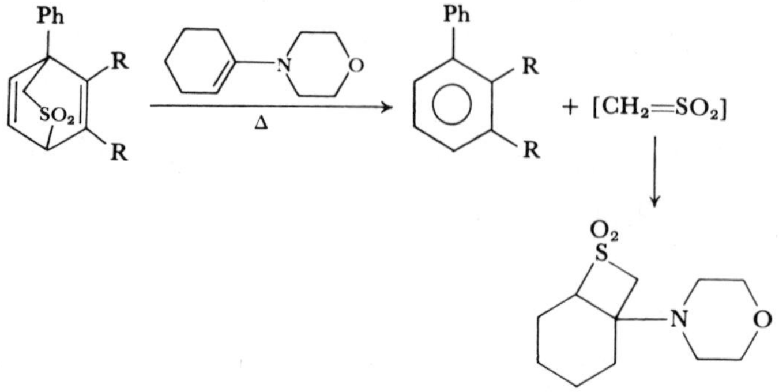

+ $[CH_2{=}SO_2]$ (7-89)

(King and Lewars, 1973)

Eq. 7-83, possible biochemical significance. This latter reaction is the mechanism proposed by Leonard for the photoreaction in *Escherichia coli* transfer-RNA between the 4-thiouridine and cytosine moieties (Bergstrom and Leonard, 1972). Additional examples akin to 1,3-dipolar addition 7-85 are considered in Section 3.4 while utilization of Eq. 7-87 in cyclopentenone syntheses is indicated in Eq. 3-74. The interesting solvent effect seen in the 1,3-dipolar addition 7-84 may be attributed to alteration in the thiocarbonyl polarity on interaction of sulfur with alcohols or acetonitrile.

7.5 Cycloelimination Reactions

Retro-ene Type Reactions

We have previously defined cycloelimination reactions as processes in which $n\sigma$-bonds and $m\pi$-bonds are transformed via an aromatic transition state into $(n - 1)\sigma$-bonds and $(m + 1)\pi$-bonds. One example of this type of reaction is illustrated in Eq. 7-90 (Giles *et al.*, 1976). This particular reaction, which can be termed a retro-ene reaction, involves a six-membered six-electron transition state (four electrons from the C—S and C—H σ bonds and two electrons from the C—C π bond). In the general terminology of Wood-

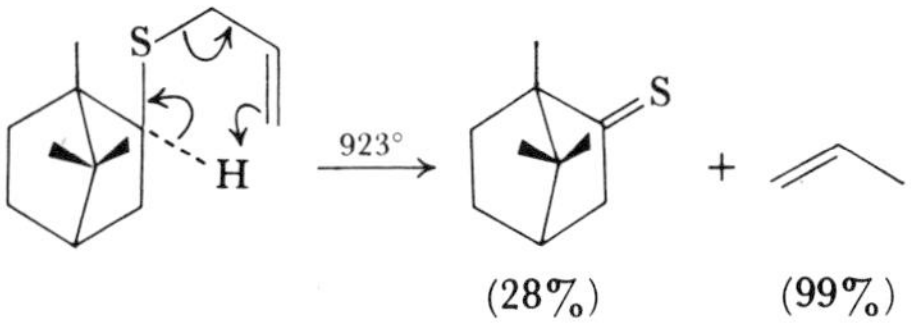

(28%) (99%) (7-90)

ward and Hoffmann, it can be regarded as a $\sigma 2_s + \sigma 2_s + \pi 2_s$ reaction. A second example of a retro-ene reaction is shown in Eq. 7-91 (Clive, 1970). Examples of the reverse process, e.g., ene ($\sigma 2_s + \pi 2_s + \pi 2_s$) reactions of organosulfur compounds, will be discussed later in this section.

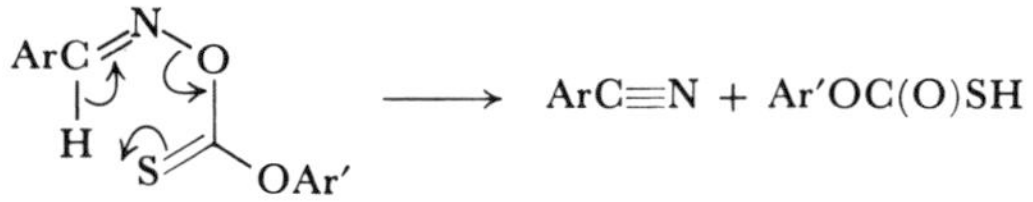

$$\longrightarrow \mathrm{ArC{\equiv}N + Ar'OC(O)SH} \qquad (7\text{-}91)$$

The sulfur counterpart of the Cope elimination, sulfoxide pyrolysis (Eq. 7-92; Kingsbury and Cram, 1960) is a cycloelimination reaction of considerable mechanistic interest and synthetic utility. This reaction, like the analogous decomposition of amine oxides, involves a five-membered six-

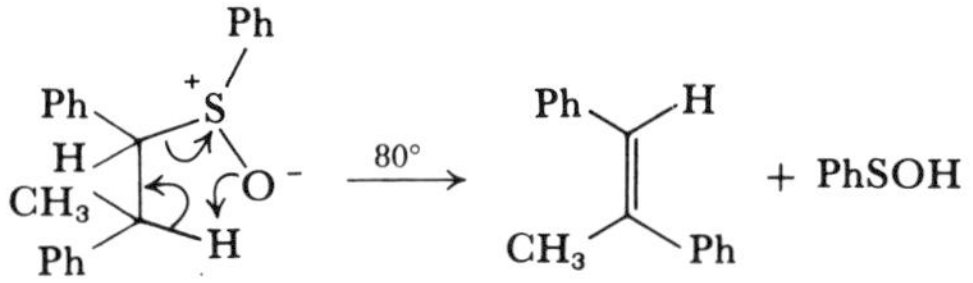

$$\xrightarrow{80^\circ} \quad + \ \mathrm{PhSOH} \qquad (7\text{-}92)$$

electron transition state (four electrons from the C—S and C—H σ bonds, two electrons as a lone pair on oxygen) and can be termed a ${}_{\sigma}2_s + {}_{\sigma}2_s + {}_{\omega}2_s$ reaction. The reaction is reversible, as nicely demonstrated with penicillin sulfoxides by deuterium exchange studies and by actual isolation of the penicillin sulfenic acid (Eq. 7-93; Chou *et al.*, 1974; Cooper *et al.*, 1973). The

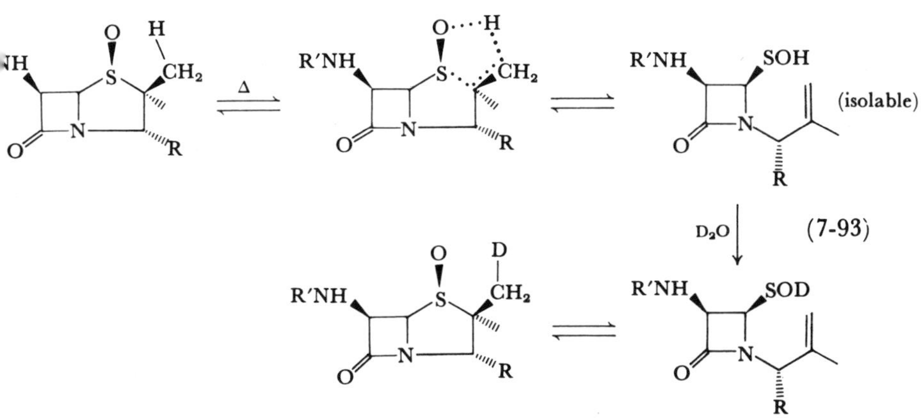

(isolable) (7-93)

addition of a sulfenic acid to an olefin can be termed a $\sigma 2_s + \pi 2_s + \omega 2_s$ process, involving the O—H σ bond, the C—C π bond and the sulfur lone pair. Alkanesulfenic acids, which can be generated at moderate temperatures from alkyl thiosulfinate esters, can be trapped in high yields with acetylenes providing a useful synthesis of α,β-unsaturated sulfoxides (Eq. 7-94; Block and O'Connor, 1974). Similar reactions can be effected with thiosulfinate **7-9** which gives an alkanethiosulfoxylic acid **7-10** (Block and O'Connor, 1974), and with sulfinimines as in Eq. 7-95 (Davis *et al.*, 1974). Thiosulfinate pyrolysis also provides a useful route to thiocarbonyl compounds, for example, **7-11** (Chou *et al.*, 1976). Some other sulfoxide pyrolyses in which the sulfenic acid intermediate undergoes novel reactions are given in Eqs. 7-96 (Baldwin *et al.*, 1971) and 7-97 (Jones *et al.*, 1975).

$$C_2H_5S(O)SCH(H)CH_3 \xrightarrow[\sigma 2_s + \sigma 2_s + \omega 2_s]{96^\circ} C_2H_5SOH + CH_3CH{=}S$$

$$C_2H_5SOH \xrightarrow{PhC{\equiv}CH} [C_2H_5\ddot{S}{-}O{-}H \cdots PhC{\equiv}CH] \xrightarrow{\sigma 2_s + \pi 2_s + \omega 2_s} C_2H_5S(O)C(Ph){=}CH_2 \quad (91\%) \qquad (7\text{-}94)$$

$$\underset{\textbf{7-9}}{(CH_3)_3CS(O)SC(CH_3)_3} \xrightarrow[PhC{\equiv}CH]{96^\circ} \underset{\textbf{7-10}}{(CH_3)_3CSSOH} + (CH_3)_2C{=}CH_2$$

$$(CH_3)_3CSSOH \xrightarrow{PhC{\equiv}CH} (CH_3)_3CSS(O)C(Ph){=}CH_2$$

$$PhS(O)N{=}CHPh \xrightarrow{\Delta} [PhSOH] + PhC{\equiv}N \qquad (7\text{-}95)$$

[PhSOH] ↓ trapped

$$\text{azetidinone (R, H, H; } SS(O)C_5H_{11}\text{-}n\text{; N-CH(CO}_2CH_3)C(CH_3){=}CH_2) \xrightarrow[C_6H_6]{\Delta} \underset{\textbf{7-11}}{\text{azetidin-2-one-3-thione (R; N-CH(CO}_2CH_3)C(CH_3){=}CH_2)}$$

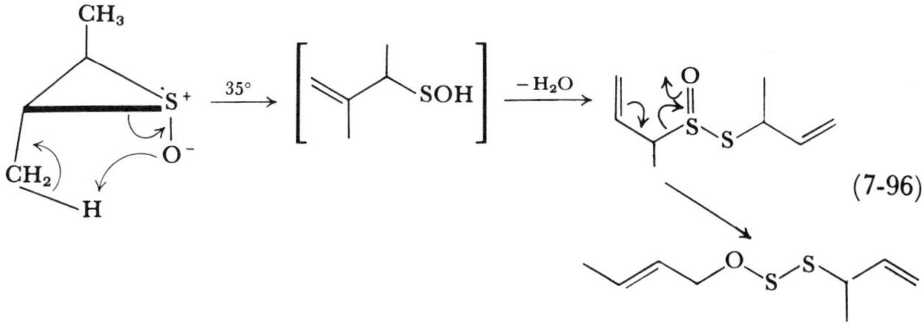

(7-96)

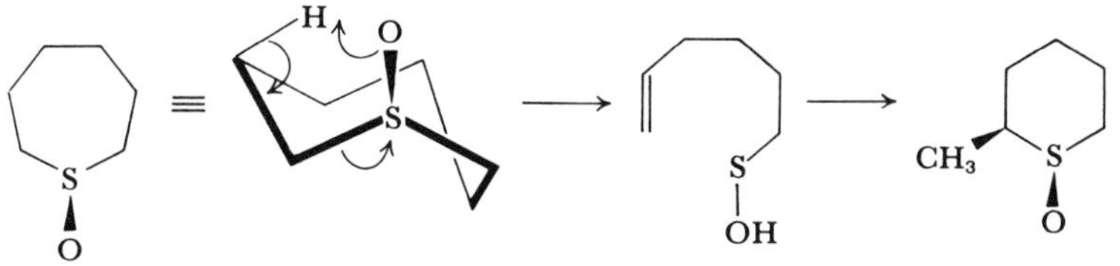

(7-97)

The chemistry associated with sulfenic acids generated by pyrolysis of sulfoxides is of interest primarily for mechanistic reasons. On the other hand, sulfoxide pyrolysis is also of great value as a method of introducing unsaturation, as may be illustrated by Schlessinger's elegant synthesis of the α-methylene lactone fungicide *dl*-avenaciolide (Eq. 7-98; Herrmann *et al.*, 1973), Trost's one-pot alkylative elimination sequence (Eq. 7-99; Trost and

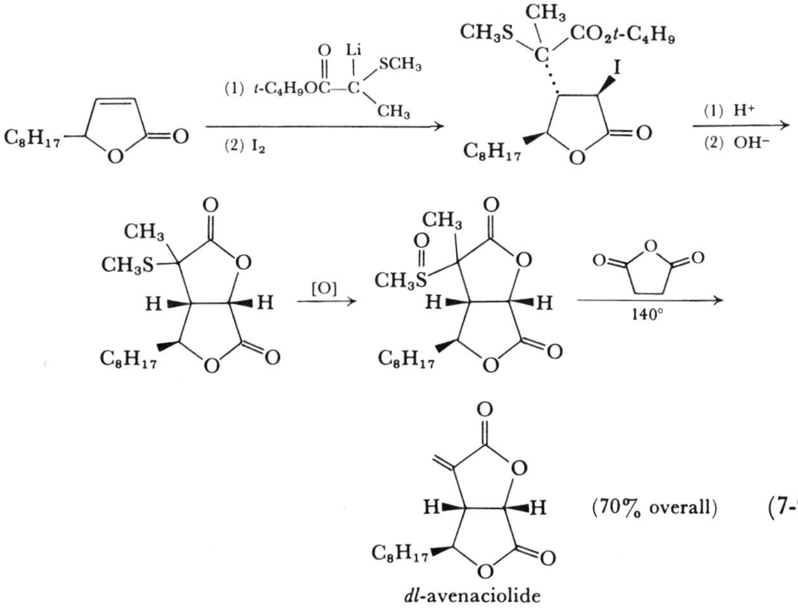

(70% overall) (7-98)

dl-avenaciolide

+ PhS(O)CHLiPh ⟶

$\xrightarrow{-\text{PhSOH}}$ (87%) (7-99)

Bridges, 1975), Jessup's preparation of 1,16-didehydrohexahelicene (Eq. 7-100; Jessup and Reiss, 1975), Grieco's approach to 5-alkyl-Δ^2-cyclopentenones (Eq. 7-101; Grieco and Pogonowski, 1975), and Nokami's route to β-keto esters (Eq. 7-102; Nokami *et al.*, 1975b). The regioselectivity of olefin formation in sulfoxide pyrolysis has been examined in a variety of acyclic and cyclic systems and is thought to be influenced by steric interactions, dipole–dipole effects, proton acidity, and double-bond stability (Trost and Leung, 1975; Jones *et al.*, 1976).

It should be noted that certain of the sulfoxide cycloelimination reactions have their counterpart in organoselenium chemistry. Thus, aryl or alkyl selenoxides undergo *syn* fragmentation under particularly mild conditions (at room temperature or lower, in some cases) (Sharpless and Lauer, 1974; Clive, 1973; Reich and Shah, 1975). Substituted aryl or alkyl selenoxides

(1) $(CH_3O)_2CH^+BF_4^-$
(2) NaH/THF
(3) $ArCO_3H$

$-CH_3SOH$, 300°

$h\nu$, I_2

(81%)

(7-100)

$$\text{2-(PhS(O))cyclopentanone} \xrightarrow[(2)\ C_4H_9I]{(1)\ R_2NLi} \text{2,2-dilithio-2-S(O)Ph... (Li, Li, S(O)Ph)} \longrightarrow \text{2-}C_4H_9\text{-5-S(O)Ph-cyclopentanone} \xrightarrow[75^\circ]{-\ PhSOH} \text{5-}C_4H_9\text{-cyclopent-2-enone} \qquad (7\text{-}101)$$

$$PhS(O)\underset{\substack{|\\ MgX}}{C}HCO_2C_2H_5 + i\text{-}C_3H_7CHO \longrightarrow$$

$$i\text{-}C_3H_7CH(OH)CH(CO_2C_2H_5)S(O)Ph \xrightarrow[\Delta]{C_6H_6} \qquad (7\text{-}102)$$

$$i\text{-}C_3H_7CH(OH){=}CHCO_2C_2H_5 \rightleftharpoons i\text{-}C_3H_7\underset{\substack{\|\\ O}}{C}CH_2CO_2C_2H_5 \ (81\%)$$

can in turn be prepared by oxidation of selenides generated via electrophilic substitution of α-selenoalkyllithium compounds or via nucleophilic substitution at selenium (i.e., nucleophilic attack of an enolate on diphenyl diselenide) (Mitchell, 1974; for a review, see Block, 1977). A typical sequence for the introduction of unsaturation employing selenoxide fragmentation is illustrated by Eq. 7-103 (Denis *et al.*, 1976).

$$\text{cyclohexanone} \xrightarrow{PhSeH} \text{1,1-bis(SePh)cyclohexane} \xrightarrow{C_4H_9Li} \text{1-Li-1-SePh-cyclohexane} \xrightarrow{ClCO_2CH_3}$$

$$\text{1-}CO_2CH_3\text{-1-SePh-cyclohexane} \xrightarrow[THF,\ 0^\circ]{H_2O_2} \text{1-}CO_2CH_3\text{-1-Se(=O)Ph-cyclohexane} \longrightarrow \text{1-}CO_2CH_3\text{-cyclohexene} \qquad (7\text{-}103)$$

Ene Reactions

Only a limited number of examples of ene reactions involving organosulfur compounds have been reported. A representative sampling would include the reaction of olefins with thioketones (Eq. 7-104; Middleton *et al.*, 1961), α-acyl thiones (Eq. 7-105; Loadman *et al.*, 1972), thioketenes (Eq. 7-106; Raasch, 1972), methylated sulfur dioxide (Eq. 7-107; Peterson *et al.*, 1975),†

$$\text{olefin (H)} + S{=}C(CF_3)_2 \xrightarrow{ene} \text{alkene–}C(CH_3)_2\text{–}S\text{–}CH(CF_3)_2 \qquad (7\text{-}104)$$

† Sulfur dioxide itself is reputed to undergo a reversible ene reaction with olefins (Rogić and Masilamani, 1977).

ene (7-105)

Ph Ph S O H Ph COPh

ene H S H CF3 F3C CF3

Diels–Alder (7-106)

CH_3F, SO_2 / SbF_5 ene R Cl CH_3O OH H

$-H^+$ (7-107)

N—Ts S N—Ts H ene HNTs NTs [2.3] SNHTs N Ts

K_2CO_3 / CH_3OH, H_2O

NHTs repeat sequence TsHN NHTs (7-108)

(63%)

(7-109)

N,N'-ditosyl sulfur diimide (Eq. 7-108; Sharpless and Hori, 1976), and methyl cyanodithioformate (Eq. 7-109; Snider *et al.*, 1976). Reactions 7-108 and 7-109 represent synthetically useful procedures for allylic amination and insertion of a functionalized carbon in an allylic C—H bond, respectively.

General References

Gilchrist, T. L., and Storr, R. C. (1972). "Organic Reactions and Orbital Symmetry." Cambridge Univ. Press, London and New York.

Gill, G. B., and Willis, M. R. (1974). "Pericyclic Reactions." Chapman and Hall, London.

Lehr, R. E., and Marchand, A. P. (1972). "Orbital Symmetry." Academic Press, New York.

Marchand, A. P., and Lehr, R. E. (ed.) (1977). "Pericyclic Reactions," Vol. I and II. Academic Press, New York.

Simmons, H. E., and Bunnett, J. F. (ed.) (1974). "Orbital Symmetry Papers." Am. Chem. Soc., Washington D.C.

Woodward, R. B., and Hoffmann, R. (1970). "The Conservation of Orbital Symmetry." Verlag Chemie, Weinheim and Academic Press, New York.

References

Aalbersberg, W. G. L., and Vollhardt, K. P. C. (1977). *J. Am. Chem. Soc.* **99**, 2792.

Anastassiou, A. G., and Chao, B. Y.-H. (1971). *Chem. Commun.* 979.

Anastassiou, A. G., Wetzel, J. C., and Chao, B. Y.-H. (1975). *J. Am. Chem. Soc.* **97**, 1124.

Baldwin, J. E., Höfle, G., and Choi, S. C. (1971). *J. Am. Chem. Soc.* **93**, 2810.

Baldwin, J. E., Lever, O. W., Jr., and Tzodikov, N. R. (1976). *J. Org. Chem.* **41**, 2312.

Barnard, D., Houseman, T. H., Porter, M., and Tidd, B. K. (1969). *Chem. Commun.* 371.

Beetz, T., Kellogg, R. M., Kiers, C. Th., and Piepenbroek, A. (1975). *J. Org. Chem.* **40**, 3308.

Bergstrom, D. E., and Leonard, N. J. (1972). *Biochemistry* **11**, 1.

Bickart, P., Carson, F. W., Jacobus, J., Miller, E. G., and Mislow, K. (1968). *J. Am. Chem. Soc.* **90**, 4869.

Block, E. (1977). *In* "Organic Compounds of Sulphur, Selenium and Tellurium" (Specialist Periodical Reports) (D. R. Hogg, ed.), Vol. IV. Chemical Society of London.

Block, E., and O'Connor, J. (1974). *J. Am. Chem. Soc.* **96**, 3929.

Block, E., Penn, R. E., Olsen, R. J., and Sherwin, P. F. (1976). *J. Am. Chem. Soc.* **98**, 1264.

Bordwell, F. G., Williams, J. M., Jr., Hoyt, E. B., Jr., and Jarvis, B. B. (1968). *J. Am. Chem. Soc.* **90**, 429.

Brandsma, L., and Verkruijsse, H. D. (1974). *Rec. Trav. Chim. Pays-Bas* **93**, 319.

Braverman, S., and Segev, D. (1974). *J. Am. Chem. Soc.* **96**, 1245. For other examples involving sulfoxylate to sulfinate rearrangement, see Büchi, G., and Freidinger, R. M. *J. Am. Chem. Soc.* **96**, 3332 (1974), and Thompson, Q. E., *J. Org. Chem.*, **30**, 2703 (1965). For other examples involving allylic and propargylic sulfinate to sulfone rearrangement, see Grieco, P. A., and Boxler, D., *Syn. Commun.* **5**, 315 (1975); Braverman, S., and Globerman, T., *Tetrahedron* **30**, 3873 (1974); Braverman, S., and Mechoulam, H., *Tetrahedron* **30**, 3883 (1974) and references therein.

Brouwer, A. C., and Bos, H. J. T. (1976). *Tetrahedron Lett.* 209.

Burger, K., Albanbauer, J., and Foag, W. (1975). *Angew. Chem. Int. Ed. English* **14**, 767.

Buter, J., Wassenaar, S., and Kellogg, R. M. (1972). *J. Org. Chem.* **37**, 4045.

Campbell, M. M., and Evgenios, D. M. (1973). *J. Chem. Soc. Perkin Trans.* **1** 2866; see also Larsson, F. C. V., Brandsma, L., and Lawesson, S.-O. *Rec. Trav. Chim. Pays-Bas* **93**, 258 (1974); Boelens, H., and Brandsma, L., *Rec. Trav. Chim. Pays-Bas* **91**, 141 (1972).

Carlsen, L., and Snyder, J. P. (1977). *Tetrahedron Lett.* 2045.

Chao, P., and Lemal, D. M. (1973). *J. Am. Chem. Soc.* **95**, 920.

Chaykovsky, M., Lin, M. H., and Rosowsky, A. (1972). *J. Org. Chem.* **37**, 2018.

Chou, T. S., Burgtorf, J. R., Ellis, A. L., Lammert, S. R., and Kukolja, S. P. (1974). *J. Am. Chem. Soc.* **96**, 1609.

Chou, T. S., Koppel, G. A., Dorman, D. E., and Paschal, J. W. (1976). *J. Am. Chem. Soc.* **98**, 7864; see also Bachi, M. D., and Vaya, J. *J. Am. Chem. Soc.* **98**, 7825 (1976).

Clive, D. L. J. (1970). *Chem. Commun.* 1014.

Clive, D. L. J. (1973). *Chem. Commun.* 695.

Cooper, R. D. G., Hatfield, L. D., and Spry, D. O. (1973). *Accounts Chem. Res.* **6**, 32.

Cope, A. C., Berchtold, G. A., and Ross, D. L. (1961). *J. Am. Chem. Soc.* **83**, 3859.

Corey, E. J., and Shulman, J. I. (1970). *J. Am. Chem. Soc.* **92**, 5522.

Corey, E. J., and Walinsky, S. W. (1972). *J. Am. Chem. Soc.* **94**, 8932.

Davis, F. A., Friedman, A. J., and Kluger, E. W. (1974). *J. Am. Chem. Soc.* **96**, 5000.

de Bruin, G. (1914). *Koninkl. Ned. Akad. Wetenschap. Proc.* **17**, 585; also see Staudinger, H., and Pfenninger, F., *Chem. Ber.* **49**, 1446 (1916).

Denis, J. N., Dumont, W., and Krief, A. (1976). *Tetrahedron Lett.* 453.

Dittmer, D. C., Chang, P. L.-F., Davis, F. A., Iwanami, M., Stamos, I. K., and Takahashi, K. (1972). *J. Org. Chem.* **37**, 1111.

Dodson, R. M., and Nelson, J. P. (1969). *Chem. Commun.* 1159.

Dodson, R. M., and Sauers, R. F. (1967). *Chem. Commun.* 1189; also see Murray, R. K., Jr., Polley, J. S., Abdel-Merguid, S., and Day, V. S., *J. Org. Chem.*, **42**, 2127 (1977).

Dodson, R. M., Srinivasan, V., Sharma, K. S., and Sauers, R. F. (1972). *J. Org. Chem.* **37**, 2367.

Evans, D. A., and Andrews, G. C. (1974). *Accounts Chem. Res.* **7**, 147; also see Evans, D. A., Crawford, T. C., Fujimoto, T. T., and Thomas, R. C., *J.Org. Chem.* **39**, 3176 (1974).

Faulkner, D. J., and Petersen, M. R. (1973). *J. Am. Chem. Soc.* **95**, 553; also see Rautenstrauch (1971).

Gassman, P. G., and Amick, D. R. (1974). *Tetrahedron Lett.* 889, 3463; see also Gassman, P. G., and Drewes, H. R., *J. Am. Chem. Soc.* **96**, 3002 (1974).

Gasteiger, J., and Huisgen, R. (1972). *J. Am. Chem. Soc.* **94**, 6541.
Giles, H. G., Marty, R. A., and de Mayo, P. (1976). *Can. J. Chem.* **54**, 537.
Grieco, P. A. (1972). *Chem. Commun.* 702.
Grieco, P. A., and Finkelhor, R. S. (1973). *J. Org. Chem.* **38**, 2245.
Grieco, P. A., and Pogonowski, C. S. (1975). *Chem. Commun.* 72; see also Bartlett, P. A. *J. Am. Chem. Soc.* **98**, 3305 (1976) and references therein.
Gundermann, K.-D., and Holtmann, P. (1966). *Angew. Chem. Int. Ed. English* **5**, 668.
Hall, C. R., and Smith, D. J. H. (1974). *Tetrahedron Lett.* 3633.
Heldeweg, R. F., and Hogeveen, H. (1976). *J. Am. Chem. Soc.* **98**, 2341.
Herrmann, J. L., Berger, M. H., and Schlessinger, R. H. (1973). *J. Am. Chem. Soc.* **95**, 7023; see also Grieco, P. A., *Synthesis* 67 (1975) for additional references.
Hoffmann, R. W., and Maak, N. (1976). *Tetrahedron Lett.* 2237.
Hoffmann, R., and Woodward, R. B. (1965). *J. Am. Chem. Soc.* **87**, 2046, 4389.
Höfle, G., and Baldwin, J. E. (1971). *J. Am. Chem. Soc.* **93**, 6307.
Jessup, P. J., and Reiss, J. A. (1975). *Tetrahedron Lett.* 1453; see also Otsubo, T., and Boekelheide, V., *Tetrahedron Lett.* 3881 (1975).
Johnson, C. R., Keiser, J. E., and Sharp, J. C. (1969). *J. Org. Chem.* **34**, 860.
Jones, D. N., Hill, D. R., and Lewton, D. A. (1975). *Tetrahedron Lett.* 2235; see also Jones, D. N., Hill, D. R., Lewton, D. A., and Sheppard, C., *J. Chem. Soc. Perkins Trans.* **1**, 1574 (1977).
Jones, D. N., Edmonds, A. C. F., and Knox, S. D. (1976). *J. Chem. Soc. Perkins Trans.* **I**, 459 and references therein.
Jung, F. (1976). *Chem. Commun.* 525.
Jung, F., Sharma, N. K., and Durst, T. (1973). *J. Am. Chem. Soc.* **95**, 3420.
Jung, F., Molin, M., Elzen, R. V. D., and Durst, T. (1974). *J. Am. Chem. Soc.* **96**, 935.
King, J. F., and Harding, D. R. K. (1976). *J. Am. Chem. Soc.* **98**, 3312.
King, J. F., and Lewars, E. G. (1973). *Can. J. Chem.* **51**, 3044.
Kingsbury, C. A., and Cram, D. J. (1960). *J. Am. Chem. Soc.* **82**, 1810.
Kondo, K., and Ojima, I. (1972). *Chem. Commun.* 62.
Kondo, K., Matsumoto, M., and Negishi, A. (1972). *Tetrahedron Lett.* 2131.
Krapcho, A. P., Silvon, M. P., Goldberg, I., and Jahngen, E. G. E. (1974). *J. Org. Chem.* **39**, 860.
Krug, R. C., and Yen, T. F. (1956). *J. Org. Chem.* **21**, 1082, 1441.
Kuckowsky, R. L., and Wilson, E. B. (1963). *J. Am. Chem. Soc.* **85**, 2028.
Kusters, W., and de Mayo, P. (1974). *J. Am. Chem. Soc.* **96**, 3502.
Kwart, H., and Schwartz, J. L. (1974). *J. Org. Chem.* **39**, 1575 and references therein.
Kwart, H., and George, T. J. (1977). *J. Am. Chem. Soc.* **99**, 5214.
Langendries, R. F. J., and de Schryver, F. C. (1972). *Tetrahedron Lett.* 4781; see also Dittmer, D. C., McCaskie, J. E., Babiarz, J. E., and Ruggeri, M. V., *J. Org. Chem.* **42**, 1910 (1977).
Lansbury, P. T., and Britt, R. W. (1976). *J. Am. Chem. Soc.* **98**, 4577.
Lawrence, A. H., Liao, C. C., de Mayo, P., and Ramamurthy, V. (1976). *J. Am. Chem. Soc.* **98**, 2219.
Leger, L., Saquet, M., Thuillier, A., and Julia, S. (1975). *J. Organometall. Chem.* **96**, 313.
Lemal, D. M., and Chao, P. (1973). *J. Am. Chem. Soc.* **95**, 922.
Loadman, M. J., Saville, B., Steer, M., and Tidd, B. K. (1972). *Chem. Commun.* 1167.
Makisumi, Y., and Sasatani, T. (1969). *Tetrahedron Lett.* 1975; see also Takano, S., Hirama, M., Araki, T., and Ogasawara, K. *J. Am. Chem. Soc.* **98**, 7084 (1976).
McGregor, S. D., and Lemal, D. M. (1966). *J. Am. Chem. Soc.* **88**, 2858.
Meyers, A. I., and Takaya, T. (1971). *Tetrahedron Lett.* 2609.

Middleton, W. J. (1965). *J. Org. Chem.* **30**, 1390, 1395.

Middleton, W. J., Howard, E. G., and Sharkey, W. H. (1961). *J. Am. Chem. Soc.* **83**, 2589; see also Lawrence, A. H., Liao, C. C., de Mayo, P., and Ramamurthy, V., *J. Am. Chem. Soc.* **98**, 2219 (1976).

Miller, J. G., Kurz, W., Untch, K. G., and Stork, G. (1974). *J. Am. Chem. Soc.* **96**, 6774.

Mitchell, R. H. (1974). *Chem. Commun.* 990.

Mock, W. L. (1966). *J. Am. Chem. Soc.* **88**, 2857.

Mock, W. L. (1970). *J. Am. Chem. Soc.* **92**, 6918.

Mock, W. L. (1975). *J. Am. Chem. Soc.* **97**, 3666, 3673.

Mock, W. L., and McCausland, J. H. (1976). *J. Org. Chem.* **41**, 242.

Morin, L., and Paquer, D. (1976). *C.R. Acad. Sci. Paris. C* **282**, 353.

Nagayama, M., Okumura, O., Noda, S., and Mori, A. (1973). *Chem. Commun.* 841.

Nakai, T., Shiono, H., and Okawara, M. (1974). *Tetrahedron Lett.* 3625; see also Hori, I., Hayashi, T., and Midorikawa, H., *Synthesis* 727 (1975).

Nakai, T., Shiono, H., and Okawara, M. (1975). *Chem. Lett.* (*Jpn.*) 249; see also Hayashi, T., and Midorikawa, H., *Synthesis* 100 (1974); Hayashi, T., *Tetrahedron Lett.* 339 (1974); for a kinetic study of this and related thio-Claisen reactions, see Nakai, T., and Ari-Izumi, A., *Tetrahedron Lett.* 2335 (1976).

Nesbitt, B. F., Beevor, P. S., Cole, R. A., Lester, R., and Poppi, R. G. (1973). *Tetrahedron* **29**, 4669.

Nieuwstad, Th. J., Kieboom, A. P. G., Breijer, A. J., Van der Linden, J., and Van Bekkum, H. (1976). *Rec. Trav. Chim. Pays-Bas* **95**, 225.

Nokami, J., Kunieda, N., and Kinoshita, M. (1975a). *Tetrahedron Lett.* 2179.

Nokami, J., Kunieda, N., and Kinoshita, M. (1975b). *Tetrahedron Lett.* 2841.

Ogura, F., Hounshell, W. D., Maryanoff, C. A., Richter, W. J., and Mislow, K. (1976). *J. Am. Chem. Soc.* **98**, 3615.

Opitz, G., and Mohl, H. R. (1969). *Angew Chem. Int. Ed. English* **8**, 73; see also Snyder, J. P., *J. Org. Chem.* **38**, 3965 (1973) for a theoretical discussion of this reaction.

Oshima, K., Takahashi, H., Yamamota, H., and Nozaki, H. (1973a). *J. Am. Chem. Soc.* **95**, 2693.

Oshima, K., Yamamoto, H., and Nozaki, H. (1973b). *J. Am. Chem. Soc.* **95**, 4446.

Paquette, L. A. (1975). *Tetrahedron* **31**, 2855.

Peterson, P. E., Brockington, R., and Dunham, M. (1975). *J. Am. Chem. Soc.* **97**, 3517.

Raasch, M. S. (1972). *J. Org. Chem.* **37**, 1347.

Ratts, K. W., and Yao, A. N. (1968). *J. Org. Chem.* **33**, 70.

Rautenstrauch, V. (1971). *Helv. Chim. Acta* **54**, 739; see also, Biellmann, J. F., and Ducep, J. B., *Tetrahedron Lett.* 33 (1971); Kreiser, W., and Wurziger, H., *Tetrahedron Lett.* 1669 (1975).

Rayner, D. R., Gordon, A. J., and Mislow, K. (1968). *J. Am. Chem. Soc.* **90**, 4854.

Reich, H. J., and Shah, S. K. (1975). *J. Am. Chem. Soc.* **97**, 3250 and earlier papers cited therein.

Rhoads, S. J., and Raulins, N. R. (1975). *In* "Organic Reactions," Vol. 22. Wiley, New York.

Rogić, M. M., and Masilamani, D. (1977). *J. Am. Chem. Soc.* **99**, 5219.

Ross, J. A., Seiders, R. P., and Lemal, D. M. (1976). *J. Am. Chem. Soc.* **98**, 4235.

Saito, S. (1968). *Tetrahedron Lett.* 4961.

Saltiel, J., and Metts, L. (1967). *J. Am. Chem. Soc.* **89**, 2232.

Sample, T. E., Jr., and Hatch, L. F. (1968). *J. Chem. Educ.* **45**, 55.

Schönberg, A., and Stephenson, A. (1933). *Chem. Ber.* **66B**, 567.

Sharpless, K. B., and Hori, T. (1976). *J. Org. Chem.* **41**, 176; see also Schönberger, N., and Kresze, G., *Justus Liebigs Ann. Chem.* 1725 (1975).

Sharpless, K. B., and Lauer, R. F. (1974). *J. Org. Chem.* **39**, 429 and earlier papers.
Snider, B. B., Hrib, N. J., and Fuzesi, L. (1976). *J. Am. Chem. Soc.* **98**, 7115.
Takahashi, H., Oshima, K., Yamamoto, H., and Nozaki, H. (1973). *J. Am. Chem. Soc.* **95**, 5803.
Tang, R., and Mislow, K. (1970), *J. Am. Chem. Soc.* **92**, 2100.
Thompson, Q. E., Crutchfield, M. M., and Dietrich, M. W. (1965). *J. Org. Chem.* **30**, 2969.
Trost, B. M., and Bridges, A. J. (1975). *J. Org. Chem.* **40**, 2014.
Trost, B. M., and Leung, K. K. (1975). *Tetrahedron Lett.* 4197.
Trost, B. M., and Stanton, J. C. (1975). *J. Am. Chem. Soc.* **97**, 4018.
Turk, S. D., and Cobb, R. L. (1967). *In* "1,4-Cycloaddition Reactions" (J. Hamer, ed.). Academic Press, New York.
van Tilborg, W. J. M., Smael, P., Visser, J. P., Kouwenhoven, C. G., and Reinhoudt, D. N. (1975). *Rec. Trav. Chim. Pays-Bas* **94**, 85.
Wiebe, H. A., Braslavsky, S., and Heicklen, J. (1972). *Can. J. Chem.* **50**, 2721.
Winter, R. E. K. (1965). *Tetrahedron Lett.* 1207.
Winterfeldt, E. (1970). *Fortschr. Chem. Forsch.* **16**, 75.
Woodward, R. B., and Hoffmann, R. (1965). *J. Am. Chem. Soc.* **87**, 395, 2511.

Appendix A | Bibliography of Recent Books on Sulfur

Ashworth, M. R. F. (1972). "The Determination of Sulphur-Containing Groups." Academic Press, New York.

Cerfontain, H. (1968). "Mechanistic Aspects in Aromatic Sulphonation and Desulphonation." Wiley (Interscience), New York.

Challenger, F. (1959). "Aspects of the Organic Chemistry of Sulphur." Butterworth, London.

Flynn, E. H. (1972). "Cephalosporins and Penicillins." Academic Press, New York.

Friedman. M., (1973). "Chemistry and Biochemistry of the Sulphydryl Group in Aminoacids, Peptides, and Proteins." Pergamon, Oxford.

Gattow, G., and Behrendt, W. (1977). "Carbon Sulfides and Their Inorganic and Complex Chemistry" (Volume 2 of "Topics in Sulfur Chemistry," A. Senning, ed.). Thieme, Stuttgart.

Gmelin (1942–present). "Handbuch der Anorganischen Chemie." Six volumes deal with inorganic sulfur chemistry; Volume 32 of the New Supplement Series (1977) deals with sulfur–nitrogen compounds. Springer-Verlag, New York.

Greenberg, D. M., ed. (1975). "Metabolic Pathways," Vol. 7, Metabolism of Sulfur Compounds. Academic Press, New York.

Hogg, D. R., ed. (1977). "Organic Compounds of Sulphur, Selenium and Tellurium," Vol. IV. The Chemical Society, London (Specialist Periodical Report).

Houben-Weyl (1955). "Methoden der Organischen Chemie," 4th ed., Vol. 9. Thieme, Stuttgart.

Janssen, M. J., ed. (1967). "Organo-Sulphur Chemistry." Wiley (Interscience), New York.

Jocelyn, P. C. (1972). "Biochemistry of the SH Group." Academic Press, New York.

Johnson, A. W. (1966). "Ylide Chemistry." Academic Press, New York.

Karchmer, J. H., ed. (1972). "The Analytical Chemistry of Sulfur and Its Compounds." Wiley (Interscience), New York.

Kharasch, N., and Meyers, C. Y., ed. (1966). "Chemistry of Organic Sulphur Compounds," Vol. 2. Pergamon, Oxford.

Kühle, E. (1973). "The Chemistry of the Sulfenic Acids." Thieme, Stuttgart.

Martin, D., and Hauthal, H. G. (1975). "Dimethyl Sulphoxide." Van Nostrand-Reinhold, Princeton, New Jersey.

Meyer, B., ed. (1965). "Elemental Sulfur." Wiley (Interscience), New York.

Meyer, B., ed. (1977). "Sulfur, Energy, and Environment." Elsevier, Amsterdam.

Miller, D. J., and Wiewiorowski, T. K., ed. (1972). "Sulfur Research Trends," ACS Adv. in Chem. Ser. No. 110. Amer. Chem. Soc., Washington, D.C.

Muth, O. H., and Oldfield, J. E., ed. (1970). "Symposium: Sulfur in Nutrition." Avi Publ., Westport, Connecticut.

Newman, A. A., ed. (1975). "Chemistry and Biochemistry of Thiocyanic Acid and its Derivatives." Academic Press, New York.

Nickless, G., ed. (1968). "Inorganic Sulphur Chemistry." Elsevier, Amsterdam.

Oae, S. (1968). "The Chemistry of Organic Sulfur Compounds." Kagaku Dozin, Tokyo.

Oae, S., ed. (1977). "Organic Chemistry of Sulfur." Plenum, New York.

Patai, S., ed. (1974). "The Chemistry of the Thiol Group," Vol. I and II. Wiley, New York.

Price, C. C., and Oae, S. (1962). "Sulfur Bonding." Ronald Press, New York.

Pryor, W. A., (1962). "Mechanisms of Sulfur Reactions." McGraw-Hill, New York.

Rao, S. M. (1971). "Xanthates and Related Compounds." Dekker, New York.

Reid, D. H., ed. (1975). "Organic Compounds of Sulphur, Selenium and Tellurium," Vol. I, 1970; Vol. II, 1973; Vol. III, 1975. The Chemical Society, London (Specialist Periodical Report).

Reid, E. E. (1958–1966). "Organic Chemistry of Bivalent Sulfur," Vol. 1–6. Chemical Publ., New York.

Roy, A. B., and Trudinger, P. A. (1970). "The Biochemistry of Inorganic Compounds of Sulphur." Cambridge Univ. Press, London and New York.

Schroeter, L. C. (1966). "Sulfur Dioxide." Pergamon, Oxford.

Senning, A., ed. (1971, 1972). "Sulfur in Organic and Inorganic Chemistry," Vol. I–III. Dekker, New York.

Senning, A., ed. (1976). "Topics in Sulfur Chemistry," Vol. 1. Thieme, Stuttgart.

Stirling, C. J. M., ed. (1975). "Organic Sulphur Chemistry." Butterworth, London (*Proc. Int. Conf. Organ. Sulphur Chem., 6th, Bangor, Wales, 1974*).

Suter, C. M. (1944). "The Organic Chemistry of Sulfur. Tetracovalent Sulfur Compounds." Wiley, New York. (Reprinted in 1971 by Gordon & Breach, New York).

Thorn, G. D., and Ludwig, R. A. (1962). "The Dithiocarbamates and Related Compounds." Elsevier, Amsterdam.

Tobolsky, A. V., ed. (1968). "Chemistry of Sulfides." Wiley (Interscience), New York.

Torchinskii, Yu. M. (1974). "Sulfhydryl and Disulfide Groups of Proteins." Consultants Bureau, New York.

Trost, B. M., and Melvin, Jr., L. S. (1975). "Sulfur Ylides." Academic Press, New York.

Appendix B | Additional Tables

Table B-1. Representative Bond Lengths for Sulfur Compounds

Bond[a]	Distance (Å) (method)[b]	Compound used	Reference[c]
H—S	1.34 (M)	CH_3SH	1
sp^3 C—$S^{2,3}$	1.80 (M)	$CH_3S(O)CH_3$, CH_3SCH_3	1
sp^3 C—S^4	1.77 (E)	$CH_3SO_2CH_3$	1
sp^2 C—$S^{2,3}$	1.78 (X)	S S	1
sp^2 C—S^4	1.73 (X)	SO_2	1
N—$S^{2,3}$	1.69 (E)	$(CH_3)_2NSN(CH_3)_2$	1
N—S^4	1.62 (X)	$(CH_3)_2N\overset{O}{\underset{O}{S}}N(CH_3)_2$	1
O—S^2	1.66 (M)	CH_3SOH	5
O—S^3	1.62 (X)	Cl, Cl, O, S=O, O	1
F—S^2	1.64 (M)	FSSF	1
F—S^6	1.57 (E)	SF_6	1
Cl—S	2.01 (M)	CH_3SCl	1
Br—S	2.24 (E)	BrSSBr	1
Si—S	2.14 (E)	$H_3SiSSiH_3$	1
S—S	2.04 (M)	CH_3SSCH_3	1
P—S	2.10 (X)	P_4S_{10}	1
O=S^3	1.49 (M)	$CH_3S(O)CH_3$	1
O=S^4	1.43 (E)	$CH_3SO_2CH_3$	1
sp^2 C=S	1.61 (M)	CH_2=S	2
sp C=S	1.57 (X)	R, =C=S (R = t-C_4H_9), R	2
N=S^4	1.53 (X)	$(CH_3)_2S(NH)_2$	2
N=S^3	1.65 (X)	p-$NO_2C_6H_4N$=$S(CH_3)_2$	3
S=S	1.88	S_2	4

Table B-2. Isotopic Abundances for S_1—S_8 Species[a]

Species present	Peak intensity relative to $P = 100$							
	$P + 1$	$P + 2$	$P + 3$	$P + 4$	$P + 5$	$P + 6$	$P + 7$	$P + 8$
S_1	0.80	4.40		0.01				
S_2	1.59	8.85	0.07	0.22				
S_3	2.40	13.27	0.21	0.63		0.01		
S_4	3.20	17.72	0.43	1.24	0.02	0.05		
S_5	4.00	22.16	0.71	2.04	0.05	0.10		
S_6	4.79	26.62	1.06	3.04	0.10	0.19		0.01
S_7	5.60	31.07	1.49	4.23	0.17	0.33	0.01	0.01
S_8	6.39	35.54	1.99	5.64	0.27	0.53	0.02	0.03

[a] Data from Beynon, J. H., Saunders, R. A., and Williams, A. E. (1968). "The Mass Spectra of Organic Molecules." Elsevier, Amsterdam.

Footnotes to Table B-1

[a] Prefix indicates hybridization state at carbon while superscript indicates coordination number of sulfur.

[b] (E) electron diffraction, (M) microwave spectroscopy, (X) x-ray crystallography.

[c] References: (1) Laur, P. H. (1972). *In* "Sulfur in Organic and Inorganic Chemistry" (A. Senning, ed.), Vol. 3. Dekker, New York. (2) Schaumann, E., Harto, S., and Adiwidjaja, G. (1976). *Angew. Chem. Int. Ed. English* **15**, 40. (3) Eliel, E. L., Koskimies, J., McPhail, A. T., and Swern, D. (1976). *J. Org. Chem.* **41**, 2137. (4) Meyer, B. (1976). *Chem. Rev.* **76**, 367. (5) Penn, R. E., Block, E., and Revelle, L. K. (1977). Unpublished results.

Table B-3. Infrared Band Positions for Organosulfur Functional Groups[a,b]

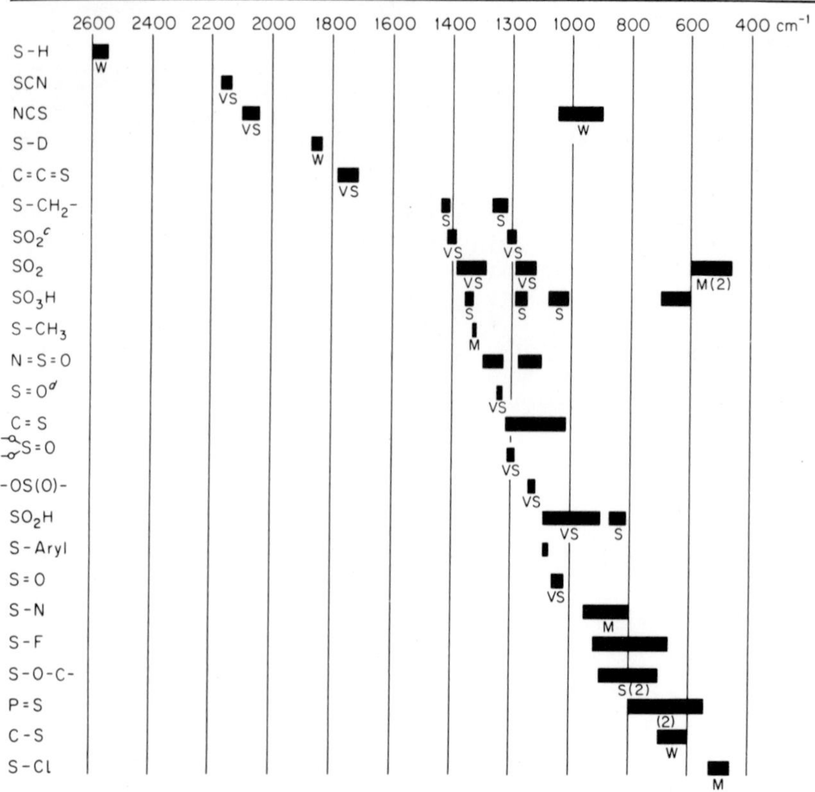

[a] Abbreviations: w = weak, m = medium, s = strong, vs = very strong.

[b] References:

Bellamy, L. J. (1958). "The Infrared Spectra of Complex Molecules." Methuen, London.

Bellamy, L. J. (1968). "Advances in Infrared Group Frequencies." Methuen, London.

Colthup, N. B., Daly, L. H., and Wiberly, S. E. (1975). "Introduction to Infrared and Raman Spectroscopy," 2nd ed. Academic Press, New York.

Elam, E. U., Rash, F. H., Dougherty, J. T., Goodlett, V. W., and Brannock, K. C. (1968). *J. Org. Chem.* **33**, 2738. (C=C=S).

Lambert, J. B., Shurvell, H. F., Verbit, L., Cooks, R. G., and Stout, G. H. (1976). "Organic Structural Analysis." Macmillan, New York.

Nakanishi, K. (1962). "Infrared Absorption Spectroscopy." Holden-Day, San Francisco, California.

Plant, D., Tarbell, D. S., and Whiteman, C. (1955). *J. Am. Chem. Soc.* **77**, 1572. (S—D)

Pouchert, C. J. (1975). "The Aldrich Library of Infrared Spectra," 2nd ed. Aldrich Chem. Co., Milwaukee, Wisconsin.

Rao, C. N. R. (1963). "Chemical Applications of Infrared Spectroscopy." Academic Press, New York.

[c] In sulfonyl fluorides or sulfate esters.

[d] In perfluoroalkyl sulfoxides.

Table B-4. ^1H NMR Chemical Shifts for Methyl Groups in Representative Organosulfur Compounds

Compound	δ[a]	Compound	δ[a]
$(CH_3)_2C(SH)_2$	1.86[m]	CH_3SO_3H	2.90[b,d]
CH_3SCH_2Ph	1.97	$CH_3S(O)(NH)C_6H_5$	2.9[b,y]
CH_3SH	2.00[ll]	CH_3SCl	2.91[h]
$(CH_3S)_2C^-{-}P^+(C_4H_9)_3$	2.04[j,cc]	$\underline{CH_3}SO_2OCH_3$	2.95
$CH_3SCH_2SCH_3$	2.05	$(CH_3)_2S(O)NC_6H_5$	2.99[bb]
CH_3SCH_3	2.06	$\underline{CH_3}S(O)SCH_3$	3.00[b,gg]
$CH_3SCH_2Li \cdot TMEDA$	2.08[j,aa]	$(CH_3)_3S^+\ ClO_4^-$	3.03[f]
$(CH_3S)_3CH$	2.16[ii]	$(CH_3)_2S(NH)_2$	3.06[b,p]
$(CH_3S)_4C$	2.17[ii]	$(CH_3S)_3C^+\ BF_4^-$	3.05[k,cc]
$\underline{CH_3}SC(O)CH_3$	2.20	CH_3SO_2OAr[jj]	3.09[b,z]
$CH_3SC(O)\underline{CH_3}$	2.24	$(CH_3)_2S(O)(NH)$	3.10[b,p]
$CH_3SCH{=}CH_2$	2.25[b]	$CH_3SO_2CF_3$	3.1[g,v]
CH_3SCH_2Cl	2.28[n]	$CH_3SO_2C_6H_5$	3.10[b]
CH_3SSCH_3	2.30	$CH_3S(O)(N(CH_3)_2)C_6H_5$	3.1[b]
$(CH_3S)_2CO$	2.32	$\underline{(CH_3)_2}S^+{-}SCH_3$	3.12[l,ee]
CF_3SCH_3	2.33[g,v]	$(CH_3S)_2CH^+\ BF_4^-$	3.13[k,cc]
$(CH_3S)_2C{=}C(SCH_3)_2$	2.33[cc]	$(CH_3)_2S^+(O)CH_2$	3.14[h,hh]
$CH_3C(S)NH_2$	2.36[b]	$\underline{CH_3}SO_2SCH_3$	3.17
$CH_3SC_6H_5$	2.43	$CH_3S(O)SO_2\underline{CH_3}$	3.17[e,o]
CH_3SSSCH_3	2.43		
$(CH_3)_2SNPh$	2.45[e,r]	$CH_3{-}\overset{+}{S}$	3.21[e,x]
$\underline{CH_3}S(O)OCH_3$	2.46		
$\underline{CH_3}S(O)N(CH_3)_2$	2.50[b,kk]	$CH_3S(O)Cl$	3.22
CH_3SCN	2.50	$(CH_3)_2S^+{-}H\quad SbCl^-$	3.27[l,dd]
$CH_3S(O)CH_3$	2.52	CH_3SF_3	3.29[e,ff]
CH_3SSCF_3	2.54[e,ff]	$CH_3SF_4CF_3$	3.3[g,v]
$\underline{CH_3}SC(S)N(CH_3)_2$	2.54	CH_3NCS	3.30[b]
$CH_3SSSSCH_3$	2.55	$\underline{CH_3}S^+(O)(OAr)N(CH_3)_2$	3.32[i,z,jj]
$CH_3SO_2S\underline{CH_3}$	2.60	$\underline{CH_3}SC(S)N(CH_3)_2$	3.35
$CH_3SO_2CH{=}CH_2$	2.62[b]	$(CH_3SO_2)_2O$	3.38[b,s]
$CH_3S(O)F$	2.62[e,ff]	$\underline{(CH_3)_2}S^+{-}OCH_3\ ClO_4^-$	3.42[f]
$CH_3S{-}O{-}C_4H_9\text{-}t$	2.63[q]	$(CH_3O)_2S{=}O$	3.45
CH_3SSSCl	2.65[n]	$(CH_3)_2S(O)(NTs)$	3.45[b,d,mm]
CH_3SCCl_3	2.66[n]	CH_3SO_2Cl	3.52
$CH_3S(O)S\underline{CH_3}$	2.67[b,gg]	$\underline{(CH_3)_2}S^+(O)C^-(CO_2CH_3)_2$	3.54[b,w]
$(CH_3)_2C{=}S$	2.67	$CH_3S(O)O\underline{CH_3}$	3.56
$CH_3S(O)N\underline{(CH_3)}_2$	2.68[b,kk]	$(CH_3O)_2SO_2$	3.80
CH_3SO_2H	2.7[b,u]	$CH_3SCH_2^+\ SbCl_6^-$	3.87[k,cc]
$(CH_3S)_2C{=}S$	2.73	$(CH_3)_2S^+{-}Cl\ SbCl_6^-$	3.88[k,dd]
CH_3SSCl	2.75[n]	$CH_3SO_2O\underline{CH_3}$	3.88
$CH_3S(O)CF_3$	2.75[g,v]	$(CH_3)_3S^+{=}O\ ClO_4^-$	3.88[f]
$\underline{CH_3}SO_2N(CH_3)_2$	2.79[b,kk]	$\underline{CH_3}S^+(O)(Ph)N(CH_3)_2$	3.95[b]
$(CH_3)_2S^+{-}C^-(CN)_2$	2.82[d]	$CH_3S(O)F_2CF_3$	4.09[e,nn]
$(CH_3)_2S^+{-}SCH_3$	2.83[l,ee]	$(CH_3)_2S^+{-}O{-}\underline{CH_3}\ ClO_4^-$	4.25[f]
$\underline{CH_3}S(O)SO_2CH_3$	2.85[e,o]	$CH_3S^+(O)(OAr)_2\ PF_6^-$	4.27[b,z,jj]
$CH_3SO_2CH_3$	2.85	$(CH_3)_2S^+(O)(OAr)$	4.29[d,jj]
$CH_3SO_2N\underline{(CH_3)}_2$	2.87[b,kk]	$(CH_3)_2S^+(O)Ph$	4.3[d,t]
$(CH_3)_2S^+{-}C^-(CO_2CH_3)_2$	2.89[w]	$\underline{CH_3}S^+(O)(OAr)N(CH_3)_2$	4.44[i,z,jj]

(continued)

Footnotes to Table B-4

[a] Unless otherwise indicated the solvent is carbon tetrachloride and the data are from Chamberlain (1971). (Chamberlain, N. F., and Reed, J. J. R. (1971). *In* "The Analytical Chemistry of Sulfur and its Compounds" (J. H. Karchmer, ed.), Part III. Wiley (Interscience), New York.)

[b] Chloroform-d.

[c] Pyridine.

[d] Dimethyl sulfoxide-d_6.

[e] No solvent indicated.

[f] Trifluoroacetic acid.

[g] Neat.

[h] 1,4-Dioxane.

[i] Acetone-d_6.

[j] Benzene-d_6.

[k] Nitromethane-d_3.

[l] Acetonitrile-d_3.

[m] SH at δ2.71. Demuynck, M., and Vialle, J. (1967). *Bull. Soc. Chim. Fr.* 1213.

[n] Pettit, G. R., Douglass, I. B., and Hill, R. A. (1964). *Can. J. Chem.* **42**, 2357.

[o] Kice, J. L., and Ikura, K. (1968). *J. Am. Chem. Soc.* **90**, 7378.

[p] Laughlin, R. G., and Yellin, W. (1967). *J. Am. Chem. Soc.* **89**, 2435.

[q] Moore, T. L., and O'Connor, D. E. (1966). *J. Org. Chem.* **31**, 3587.

[r] Claus, P. and Vycudilik, W. (1968). *Tetrahedron Lett.* 3607.

[s] Robinson, E. A., and Silberber, V. (1966). *Can. J. Chem.* **44**, 1437.

[t] Ryoke, K., Minato, H., and Kobayashi, M. (1976). *Bull. Chem. Soc. Jpn.* **49**, 1455.

[u] Wudl, F., Lightner, D. A., and Cram, D. J. (1967). *J. Am. Chem. Soc.* **89**, 4099.

[v] Yu, S.-L., Sauer, D. T., and Shreeve, J. M. (1974). *Inorg. Chem.* **13**, 484.

[w] Ando, W., Yagihara, T., Tozune, S., Imai, I., Suzuki, J., Toyama, T., Nakaido, S., and Migita, T. (1972). *J. Org. Chem.* **37**, 1721.

[x] Acheson, R. M., and Harrison, D. R. (1970). *J. Chem. Soc.* (*C*) 1764.

[y] Johnson, C. R., Haake, M., and Schroeck, C. W. (1970). *J. Am. Chem. Soc.* **92**, 6594.

[z] Chalkley, G. R., Snodin, D. J. Stevens, G., and Whiting, M. C. (1970). *J. Chem. Soc.* (*C*) 682.

[aa] Peterson, D. J. (1967). *J. Org. Chem.* **32**, 1717.

[bb] Heintzelman, R. W., Bailey, R. B., and Swern, D. (1976). *J. Org. Chem.* **41**, 2207.

[cc] Walinsky, S. W. (1971). *Ph.D. Thesis*, Pennsylvania State Univ.; the CH in $(CH_3S)_2CH^+$ appears at δ11.23 while the CH_2 in $CH_3SCH_2{}^+$ appears at δ5.78.

[dd] Hansen, D. W. Jr., and Olofson, R. A. (1971). *Tetrahedron* **27**, 4221.

[ee] Kice, J. L., and Favstritsky, N. A. (1969). *J. Am. Chem. Soc.* **91**, 1751.

[ff] Gombler, W., and Budenz, R. (1976). *J. Fluorine Chem.* **7**, 115.

[gg] Block, E., and O'Connor, J. (1974). *J. Am. Chem. Soc.* **96**, 3921.

[hh] Corey, E. J., and Chaykovsky, M. (1965). *J. Am. Chem. Soc.* **87**, 1353.

[ii] Seebach, D., Geiss, K.-H., Beck, A. K. Graf, B., and Daum, H. (1972). *Chem. Ber.* **105**, 3280.

[jj] Ar = *p*-tolyl.

[kk] Moriarty, R. M. (1965). *J. Org. Chem.* **30**, 600.

[ll] The SH proton appears at δ0.92 in C_6H_{12}, δ1.04 in CCl_4, δ1.21 in $CHCl_3$, δ1.23 in $C_2H_5OC_2H_5$, and δ1.38, neat.

[mm] Heintzelman, R. W., and Swern, D. (1976). *Synthesis* 731.

[nn] Sprenger, G. H., and Cowley, A. H. (1976). *J. Fluorine Chem.* **7**, 333.

Table B-5. Ultraviolet Spectra of Organosulfur Compounds

Compound[a]	$^{\lambda}max(\varepsilon)$[b]	Compound[a]	$^{\lambda}max(\varepsilon)$[b]
$C_2H_5SCH_3$	229(139)	PhSCH$_3$	236(10,000)
S	229(183)	PhSPh	201(38,000), 231(6500), 250(12,000), 274(5600)
S	239(54)	S	231(7100)
S	242(43)[g]	$CH_2{=}CHSCH(CH_3)_2$	230(6000)
S	257(40)	$CH_2{=}CHSCH{=}CH_2$	240(8350), 255(7600)
S	270(32)	$CH_3COCH_2SC_2H_5$	243(450), 299(260)
S	277(43)	S, O	230(640),[h] 291(21)
S	278(14)	S, O	238(2600)
CF_3SCF_3	210(7)	$PhCOCH_2SC_2H_5$	278(2300), 290(1600), 300(630), 330(450), 338(500)
		i-$C_4H_9C(O)SC_2H_5$	233(4100), 276(690)
		$CH_2(SCH_3)_2$	235(540)
		S, S, S	240(135)

(continued)

Table B-5. *(Continued)*

Compound[a]	$\lambda_{max}(\varepsilon)$[b]
S S (five-membered ring)	207(1400), 247(370)
S S (four-membered ring)	216(860), 293(38), 311(sh, 20)
C_4H_9SH	233(*ca.* 150)
CF_3SH	218(46)
PhSH[i]	236(10,000)
CH_3SSCH_3	201(2200), 254(400)
CF_3SSCF_3	237(360)
t-C_4H_9SS-t-C_4H_9	201(4000), 222(sh, 460)
PhSSPh	<215(16,900), 242(17,400), 320(783)
S—S (five-membered ring)	330 (—)
S—S (six-membered ring)	202(1300), 211(1150), 286(295)
S—S (seven-membered ring)	192(2300), 200(2300), 206(1900), 259(467)
$CH_3S(O)SCH_3$	<215(1600), 246(2080), 320(10)
$C_4H_9S(O)CH{=}CH_2$	202(2300),[i] 249(2600)
$CH_2{=}CHS(O)CH{=}CH_2$	208(3950),[i] 236(2720)
$PhS(O)CH_3$	238(4100), 256(2800)
PhS(O)Ph	202(33,000), 233 (14,000), 265(2100)
$(Alkyl)_2SO_2$	<180(—)
$CH_2{=}CHSO_2CH{=}CH_2$	209(3200)
$PhSO_2Ph$	201(38,000), 235 (15,000), 260(1700), 266(2100), 274(1400)
$C_6H_{11}COC^-HS^+(O)(CH_3)_2$	252(13,600)
$C_3H_7C(S)C_3H_7$	213(5000), 231 (6200), 492(7.9)[f]
t-$C_4H_9C(S)t$-C_4H_9	237(7900),[f] 536(8.9)
	210(3700), 246(1100),[f] 481(10.5)
$CF_3C(S)CF_3$	308(13.6), 313(14.4), 318(14), 580(12.7)
PhC(S)Ph	235(~10,000), 315(15,100), 595(177)
$CH_3C(S)OC_2H_5$	250(—)
PhC(S)OPh	289(1500), 433(135)
$CF_3SC(S)CF_3$	532(11.5)

$CH_3SO_2SCH_3$	<215(180), 280(26)	$(CF_3S)_2C{=}S$	259(4900), 299 (6400), 496(16)
$CH_3SO_2S(O)CH_3$	240–250(1650)	$(CH_3S)_2C{=}S$	238(3780), 302(16900), 430(28)
$CH_3SO_2SO_2CH_3$	<215(3580), 280(1)		292(10,600), 311 (12,500), 460(69)
PhS(O)SPh	<215(16,600), 224 (16,100), 284(7400)		230(~16,000), 315(~32)
$PhSO_2SPh$	<215(15,000), 228 (14,000), 270(3900), 276(3520), 320(181)	$PhCH_2N{=}C{=}S$	251.5(1520)
$PhSO_2SO_2Ph$	238(25,500) 257(10,700)	$(CF_3)_2C{=}C{=}S$	239(5590), 503(8.5)[c]
	244(—)	$(t\text{-}C_4H_9)_2C{=}C{=}S$	211(32,400), 239(4170),[d] 570(7.8)
CF_3SSSCF_3	238(1500)	$(t\text{-}C_4H_9)_2C{=}C{=}S{=}O$	210(52,500), 251(13,500),[d] 315(759), 330(631)
CF_3SCl	324(12)		
$t\text{-}C_4H_9SOC_2H_5$	266(71)	$(t\text{-}C_4H_9)_2C{=}S{=}O$	264(9000)[e]
PhSCN	230(5600),[j] 260(400)	S_8	277(794)[k]
		SO_2	290(250)
$PhCH_2SCN$	260(470)	SO_3	290(—)
$C_4H_9S(O)C_4H_9$	207(3800),[i] 222(1390)		

[a] Unless otherwise indicated, the data have been taken from E. Block, *Quart. Rep. Sulfur Chem.* **4**, 237 (1969) and the solvent is ethanol.

[b] In nanometers.

[c] Raasch, M. S. (1970). *J. Org. Chem.* **35**, 3470; solvent is isooctane.

[d] Elam, E. U., Rash, F. H., Dougherty, J. T. Goodlett, V. W., and Brannock, K. C. (1968). *J. Org. Chem.* **33**, 2738; solvent is cyclohexane.

[e] Back, T. G., Barton, D. H. R., Britten-Kelly, M. R., and Guziec, F. S., Jr. (1975). *Chem. Commun.* 539.

[f] Ohno, A., Nakamura, K., Nakazima, Y., and Oka, S. (1975). *Bull. Chem. Soc. Jpn.* **48**, 2403.

[g] In isooctane.

[h] In CF_3Cl.

[i] In cyclohexane.

[j] In water.

[k] In chloroform.

Table B-6. Bond Dissociation Energies for Some Organosulfur Compounds[a]

Bond	Dissociation energy (kcal/mole)[b]	Bond	Dissociation energy (kcal/mole)[b]
$CH_2{=}S$	129 ± 5	AlkylSS—H	70
SC=S	103.4	$CH_3S{-}SCH_3$	74
$CH_3NC{=}S$	71	$C_6H_5S{-}SC_6H_5$	55
$CH_3S{-}CN$	97	$HSS{-}SCH_3$	54
$CH_3{-}SCH_3$	77	$AlkylS_2{-}S_2Alkyl$	33.6
$CH_3{-}SH$	75	$(CH_3)_2S{=}S$[d]	[53 ± 4]
$C_2H_5{-}SCH_3$	74	$PhSO_2{-}SO_2Ph$	41
$CH_3S(O)_2{-}CH_3$	68	PhS—S(O)Ph	36
CH_3—SS-alkyl	57	$PhSO_2{-}S(O)Ph$	28
$CH_3S(O){-}CH_3$	[55][c]	$(CH_3O)_2S{=}O$	[116 ± 6]
$CH_3S(O)_2$-allyl	55	$(CH_3)_2S(O){=}O$	112
$CH_3S(O)_2$-benzyl	56	$(CH_3)_2S{=}O$	86.6
$CH_3S{-}H$	92	$CH_3S{-}Cl$	[70 ± 3][c]
$C_6H_5S{-}H$	82	$CH_3S{-}NO$	[25][c]

[a] S. W. Benson, *Chem. Rev.*, in press.
[b] Unless otherwise indicated, ±2 kcal/mole.
[c] Estimated value.
[d] Hypothetical molecule.

Table B-7. Raman Frequencies for Some Organosulfur Functional Groups[a]

Functional group	Range (cm^{-1}) and intensity[b]	Assignment
S—H	2590–2560 (vs)	S—H stretch
	850–820 (vs)	S—H in plane
	700–600 (vs)	C—SH stretch
	340–320 (vs)	S—H out of plane
S—S	550–430 (vs)	S—S stretch
C—S—C	705–570 (s)	C—S stretch
Thiophenes	740–680 (vs)	C—S—C stretch
	570–430 (s)	ring deformation
C—S(O)—C	1050–1010 (s)	S=O
$C{-}S(O)_2{-}C$	1280–1260 (m)	SO_2 antisymmetric stretch
	1150–1110 (s)	SO_2 symmetric stretch
	610–545 (s)	SO_2 scissoring
$C{-}SO_2NH_2$	1155–1135 (vs)	SO_2 stretch
$C{-}SO_2Cl$	1230–1200 (m)	S=O stretch
C—S—C≡N	650–600 (s)	C—S stretch
O—C(S)—S (xanthates)	670–620 (vs)	C=S stretch
	480–450 (vs)	C—S stretch

[a] Lambert, J. B., Shurvell, H. F., Verbit, L., Cooks, R. G., and Stout, G. H. (1976). "Organic Structural Analysis." Macmillan, New York.
[b] Abbreviations: m = medium, s = strong, vs = very strong.

Index

Numbers followed by t refer to items that occur in tables.

E

F

G

H

S

ORGANIC CHEMISTRY

A SERIES OF MONOGRAPHS

EDITOR

HARRY H. WASSERMAN
Department of Chemistry
Yale University
New Haven, Connecticut

1. Wolfgang Kirmse. CARBENE CHEMISTRY, 1964; 2nd Edition, 1971
2. Brandes H. Smith. BRIDGED AROMATIC COMPOUNDS, 1964
3. Michael Hanack. CONFORMATION THEORY, 1965
4. Donald J. Cram. FUNDAMENTALS OF CARBANION CHEMISTRY, 1965
5. Kenneth B. Wiberg (Editor). OXIDATION IN ORGANIC CHEMISTRY, PART A, 1965; Walter S. Trahanovsky (Editor). OXIDATION IN ORGANIC CHEMISTRY, PART B, 1973; PART C, 1978
6. R. F. Hudson. STRUCTURE AND MECHANISM IN ORGANO-PHOSPHORUS CHEMISTRY, 1965
7. A. William Johnson. YLID CHEMISTRY, 1966
8. Jan Hamer (Editor). 1,4-CYCLOADDITION REACTIONS, 1967
9. Henri Ulrich. CYCLOADDITION REACTIONS OF HETEROCUMULENES, 1967
10. M. P. Cava and M. J. Mitchell. CYCLOBUTADIENE AND RELATED COMPOUNDS, 1967
11. Reinhard W. Hoffman. DEHYDROBENZENE AND CYCLOALKYNES, 1967
12. Stanley R. Sandler and Wolf Karo. ORGANIC FUNCTIONAL GROUP PREPARATIONS, VOLUME I, 1968; VOLUME II, 1971; VOLUME III, 1972
13. Robert J. Cotter and Markus Matzner. RING-FORMING POLYMERIZATIONS, PART A, 1969; PART B, 1; B, 2, 1972
14. R. H. DeWolfe. CARBOXYLIC ORTHO ACID DERIVATIVES, 1970
15. R. Foster. ORGANIC CHARGE-TRANSFER COMPLEXES, 1969
16. James P. Snyder (Editor). NONBENZENOID AROMATICS, VOLUME I, 1969; VOLUME II, 1971
17. C. H. Rochester. ACIDITY FUNCTIONS, 1970

18. Richard J. Sundberg. The Chemistry of Indoles, 1970

19. A. R. Katritzky and J. M. Lagowski. Chemistry of the Heterocyclic *N*-Oxides, 1970

20. Ivar Ugi (Editor). Isonitrile Chemistry, 1971

21. G. Chiurdoglu (Editor). Conformational Analysis, 1971

22. Gottfried Schill. Catenanes, Rotaxanes, and Knots, 1971

23. M. Liler. Reaction Mechanisms in Sulphuric Acid and Other Strong Acid Solutions, 1971

24. J. B. Stothers. Carbon-13 NMR Spectroscopy, 1972

25. Maurice Shamma. The Isoquinoline Alkaloids: Chemistry and Pharmacology, 1972

26. Samuel P. McManus (Editor). Organic Reactive Intermediates, 1973

27. H.C. Van der Plas. Ring Transformations of Heterocycles, Volumes 1 and 2, 1973

28. Paul N. Rylander. Organic Syntheses with Noble Metal Catalysts, 1973

29. Stanley R. Sandler and Wolf Karo. Polymer Syntheses, Volume I, 1974; Volume II, 1977

30. Robert T. Blickenstaff, Anil C. Ghosh, and Gordon C. Wolf. Total Synthesis of Steroids, 1974

31. Barry M. Trost and Lawrence S. Melvin, Jr. Sulfur Ylides: Emerging Synthetic Intermediates, 1975

32. Sidney D. Ross, Manuel Finkelstein, and Eric J. Rudd. Anodic Oxidation, 1975

33. Howard Alper (Editor). Transition Metal Organometallics in Organic Synthesis, Volume 1, 1976; Volume II, 1978

34. R. A. Jones and G. P. Bean. The Chemistry of Pyrroles, 1976

35. Alan P. Marchand and Roland E. Lehr (Editors). Pericyclic Reactions, Volume I, 1977; Volume II, 1977

36. Pierre Crabbé (Editor). Prostaglandin Research, 1977

37. Eric Block, Reactions of Organosulfur Compounds, 1978

38. Arthur Greenburg and Joel Liebman, Strained Organic Molecules, 1978

39. Philip S. Bailey, Ozonation in Organic Chemistry, Vol. 1, 1978

A B C D E F G H I J
8 9 0 1 2 3 4 5